# Philadelphia and Reading Railroad.

## PASSENGER TRAIN TIME TABLE.

Leave Philadelphia from the Depot, Broad and Callowhill Street, at 7½ A. M., and 3½ P. M. Daily Except Sundays, when an Excursion Train leaves at 7½ A. M. Returning leaves Pottsville at 4 P. M.

| UP TRAINS. | | | DOWN TRAINS. | | |
|---|---|---|---|---|---|
| STATIONS. | *Exp's Morn.* | *Way Aft'n.* | STATIONS. | *Wag Morn.* | *Exp's Aft'n.* |
| Leaves PHILADELPHIA, | 7.30 | 3.30 | Leaves POTTSVILLE, | 7.30 | 3.30 |
| Passes Schl. Viaduct, | — | 3.41 | Passes MT. CARBON, | 7.37 | 3.37 |
| " Manayunk, | — | 3.50 | " SCHL. HAVEN, | 7.46 | 3.45 |
| " Conshohocken, | — | 4.05 | " Orwigsburg, | 7.57 | — |
| " Norristown, | — | 4.12 | " Auburn, | 8.05 | — |
| " Port Kennedy, | — | 4.21 | " PORT CLINTON, | 8.20 | 4.10 |
| " Valley Forge, | — | 4.26 | " Hamburg, | 8.30 | — |
| " PHŒNIXVILLE, | 8.31 | 4.41 | " Mohrsville, | 8.48 | — |
| " Royer's Ford, | — | 4.51 | " Althouse's, | 8.53 | — |
| " Limerick, | — | 4.56 | " READING, | 9.10 | 4.51 |
| " POTTSTOWN, | 8.58 | 5.13 | " Birdsboro, | 9.32 | — |
| " Douglassville, | — | 5.22 | " Douglassville, | 9.41 | — |
| " Birdsboro, | — | 5.33 | " POTTSTOWN, | 9.51 | 5.30 |
| " READING, | 9.34 | 6.00 | " Limerick, | 10.04 | — |
| " Althouse's, | — | 6.25 | " Royer's Ford, | 10.08 | — |
| " Mohrsville, | — | 6.30 | " PHŒNIXVILLE, | 10.18 | 5.56 |
| " Hamburg, | — | 6.48 | " Valley Forge, | 10.29 | — |
| " PORT CLINTON, | 10.17 | 6.58 | " Port Kennedy, | 10.34 | — |
| " Auburn, | — | 7.11 | " Norristown, | 10.44 | — |
| " Orwigsburg, | — | 7.19 | " Conshohocken, | 10.51 | — |
| " SCHL. HAVEN, | 10.43 | 7.27 | " Manayunk, | 11.06 | — |
| " MOUNT CARBON, | 10.52 | 7.36 | " Schl. Viaduct. | 11.18 | — |
| Arrives at POTTSVILLE, | 11.00 | 7.45 | Arrives at PHILAD'A. | 11.30 | 7.00 |

## STAGE CONNEXIONS.

At PHŒNIXVILLE, with Express and Way Trains, for Yellow Springs, &c.
At POTTSTOWN, with Express Trains, for Boyerstown, Allentown, &c.
At READING, with Express Trains, for Lebanon, Harrisburg, Bernville, Jonestown, &c.
At POTTSVILLE, with Express Trains, for Northumberland, Sunbury, Danville, Catawissa, &c.

## RAILROAD CONNEXIONS.

At PORT CLINTON, to Tamaqua, thence by Stage to Mauch Chunk, Wilkesbarre, Lackawanna, Hazelton, &c.
At SCHUYLKILL HAVEN, to Minersville, Tremont, &c.
At MOUNT CARBON, to Tuscarora, Middleport, &c.

# Philadelphia and Reading Railroad.

## PASSENGER FARES AND DISTANCES.

| UP TRAINS. | | | | DOWN TRAINS. | | | |
|---|---|---|---|---|---|---|---|
| Dist. | From Philad'a to | FARES. No. 1 | No. 2 | Dist. | From Pottsville to | FARES. No. 1 | No. 2 |
| 3½ | Schl. Viaduct, | .15 | .10 | 1 | MOUNT CARBON, | .05 | .05 |
| 7 | Manayunk, | .20 | .15 | 4 | SCHL. HAVEN, | .15 | .10 |
| 13½ | Conshohocken, | .30 | .25 | 7 | Orwigsburg, | .20 | .15 |
| 17 | Norristown, | .40 | .30 | 10 | Auburn, | .30 | .25 |
| 21½ | Port Kennedy, | .65 | .50 | 15 | PORT CLINTON, | .45 | .35 |
| 23½ | Valley Forge, | .70 | .60 | 18 | Hamburg, | .55 | .45 |
| 27½ | PHŒNIXVILLE, | .80 | .65 | 25 | Mohrsville, | .75 | .60 |
| 32 | Royer's Ford, | .95 | .80 | 27 | Althouse's, | .80 | .65 |
| 34 | Limerick, | 1.05 | .85 | 35 | READING, | 1.05 | .85 |
| 40 | POTTSTOWN, | 1.20 | 1.00 | 44 | Birdsboro, | 1.30 | 1.10 |
| 44½ | Douglassville, | 1.35 | 1.10 | 48½ | Douglassville, | 1.45 | 1.20 |
| 49 | Birdsboro, | 1.50 | 1.25 | 53 | POTTSTOWN, | 1.60 | 1.30 |
| 58 | READING, | 1.75 | 1.45 | 59 | Limerick, | 1.75 | 1.45 |
| 66 | Althouse's, | 2.00 | 1.65 | 61 | Royer's Ford, | 1.80 | 1.50 |
| 68 | Mohrsville, | 2.05 | 1.70 | 65½ | PHŒNIXVILLE, | 1.95 | 1.65 |
| 75 | Hamburg, | 2.25 | 1.90 | 69½ | Valley Forge, | 2.05 | 1.70 |
| 78 | PORT CLINTON, | 2.35 | 1.95 | 71½ | Port Kennedy, | 2.10 | 1.75 |
| 83 | Auburn, | 2.50 | 2.10 | 76 | Norristown, | 2.35 | 1.95 |
| 86 | Orwigsburg, | 2.60 | 2.15 | 79½ | Conshohocken, | 2.45 | 2.00 |
| 89 | SCHL. HAVEN, | 2.70 | 2.20 | 86 | Manayunk, | 2.55 | 2.15 |
| 92 | MOUNT CARBON, | 2.75 | 2.25 | 89½ | Schl. Viaduct. | 2.65 | 2.20 |
| 93 | POTTSVILLE, | 2.75 | 2.25 | 93 | PHILADELPHIA. | 2.75 | 2.25 |

Way Trains stop at all the points stated: EXPRESS Trains only at those Stations in SMALL CAPITALS, and *positively* at no others.

All Passengers will purchase their tickets before entering the cars.

Fifty pounds of Baggage are allowed each Passenger.

Passengers are strictly forbidden to stand outside, on the Platforms, while the cars are in motion.

Passengers waiting for Way Trains, at Way Points where there is no Railroad Agent, will signal the approaching Trains, otherwise they will not stop.

PORT RICHMOND.

THE

# PICTORIAL SKETCH-BOOK

OF

# PENNSYLVANIA.

OR

## ITS SCENERY, INTERNAL IMPROVEMENTS, RESOURCES, AND AGRICULTURE,

POPULARLY DESCRIBED,

BY

**ELI BOWEN.**

AUTHOR OF THE "U. S. POST-OFFICE GUIDE," AND LATE OF THE GENERAL POST-OFFICE.

Illustrated with over Two Hundred Engravings,

AND A COLORED MAP.

PHILADELPHIA:

WILLIS P. HAZARD,

178 CHESTNUT STREET.

1852.

Stereotyped by SLOTE & MOONEY, Philadelphia.
C. SHERMAN, Printer.

TO JOHN TUCKER, Esq.,

PRESIDENT OF THE READING RAILROAD, THE SCHUYLKILL VALLEY

RAILROAD, ETC. ETC.,

THESE SKETCHES OF THE ANTHRACITE COAL REGIONS, AND

THEIR PRINCIPAL OUTLET,

ARE INSCRIBED

AS A SLIGHT INTIMATION OF THE RESPECT WHICH IS

ENTERTAINED FOR HIM BY THE THOUSANDS DIRECTLY AND INDIRECTLY CONNECTED

WITH THE COAL TRADE.

# A WORD BEFORE WE GO.

In all parts of Europe the traveller is supplied with Guide-books, detailing, for his special information and satisfaction, the leading features of all objects of interest on his route. There is not an antiquated castle, a battle-field, a mountain, or a river, but has its peculiar points revealed for the entertainment of the stranger, as he rambles along from place to place. No doubt this materially adds to the interest and subsequent *value* of travel; and probably constitutes one of the paramount attractions of a tour in Europe, since all its incidents are thus permanently impressed on the mind.

In the United States no such conveniences exist; and this is probably one reason why foreigners generally misunderstand and misrepresent us —they are not *sufficiently informed* to give a correct estimate of our resources, peculiarities, and institutions. They hastily pass over our railways and rivers, and, for the want of suitable printed-guides, return as profoundly ignorant of the routes traversed as they were at the starting-point—for seeing is not *understanding.*

In her physical aspect and resources, Pennsylvania is pre-eminently the most interesting State in the Union—yet, for the want of *popular descriptions* and references, her real character is comparatively obscured from the public view. The most intelligent individual may make the tour from the Delaware to the Ohio by railroad, and yet be unable to identify one-half the towns, or mountains, or streams, or otherwise explain correctly the prominent local characteristics of the route traversed. Thousands of persons, of fortune and leisure, owing to this evil, are intimidated from travelling; while many proceed direct to Europe, before visiting the objects of interest in their own immediate land.

It was as much with the hope of converting our time to a useful purpose, as receiving a reasonable compensation for it, that we undertook to sketch, in a sprightly and popular way, some of the prominent features of our time-honoured Commonwealth. If we have collected together, in tolerable order, a mass of matter that will relieve, to some extent, the fatigue and monotony of travel, our main object has been attained.

We may add, that over seventeen hundred dollars have been expended for pictorial illustrations, some of which we can point to as fair specimens of the art. During the particular time we were engaged in the preparation of these pages, however, an unusual activity prevailed among our best wood engravers, in consequence of large orders from the Government. We were, therefore, in several instances, forced to employ artists of ordinary talent—though, upon the whole, we think the reader will find little to complain of under this head.

The matter is, what it purports to be, off-hand, and no particular credit is claimed or expected for it. We have profited from the works of others to a greater extent than we should, had our time been less limited. Our acknowledgments are due to the works of the late Prof. Richardson, and to those of Mr. Day, Mr. Trego, and others, from which the matter not strictly original has been mainly extracted. With these explanations our work is done.

E. B.

N. B.—It is proper to add, that not having corrected the latter portion of this work as it was passing through the press, some errors appear which would not otherwise have occurred.

# PART I.

## Valley of the Schuylkill.

### PHILADELPHIA TO POTTSVILLE.

Starting from the Depot—Preston Retreat—Girard College—Fairmount WaterWorks—Lager Beer Establishments—Railroad Bridge—Ice Houses—Tom Moore's Cottage—Laurel Hill Cemetery—Falls of the Schuylkill—Port Richmond—The Wissahickon—Norristown Railroad—Manayunk—The Factory System—Limestone Region—Spring Mill—The Copper Region—The Lead Region—Scenery of the Schuylkill—Conshehocken—Norristown—Montgomery County—Bridgeport—Valley Forge—Historical Associations—Phœnixville—Yellow Springs—Tunnel and Bridge—Royer's Ford—Limerick—Pottstown—Berk's County—Douglassville—Birdsboro'—Reading—Union Canal—Schuylkill Navigation Company—The Reading Railroad Company—Dragon's Cave—Revolutionary Incidents—Hamburg—Port Clinton—Auburn—Orwigsburg—Schuylkill Haven—Mine Hill Railroad—Head's Mansion House—Pottsville—Speculating Scenes—Swatara Falls—Tamaqua—Summit Hill—Gravitation Railroad—Mauch Chunk.

# PART II.

## The Anthracite Coal Region.

Coal, its Nature and Origin—Fossils—Coal Vegetation—The Strata of the Earth—The Dip—Valleys—Discovery of Anthracite Coal—Geographical Position, Dimensions and Structure of the Coal Basins—Geological

## PART III.

## Wyoming.



## PART IV.

## Philadelphia to Pittsburg.

# PENNSYLVANIA.

BY PHILIP FRENEAU.

SPREAD with stupendous hills, far from the main,
Fair Pennsylvania holds her golden reign;
In fertile fields her wheaten harvest grows,
Charged with its freights her favorite *Delaware* flows;
From ERIE's lake her soil with plenty teems
To where the *Schuylkill* rolls his limpid streams—
Sweet stream! what pencil can thy beauties tell—
Where, wandering downward through the woody vale,
Thy varying scenes to rural bliss invite,
To health and pleasure add a new delight.
Here *Juniata*, too, allures the swain,
And gay *Cadorus* roves along the plain;
*Swatara*, tumbling from the distant hill,
Steals through the waste, to turn the industrious mill—
Where'er those floods through groves or mountain stray,
That God of nature still directs the way;
With fondest care has traced each river's bed,
And mighty streams thro' mighty forests led;
Bade agriculture thus export her freight,
The strength and glory of this favored STATE.

She, famed for science, arts, and polished men,
Admires her **FRANKLIN**, but adores her **PENN**,

Who wandering here, made barren forests bloom,
And the new soil a happier robe assume:
He planned no schemes that virtue disapproves,
He robbed no Indian of his native groves,
But, just to all, beheld his tribes increase,
Did what he could to bind the world in peace,
And, far retreating from a selfish band,
Bade Freedom flourish in this foreign land.

Gay towns unnumbered shine through all her plains,
Here every art its happiest height attains:
The graceful ship, on nice proportions planned,
Here finds perfection from the builder's hand,
To distant worlds commercial visits pays,
Or war's bold thunder o'er the deep conveys.

# OFF-HAND SKETCHES.

## PART I.

### The Valley of the Schuylkill.

Let us, since life can little else supply,
Than just to look around us, and to die,
Expatiate free o'er all this scene of man—
A mighty maze, but not without a plan;
A wild, where weeds and flowers promiscuous shoot,
Or garden, tempting with forbidden fruit.
Together let us roam this ample field—
Try what the open, what the covert yield;
Eye Nature's walks—shoot folly as it flies,
And catch the manners, living, as they rise;
Laugh where we must—be candid where we can,
But always vindicate the ways of God to man!

FROM Philadelphia to Pottsville, Tamaqua and Mauch Chunk, thence to Wilkesbarre, in Wyoming;—this is the journey before us. Having seated ourselves in the comfortable cars of the Reading Railroad Company, the first object which arrests our attention, after leaving the depot at Broad near Callowhill street, is the Preston Retreat, a fine marble building on our right; we then catch glimpses of the Eastern Penitentiary, which served as a model for European Institutions of a like character, and of Girard College, the finest building of the kind in the United States, and one of the finest in the world.

GIRARD COLLEGE.

Shortly after which we see, on our left, the Fairmount Water-works, and although a notice of it is not strictly within the range of this work, it may nevertheless prove interesting to many to learn something of its leading features, especially as it was the first establishment of the kind ever erected in the United States; and, in point of boldness of conception and romantic profile, probably inferior to none in any quarter of the globe.

The first water-works were commenced in 1799. A steam-engine was placed in Chestnut street, near the Schuylkill, by means of which the water was elevated to a basin in Penn square, and from thence distributed to the city in wooden pipes. The quantity of water thus obtained was soon found to be entirely too small to supply the increasing demand, and the works were abandoned in 1815, after nearly $700,000 had been spent upon them. In 1816 the works at Fairmount were commenced, the water being again raised by steam to an elevated reservoir. Steam was found too expensive, and arrangements were adopted in 1818, by which the water-power of the river was applied. A dam was erected in a diagonal course across the river, securing a head of water nearly thirty feet in depth, and conducted to the mill-houses, on the eastern side of the stream, as represented in the engraving on the opposite page. Here the water

FAIRMOUNT WATER-WORKS.

is forced up to the reservoir, elevated about one hundred feet above the level of the river, and fifty feet above the highest ground in the city. The reservoir, when full, contains twelve feet of water, and is capable of holding over twenty-two millions of gallons. There are eighty-three miles of water-pipe laid down in the city, exclusive of the works of Spring Garden and the Northern Liberties, which probably have an equal extent in the adjoining districts of the city proper. The daily average consumption of water, from these works, is nearly five million gallons. Their total cost was $1,615,169, and they were designed and executed by the late Frederick Graff, to whose memory a handsome monument is erected in the grounds fronting on the Schuylkill, from a design by his son.

The comparison between the present works and the old steam-works, is greatly to the advantage of the former. It was not possible, with the steam-engines, to raise one million two hundred and fifty thousand gallons per day—whereas, the present works, with only three wheels, can readily raise three times this amount, without any increase of expense. But if the same quantity were required to be raised by additional steam-engines, the annual expense would probably be at least $75,000. In other words, the expense of raising three hundred and seventy-five thousand gallons per day, by steam, would be $206—by water, it is only $4. In this estimate, the first cost of the steam-engines or of the water-power is not considered.

These works are eminently worth a visit from the stranger. They are delightfully situated, and present a view, in connection with surrounding objects, of rare beauty and spirit. The wire-bridge, stretching across the Schuylkill, is also an interesting object, and is probably one of the most complete structures of the kind, as we believe it was one of the first, ever erected in the United States. The accompanying sketch exhibits a faithful view of the Water-works and Wire-bridge, as seen from the opposite side of the river.

Within the suburbs of the city, scattered along the rail-road, several "lager beer" establishments will be noticed. These breweries are all of very recent origin, and lager beer is, to many, an unknown beverage. It is a German drink, of which they are very fond, and is similar in taste and appearance to porter, but is said to have none of its deleterious qualities. It is a weak, bitter, but not unpleasant beer, containing an abundance of hops. It derives its peculiar value and flavor from storage in vaults, as the word "lager" sufficiently implies. The longer it is stored, the finer becomes its quality. The vicinity of Fairmount has lately become the fountain-head of this description of manufacture, and it is consequently a favorite resort for Germans, who, ranged around their little square tables, with cigars, pipes, newspapers and glass-mugs of *lager*,

> Mingle o'er their friendly bowl,
> The feast of reason and the flow of soul.

As we pass Fairmount, the river Schuylkill, with its green banks, soft verdure, and rich foliage, is brought into view—the rail-road, until it crosses the bridge, diverging along its eastern bank. Here a number of spacious warehouses will be noticed, built directly along the water's edge, and affording access for carts by means of scaffolding erected to their upper stories. These are *ice-houses*. They are built with much care, the walls filled in with tan-bark to exclude the air, and capable of storing an immense quantity of ice. The manner of collecting and storing the ice is very simple, and is fully illustrated in the accompanying engraving.

Of late years, the storage and shipment of ice has become a very considerable item of trade. A large quantity is required for the consumption of the city, but in addition to this, no inconsiderable amount is shipped to the South, as well as to foreign countries where the climate forbids its production. Vessels freighted with ice always obtain a return load, and thereby a judicious exchange of local commodities is effected with points where, under other circumstances, our trade would probably be less extensive, and our communication

CUTTING ICE ON THE SCHUYLKILL.

less frequent. Our eastern neighbors, always the first in the market with their "notions," have now a serious competitor in Philadelphia in this branch of commerce.

In seasons of scarcity, ice is brought down the Schuylkill, in the spring, from the mountain regions of Schuylkill county, where, the climate being somewhat colder, and the streams less impregnated with sediment, it attains a good thickness, as well as a pure and transparent quality. On the Schuylkill, it usually attains a thickness of from four to twelve or more inches, and is probably unrivalled for the purity of its mineral composition, and freedom from foreign and deleterious substances. Its color varies from snowy opaqueness to translucency, and sometimes to the most beautiful watery transparency.

As we pass over the splendid rail-road bridge, a very interesting object presents itself. A beautiful little island, overgrown with tall and slender trees, nestles in the midst of the river, and immediately in front of it, on the western shore, is situated an unique cottage, built of stone, and apparently of some antiquity. In front of the cottage are two old trees, wrinkled and gnarled, like the furrows in an old man's face. This cottage is now a rough and dilapidated affair, but it was once the temporary residence of the late Thomas Moore, the celebrated Irish poet. It bears the rather unpoetic name of "Pig's Eye," but to many is known as Tom Moore's cottage. We entered the house while our friend Brightly was sketching it, and found it indeed a relic of the past. The ceilings, which have never been plastered, reveal the rough joists, now blackened with smoke and greasy rust, while the occupants complained of the condition of the roof, which leaks badly. The cottage appeared otherwise warm and comfortable, as answering the humble pretensions of the lessee. But we thought there was some reason for his complaint against the landlord, who failed to put it in tenantable order, after receiving two months' rent in advance. If properly fitted up, it might still serve as the abode of the muse,—but, alas! it can never again return to the scenes of its former glory. It is about forty-five years since Mr. Moore visited this country; and the changes which have been made during this time, are probably no less striking elsewhere than on this spot. At that time, this little cot was surrounded with a comparative wilderness, the abode of merry warblers and of wild-flowers;—the Schuylkill yet flowed in undisturbed tranquillity, and its peaceful shores were fragrant with the rich profusion of its foliage. It was a spot well calculated to tempt the poet from the noisy scenes of the town, and no less calculated to lend inspiration to the harp which has given such celebrity to his melodies.

"I went to America," (says the poet, after his return to Europe,) "with prepossessions by no means unfavorable, and indeed rather indulged in many of those illusive ideas with respect to the purity of the government and the primitive happiness of the people, which I had early imbibed in my native country, where, unfortunately, discontent at home enhances every distant temptation, and the western world has long been looked to as a retreat for real or imaginary oppression; as, in short, the Elysian Atlantis, where persecuted patriots might find their visions realized, and be welcomed by kindred spirits to liberty and repose. In all these flattering expectations I found myself completely disappointed, and felt inclined to say to America, as Horace says to his mistress, "*intentata nites.*" Brissot, in the preface to his travels, observes that 'freedom in that country is carried to so high a degree as to border upon a state of nature;' and there certainly is a close approximation to savage life, not only in the liberty which they enjoy, but in the violence of party spirit, and of private animosity which results from it. . . . . . The rude familiarity of the lower orders, and indeed the unpolished state of society in general, would neither surprise nor disgust if they seemed to flow from that simplicity of character, that honest ignorance of the gloss of refinement, which may be looked for in a new and inexperienced people. But, when we find them arrived at maturity in most of the vices, and all the pride of civilization, while they are still so far removed from its higher and

TOM MOORE'S COTTAGE.

better characteristics, it is impossible not to feel that this youthful decay, this crude anticipation of the natural period of corruption, must repress every sanguine-hope of the future energy and greatness of America."

During his brief sojourn on the green banks of the Schuylkill, the poet produced several choice effusions; but it is to be regretted that these gems are associated with so much that, for his own high reputation, had better been "left unsung." In his poem addressed to the Hon. W. R. Spencer, he speaks thus disparagingly of us:

All that creation's varying mass assumes
Of grand or lovely, here aspires and blooms;
Bold rise the mountains, rich the gardens glow,
Bright lakes expand, and conquering rivers flow;
But mind, immortal mind, without whose ray
This world 's a wilderness and man but clay;
Mind, mind alone in barren, still repose,
Nor blooms, nor rises, nor expands, nor flows.
Take Christians, Mohawks, Democrats, and all—
From the rude wigwam to the Congress hall—
From man the savage—whether slaved or free,—
To man the civilized, less tame than he,—
'Tis one dull chaos, one unfertile strife
Betwixt half-polished, and half-barbarous life;
Where every ill the ancient world could brew
Is mixed with every grossness of the new,—
Where all corrupts, though little can entice,
And naught is known of luxury but its vice.

In his sweeping denunciations of the American character, he spares only the "sacred few" whom he met in Philadelphia:

Yet, yet forgive me, oh ye sacred few,
Whom late by Delaware's green banks I knew;
Whom, known and loved through many a social eve,
'Twas bliss to live with, and 'twas pain to leave.

* * * * * *

Believe me, Spencer, while I winged the hours
Where Schuylkill winds his way through banks of flowers,
Though few the days, the happy evenings few,
So warm with heart, so rich with mind they flew,
That my charmed soul forgot its wish to roam,
And rested there, as in a dream of home.

The following lines purport to have been written on leaving Philadelphia:

Alone by the Schuylkill a wanderer roved,
And bright were its flowery banks to his eye;
But far, very far were the friends that he loved,
And he gazed on its flowery banks with a sigh.
Oh Nature, though blessed and bright are thy rays,
O 'er the brow of creation enchantingly thrown,
Yet faint are they all to the lustre that plays
In a smile from the heart that is fondly our own.
Nor long did the soul of the stranger remain
Unblessed by the smile he had languished to meet;
Though scarce did he hope it would soothe him again,
Till the threshold of home had been pressed by his feet.
But the lays of his boyhood had stolen to their ear,
And they loved what they knew of so humble a name;
And they told him, with flattery welcome and dear,
That they found in his heart something better than fame.
Nor did woman—oh woman! whose form and whose soul
Are the spell and the light of each path we pursue;
Whether sunned in the tropics or chilled at the pole,
If woman be there, there is happiness too;—
Nor did she her enamoring magic deny;—
That magic his heart had relinquished so long,—
Like eyes he had loved was *her* eloquent eye,
Like them did it soften and weep at his song.
Oh, blessed be the tear, and in memory oft,
May its sparkle be shed o 'er the wanderer's dream;
Thrice blessed be that eye, and may passion as soft,
As free from a pang, ever mellow its beam!
The stranger is gone—but he will not forget,
When at home he shall talk of the toils he has known,
To tell, with a sigh, what endearments he met,
As he strayed by the banks of the Schuylkill alone!

It was also during his lonely rambles on the banks of the Schuylkill that the following beautiful ballad stanzas were written—most probably while contemplating some neighboring cottage:

I knew by the smoke, that so gracefully curled
Above the green elms, that a cottage was near;
And I said, "If there 's peace to be found in the world,
A heart that was humble might hope for it here!"

THE SCHUYLKILL FROM LAUREL HILL LANDING.

It was noon, and on flowers that languished around
  In silence reposed the voluptuous bee;
Every leaf was at rest, and I heard not a sound
  But the woodpecker tapping the hollow beech-tree.
And "Here, in this lone little wood," I exclaimed,
  "With a maid who was lovely to soul and to eye,
Who would blush when I praised her, and weep if I blamed,
  How blest could I live, and how calm could I die!"
By the shade of yon sumach, whose red berry dips
  In the gush of the fountain, how sweet to recline,
And to know that I sighed upon innocent lips,
  Which had never been sighed on by any but mine!

Whatever may be thought of the justness of Mr. Moore's estimate of our country forty-five years ago, it hardly needs comment now. The poet, then young and inexperienced, lived long enough to form different and more correct opinions. It is but a few months since he died, after lingering, for a considerable time, in a melancholy and imbecile state of mind. Whatever his sentiments may have been, subsequently to his visit to this country, as to the state of American civilization, literature, and the arts, is now perfectly immaterial;—for, as a nation and a people, we have lived long enough to *learn a little*, and have not been without opportunities of illustrating our progress. We have paid our respects to old England in various ways, and at sundry times;—and there can be no doubt but that she *knows us*. Whatever our progress *is*, she finds it no child's play to keep up with us, whether on *land* or *sea*. As for poor Ireland—she, too, has heard from us, and whether we be "savages," "democrats," or "poets," she probably has a correct idea of the extent of our *productive resources*, if not of our *benevolence*. The spirit that can prompt generous feelings in one case, can supply it in all cases. No matter what the bard thought of *us*, *we* had a good opinion of him; and the day will never dawn when American hearts will cease to beat to his happy strains.

After leaving the cottage, we pass on to the Falls of Schuylkill, some six miles from the city. On our right, on the other or eastern side of the river, is Laurel Hill Cemetery, one of the most lovely and inviting spots of the kind in this country. So popular has this necropolis of the dead become, that the company has been obliged to increase its area, and several adjacent tracts of land have accordingly been added to it. Many strangers own lots in this beautiful Cemetery, and some of its handsomest tombs and monuments have been erected over the remains of eminent men who served their country in important public capacities. We give an accurate view of a portion of the grounds from an original sketch just taken, as also a view of the grand entrance.

ENTRANCE TO LAUREL HILL.

The Falls of the Schuylkill were so called, because, in former times, before the erection of the Fairmount dam, they were quite perceptible, but have since entirely disappeared.

The Reading Railroad makes a fork at this point—one branch crossing the river by a splendid bridge, and extending to Port Richmond on the Delaware; the other branch extending to the city, over which we have just passed. The road to Port Richmond is about five miles in length, and it is at this place that the great bulk of the coal brought down by the Reading Railroad is shipped. The facilities for this purpose are of the most extensive and admirable character. The wharves are extended a considerable distance into the river, over

SCENE IN LAUREL HILL.

PORT RICHMOND.

which the railroad is prolonged into numerous lateral branches, supported on strong tressel-works. The loaded cars are hauled to the water's edge, where large apartments are erected for the storage of the coal. These apartments lie under the tressel-works, the bottoms of which descend, with a slight inclination, over the water's edge. The contents of the cars are discharged from the bottom, (being constructed expressly for that purpose,) and the coal falls directly into the proper apartments below, assigned for the different sizes and qualities. A vessel, therefore, to be loaded, has merely to be drawn up to the wharf, under the projecting spout of the coal apartments, when a wicket is raised, and the coal issues out in one continuous stream. The operation of unloading the car, and of loading a vessel, is consequently very simple; yet the contrivance, in its original conception, is one of great practical merit, saving annually, as it does, a large amount of money and time. The engraving illustrates the process just described, at the same time that it conveys an idea of the extent of the business of shipping the coal at Port Richmond. The Reading Railroad, after many years of hard struggling, has laid down a foundation for future success as broad, and practical, and comprehensive, as it was possible for human industry and ingenuity to devise. The earnings of the company, amidst all its former embarrassments, were, in a great measure, necessary to its complete equipment. To make it *productive*, accommodations corresponding with the stupendous trade of the road had to be provided; and this, too, in the midst of its darkest and most trying history. But the improvements are now made and completed, and stand forth as shining monuments to the energy and well-directed management of the road.

On our return to the Schuylkill, we shall diverge into the city, and "see what is to be seen" on the Philadelphia and Norristown Railroad, which, on the opposite shore of the river, runs parallel with the Reading Railroad from the Falls to Norristown, and embraces nearly every object of interest between those two places. The first object that strikes us, in connection with this road, is a new, elegant, and imposing one, viz.: the depot situated at the corner of Ninth and Green streets. This handsome edifice has just been completed, at a cost of some $10,000. It is, in many points of view, a model of architectural skill—combining the practical with the ornamental, at the lowest possible cost. The business of this road, extending from

PHILADELPHIA AND NORRISTOWN RAILROAD DEPOT.

Philadelphia to Norristown, with a branch to Germantown, is rapidly increasing, and has been the instrument of scattering along the route it traverses an active, intelligent, and enterprising population. The trade, of course, is mainly local, including the conveyance of passengers. Many of the business men of Philadelphia have summer residences in the vicinity of the road, while others permanently reside in the country. These, added to the ordinary movements of the dense population along the route, make the conveyance of passengers an important item, which must annually increase with the progressive increase of business. The road, a short distance from the city, passes over the Port Richmond branch of the Reading Railroad, and soon after appears at the point from which we diverged, viz: the Falls of Schuylkill, a view of which is annexed. The extensive buildings lying at the western end of the village, between the railroad and the river, comprise the chemical works of Messrs. Weightman, Harrison & Co. The greater portion of the population is supported by these large and splendid works, the proprietors of which have an establishment, equally extensive, in the city. Philadelphia is justly distinguished for its chemical productions, and the firm above mentioned probably stands at the head of this description of manufacture—one of the most complicated and arduous, we may add, that human industry and capital could embark in.

FALLS OF SCHUYLKILL CHEMICAL WORKS.

The vicinity of the Falls is much frequented, in the summer time, by the citizens of the town. They ride out here to obtain an airing. The romantic and picturesque Wissahickon empties into the Schuylkill a short distance above the village, and this is the principal source of attraction. Its banks are bold and rocky, overgrown with stately trees, whose shade affords a cool retreat from the heat, and dust, and parched and sultry avenues of the city. There are several hotels, or places of refreshment, both in the village and on the Wissahickon, and there is no lack of *material* to gratify or amuse the visitor. The drive from the city is very refreshing—the road being remarkably smooth, and studded all along with handsome cottages and tasteful scenery, as well as objects of historical and general interest. It is

THE HIGH BRIDGE ACROSS THE WISSAHICKON.

customary to enjoy the ride late in the afternoon, before dusk, while many drive out to partake of the celebrated "catfish and coffee," and return by "the light of the moon." Riding by horse-back, both for ladies and gentlemen, is in these days one of the requisites of a polite education;—and the taste for the exercise is indulged to the fullest extent—though there is still a corresponding number of vehicles, some of them splendid equipages, to be met on the road. Pic-nic parties are very frequent in this quarter, and the arrangements of the Norristown Railroad are no less complete for their accommodation than the attractions of the grounds.

The engraving on page 41 exhibits a view of the Norristown Railroad bridge across the Wissahickon, near its junction with the Schuylkill. The bridge is a fine specimen of architecture, and viewed in connection with the adjacent scenery, is probably one of the most picturesque scenes to be found in this quarter of the State. It stands seventy-five feet above the level of the water, and is about three hundred feet in length. The entire route of this rail-road, from Philadelphia to Norristown, is full of beautiful and varied scenery, nearly every inch of which is identified, in some way or other, with historical associations more or less interesting. From Fairmount to Manayunk, there is a succession of smiling villas, handsome grounds, and unique cottages,—while the hum and rattle of the loom and the shuttle, the clinking of the hammer, the grit of the saw-mill, the steam and blaze of the numerous iron works and manufactories, no less than the general life and bustle of the way-side, keep the visitor continually on the *qui vive* of excitement.

About one mile beyond the Wissahickon is MANAYUNK, situated on the east side of the river. It is the seat of very extensive and varied manufactures — embracing cotton and woollen factories, flour and paper mills, furnaces, machine shops, &c. The town owes its origin and onward progress entirely to the facilities afforded by the Railroad, and the Schuylkill canal, which passes directly through the principal street, and supplies the water-power for all its manufacturing establishments. Some of these establishments are among the oldest in the United States, having been commenced in 1819, upon the completion of the canal, and when the present site of the place was overgrown with trees and wild bushes. At that time, Manayunk, with some other points lower down, was an excellent spot for shad-fishing;—but since the erection of the dams in the Schuylkill, this splendid fish has ceased its periodical visits to these waters, and the business, once very considerable, is now entirely discontinued.

The present population of Manayunk is probably about seven thousand—almost every person, of both sexes, being engaged in the industrial interest of the place. It was our desire to have presented a view of the interior of a cotton factory, and for that purpose, in company with our artist, we waited on one of the principal factors in that place. Without deigning to see us, he refused to give us admission, and the refusal was couched in terms so sharp and contemptu-

MANAYUNK.

ous, that it naturally suggested a few thoughts as to the moral and social tendencies of the factory system in our country. We never before realized to the extent we did on this occasion, the haughty and austere manner, the cold, biting dignity, which a commanding position over two or three hundred poor operatives, is calculated to impress on some individuals. While we are free to admit the advantages which these establishments are capable of conferring, it is not to be disguised that, in the hands of some men, they may be converted into engines of great social oppression. The spirit of enterprise which induces our citizens to make large investments in the industrial pursuits, cannot be too highly admired and extolled; but the motives which subsequently turn some of them into uncouth and selfish nabobs, are altogether unworthy the character of a gentleman or a republican.

The most prominent evils attending the factory system in this country, are the natural results of *capital*, combined with a *speculative purpose*, to which the factory is made subservient and subordinate. Thus, an individual with a capital of one hundred thousand dollars, purchases a favorable location for a factory—no matter what kind—which is to employ from one to three hundred operatives. The location we will suppose to comprise one hundred acres of land, for which five thousand dollars are paid cash, and the remainder upon a credit of twelve months. The factory buildings are now commenced, which will absorb fifteen thousand dollars—one-half to be paid cash, the remainder upon credit. In the meantime, the tract of one hundred acres is laid off into town lots, and twenty or thirty tenements erected, at a cost of four thousand dollars—one-half cash. Thus far about $28,000 have been expended, of which one-half is credit. From ten to thirty thousand dollars are yet required to equip the factory with the necessary machinery,—say $20,000, and we have a total expenditure of upwards of $50,000. There now remains a capital of $50,000 additional to purchase stock, and as a fund to carry on the mill, which is set in *operation at high wages* for the operatives, and under favourable auspices to attract mechanics, labourers and tradesmen to the village. A brisk demand for houses and lots ensues, and the greater the demand the greater becomes the value—*ergo,* in a short time probably more than one hundred thousand dollars will have been realized from the sales and rents of houses and lots, and thus, without reference to the immediate productions of the factory, a handsome fortune has been realized from a comparatively small investment. As soon as this manœuvre is terminated, and when the village is filled to repletion, another card is to be played. The factory cannot be conducted with profit under high rates of wages—the prices of labor must be reduced, or the works suspend operation. Here ensues a panic—a general stagnation of all the affairs of the village. Many will sell out their houses and lots at a sacrifice, and move away; others will seek new employments, while, in the meantime, some will work on at *reduced wages.* The speculator now again makes his appearance, and it is not long before a large number of lots are in his possession, and thus, probably to a less extent than before, the same speculation is acted over.

We would not create the inference that our factory system generally is under the influence of such speculative movements; but we mention this as one of the evils which have hitherto surrounded it, and which have, in a great measure, crippled its operations, and raised an amount of political opposition which could not, under other circumstances, exist. But, independent of this, there are other objections, arising from the centripetal agency of such establishments in attracting around them the necessary operatives, always constituting a population more or less numerous. As this population is solely dependent upon the "lord of the loom," it is liable to be thrown into idleness at his whim or caprice, and thus a general depreciation of their condition and prospects ensues, while frequently industrious and virtuous families are exposed to want and suffering. Labor is not only degraded by such means, but it is robbed of its just reward, and, as a natural consequence, children of both sexes are driven, by the necessities of their condition, into the factory, where they earn a bare subsistence at the same time that they secure premature graves. It is impossible to contemplate the condition of these operatives without arriving at the conclusion that there is something morally wrong in the system, as well as socially inconsistent with the spirit of our institutions.

ANTHRACITE FURNACE ABOVE MANAYUNK.

A short distance above Manayunk is one of the most extensive anthracite furnaces in Pennsylvania. It is situated on the east bank of the Schuylkill, and presents a spirited scene when viewed from the Reading Railroad, on the opposite side of the river, especially in the evening, when the flames issue from the chimneys illuminating the whole establishment in their red glare. These works have been in operation for several years, notwithstanding the recent depression of the iron market, which prostrated a large number of furnaces in the valley of the Schuylkill. As we shall elsewhere make some remarks in reference to the manufacture of iron, we must forbear touching upon that point in this connection.

With the exception of the natural scenery along the Schuylkill, there is little of interest or importance until we reach the great limestone region which traverses a large area of the south-eastern counties of Pennsylvania. This limestone formation is a continuation of the great valley of Chester county, and constitutes, by far, its most important feature. It occupies a position in the stratified primary group, and teems with narrow belts and valleys, adapting it for successful culture as well as for excavation. The largest beds of limestone are met above Spring Mill, and alternate in subordinate ridges and valleys of denudation for some distance above Norristown, where the hills of the Mine Ridge, somewhat flattened down, rise through and cut off the basin. The limestone is not uniform in quality, but the lime produced from it is, upon the whole, highly esteemed, and probably the best argument in its favor is the immense quantity annually produced and shipped for the supply of Philadelphia and

LIME KILNS NEAR SPRING MILL.

other points more distant. At various points along the Schuylkill, especially near Spring Mill, Conshehocken, and Port Kennedy, there are very extensive quarries, where kilns have been erected for burning the stone—the canal and railroads, on both side of the river, affording excellent facilities for transporting the lime to market, as well as for supplying the kilns with fuel. The lime kilns are large and substantial, but erected without much regard to ornament. They are generally placed on the slope of a hill, so as to allow the limestone to be hauled to and thrown in at the top. The stones, as placed in the kiln, form an arch over the hearth, with sufficient space between the alternate layers of stone and coal to permit the heat to penetrate and decompose them. The stones are thus thoroughly roasted, and in due time crumble into powder or small white particles, in which state the lime is in a marketable condition.

Spring Mill receives its name from a superb spring, which bubbles up with great force in the midst of a beautiful grove of trees. The water is as clear as crystal, through which the pebbles at the bottom, some thirty feet, can be distinctly seen. It is quite cold in the warmest weather.

Limestone is the general name applied to all massive varieties of carbonate of lime, that form beds of great extent, or mountains. Calcspar is carbonate of lime in its purest state. It is generally transparent or translucent, the faces of the crystals sometimes very brilliant, but the bases of the hexagonal prism are always opaque. Its color varies, sometimes perfectly colorless, often of a topaz or honey yellow, and sometimes grey or reddish. Exposed to the blowpipe, carbonate of lime does not swell nor fall to powder, but becomes white and caustic—it is then *quick-lime;* some varieties are phosphorescent when heated, and shine with a pale yellow light. *Satin-spar* consists of fine parallel fibres, either straight or waved, and has a silky lustre; it fills small veins in limestone rock, the fibres laying across the vein. There is a particular kind of limestone containing a large proportion of bitumen, which, when rubbed or scratched by any hard substance, or slightly heated, gives out a strong fetid odor. *Chalk* is carbonate of lime of an earthy texture. It forms the cliffs along the south-east coast of England, which acquired for that island the name of Albion. Chalk formations are not often met with in the United States, though it probably exists to some extent. *Rockmilk* resembles chalk, but it is much more tender. It is found in the clefts of mountains, where it is deposited by water containing calcareous particles. *Stalactites* are sometimes transparent, and have the crystalline structure of calcspar; sometimes they consist of parallel layers of different shades of color. This rock is often employed for vases and slabs, under the names of *alabaster* and *onyx marble. Stalactites* are constantly forming in nearly all rich limestone formations of a cavernous structure. In the district of Port Kennedy, a few years ago, an extensive cavern was reached, where the process of the accumulation of stalactitic matter was illustrated. They are produced from the drippings of minute particles of calcareous matter, from water which percolates through the roof or sides of the rocks. When a small quantity of moisture arrives at the inner surface of the roof, before a drop is formed sufficiently large to fall by its own weight, a portion of it evaporates, and a rim-shaped film of solid matter is left adhering to the rock. Every succeeding drop increases the thickness of this film, until at length a slender tube is formed, which is constantly increased in thickness as well as in length. In general, the interior is quickly filled up, and becomes perfectly solid; but sometimes the stalactites are hollow throughout a great part of their length. At Port Kennedy, where the process of formation had been interrupted while yet in its incipient stages, the stalactites coated the interior rocks with irregular thin fibres, in some cases forming conical arches, with borders of variegated color, and in others forming pyramids on the floor. The cavern was an object of considerable curiosity during the brief period it was open to visitors, and its numerous chambers presented an aggregate area fully equal to many of our largest public buildings. A concert was held in one of its largest *saloons*, on the fourth of July, 1846, at which several hundred persons were present.

The limestone formations of this State, at numerous points, afford several varieties of superior marble. The eastern portion of the state, drained by the Schuylkill, is particularly rich in this valuable mineral, and finds a cheap and easy outlet to market. Much of the marble used for building purposes, as well as for monuments and articles of furniture, is obtained within a range of from ten to twenty-five miles around Philadelphia. The extensive buildings comprising the Girard College, were in part constructed from marble obtained in this neighborhood. There are several productive quarries in Lancaster and other counties; but those of Chester and Montgomery are the most extensive and abundant. Some of these quarries are over one hundred and fifty feet in depth, and powerful levers are used to hoist the massive pieces from their beds. At Conshehocken there is an extensive marble mill, where the rough pieces are sawed into patterns to suit the demands of the market. It is worthy of remark, that the edges of our limestone basins usually afford a marble of conglomerated character, beautifully variegated in color, similar to a variety of the Potomac marble, or to that constituting the interior pillars of the House of Representatives at Washington. This marble is a sedimentary deposit, the various pebbles being cemented together by the calcareous matter of which it is composed. Though extremely hard, it is, in some places, susceptible of the finest polish, and the reflection of the atoms upon the polished surface, at first glance, gives the impression of *roughness*, which is only dispelled by rubbing your hand upon it. A fine deposit of this peculiar rock lies near Bainbridge, in Lancaster county; also near Reading, in Berks county, while it is elsewhere met with along the borders of our limestone beds, in the vicinity of slate and shale.

This extensive region of limestone, which occupies, in numerous distinct belts or basins, a large portion of the area of what is termed the Atlantic slope is also associated with several useful metals, as the ores of copper, iron, lead, chrome, &c. The region of copper is principally north of the Mine Ridge—(the first chain of elevation met with proceeding in a north-west course,) and outcrops at various points in Pennsylvania, between New Jersey and Maryland, which States it also penetrates. The most extensive deposit is probably in Adams county, where ample preparations for mining have recently been made, in the immediate vicinity of Gettysburg. Mining exploitations were also conducted, until lately, in the vicinity of Pottstown, but the ore was not found to be sufficiently productive to justify the continuation of the enterprize. More recently, operations have been commenced near Valley Forge, and the probability is, that they will prove successful. In various other points attention has been directed to this vast mineral formation, and the time is evidently not far distant, when the eastern portion of Pennsylvania will be as noted for its mines of copper and lead, as other parts of the State now are for their inexhaustible and extraordinary deposits of coal and iron.

*Native Copper.*—Like most of the native metals, it crystallizes in the octahedral system; but perfect crystals are seldom met with. It occurs sometimes in very large masses, but most frequently in branching and leaf-like forms, scattered among the veinstone, or penetrating it; and the surface of these ramifications is often thinly coated with green carbonate of copper, or tarnished with a brown color. In general it is very nearly pure copper, and has the color, hardness, and malleability of the refined metal, as we are accustomed to see it—sometimes it contains a minute proportion of silver.—(Varley's Mineralogy) Lake Superior is the most extensive region in the world for the production of native copper. In some parts of that region, the copper is penetrated by threads of pure silver, and grains of the same metal are scattered through it—a circumstance which has never been observed elsewhere. Its softness and ready solubility in every kind of acid, and in ammonia, distinguish copper from the few metallic minerals which at all resemble it. Copper is one of the metals that has been known and worked from the earliest period;—alloyed with tin, its hardness is much increased; and this alloy proved the various kinds of bronze of which armor, weapons, knives, and other tools were manufactured by the former inhabitants of both the old and the new continent. Axes and knives from the tombs of the ancient Peruvians and Mexicans, chisels found in the quarries of Egypt, and Roman and Gaulish swords of great antiquity, have been analyzed, and found to contain from 75 to 96 per cent. of copper alloyed with tin.—(*Ib.*) In the mines of Lake Superior tools and implements, and marks of previous workings, have been found, which can only be attributed to a race far anterior to the present era of the human family. The mines of New Jersey were worked by the first settlers long anterior to the Revolution.

*Ruby Copper.*—(*Red Oxyde of Copper.*)—This substance is of a fine crimson color, sometimes almost black, with vitreous lustre, ranging from semi-transparent to nearly opaque. It is brittle, and about as hard as fluorspar, with a specific gravity of 5·6. It is often intermixed with native copper, but seldom with the other ores of this metal. *Tile ore* is a variety which is intermixed with oxyde of iron and other impurities, and forms thin flattish masses, something like dark colored tiles.

*Black Oxyde of Copper*—which is more oxydated than the preceding species, occurs in the form of a fine black powder, or in small masses of an earthy texture, with some other copper ores. Both this and ruby copper are easily reduced on charcoal to metallic copper.

*Sulphurets of Copper.*—There are several combinations of sulphur with copper, some of which are valuable ores. *Copper Glance*—or vitreous copper ore—has a dark steel, gray color, and when freshly broken, a perfectly metallic lustre; but the exterior is often black and dull. It is most generally found in masses without any regular form, or filling small veins. This is the richest of all the sulphurets of copper, affording 75 per cent. of metal, and being in general very free from any other. It has been met with in some of the Cornish copper mines, but only in small quantity—but in the Ural mountains it is an object of extensive exploitation, occurring there in nodules of various sizes, disseminated in veins of clay and gravel.

*Variegated Copper Ore.*—This was long considered to be the same substance as copper pyrites, of which the surface was tarnished; but it differs from it in containing less iron and sulphur, affording about 60 per cent. of copper, while copper pyrites does not yield more than 33 or 34. It is softer than the latter, and the color much redder—and it is less easily fusible than copper glance.

*Copper Pyrites.*—(*Yellow Copper Ore*)—is the most abundant of the English copper mines. Its color is that of brass, and its lustre perfectly metallic and shining, particularly when fresh broken. It is easily scratched by a knife; differing, in this character, from iron pyrites, which is much harder. Groups of small crystals are often sprinkled over other substances, as quartz, calcspar, fluorspar, galena, and blende. When pure, copper pyrites consist of sulphur 35·87, copper 34·40, and iron 30·47. Copper pyrites form veins in granite, slate, and other rocks, sometimes filling them entirely, sometimes distributed in irregular masses varying in size, and occasionally weighing some hundreds of pounds.

*Gray Copper Ore.*—The composition of this ore varies exceedingly in different localities; but it still presents nearly the same appearance—a light gray metallic substance. It consists principally of sulphuret of copper, antimony, and iron, with arsenic, zinc, or silver, and sometimes with all these metals—the proportion of the latter, in some instances, amounts to seventeen per cent., when the ore is worked for the sake of the silver as well as the copper.

In the same region traversed by the limestone,—(or rather in the valley above the Mine Ridge—) lead is found outcropping at different points. The ores are of various kinds, and in some cases comprise *galena*, with a plentiful mixture of silver. Extensive mining operations have lately been commenced in the vicinity of Phœnixville, Valley Forge, and Perkiomen; while the veins of the basin have been traced, here and there, over portions of the counties of Bucks, Montgomery, Chester, and Lancaster. The ore is, for the most part, pyromorphite—(phosphate of lead) a beautiful mineral, but not very abundant at any one district, though it appears to be plenty here. Its color rises from bright grass-green to yellow, orange, brown, and sometimes a dull violet. Most specimens contain nine or ten per cent. of the chlouret of lead—sometimes arseniate, and those of an orange color, chromate of lead. The chromates are found in great abundance in the Mine Ridge, in Lancaster county, of which we shall hereafter take occasion to speak. The ores are found in other spots in various stages of combination with other substances. In Bucks county, plumbago has long been mined to a considerable extent, and there is every reason to suppose that this interesting mineral, like the others, occupies a large portion of the peculiar formation in which it is found.

*Native Lead* is of rare occurrence. It has been observed in small grains and laminæ in some volcanic products, and, also, in some specimens of galena. It may be distinguished by its softness and sectility from the galena, which is hard and brittle. Minium (red oxyde of lead) is a red substance, occuring in the form of powder in some veins of galena and calamine. It is the same substance as the red lead used in painting; but for this and other purposes it is prepared artificially.

*Yellow Oxyde* of lead, which is less oxydized than the preceding ore, occurs sometimes as a powder, sometimes in solid masses, not unlike yellow jasper; its great weight distinguishes it from all other minerals which otherwise resemble it.

*Arseniate of Lead* takes the same forms as the phosphate of lead, but the prisms are often swelled in the middle, so as to have something of a *bowl shape;* color, generally pale yellow brown, and lustre often silky.

*Carbonate of Lead*—(White Lead ore.)—This is the same substance as the white lead of commerce, which is prepared artificially. It is abundant in some lead mines, filling large veins or intermixed with the other ores, either compact, earthy, or crystalline. Its crystals are derived from a light rhombic prism, differing very little from that of arthente, and are often grouped so as to form a cross or star. It is the only mineral which equals the diamond in brilliancy; its lustre, when pure and transparent, being adamantine. When fused by the blow-pipe on charcoal, a bead of lead is obtained; or if dissolved in nitric acid, upon immersing a piece of zinc (the surface being quite clean) metallic lead will quickly be precipitated on it in brilliant laminæ. When massive it has sometimes a crystalline structure, splitting readily into large flakes, with a brilliant surface. An earthy variety resembling chalk in its appearance is common in some lead mines. But it is a mineral met with not only in lead mines: it is frequently found with copper pyrites and blende. In these situations it often forms columnar masses, with a silky lustre. The sulpho-carbonate of lead, which is this substance combined with about one-third of its weight of sulphate of lead, resembles it in brilliancy and weight, but when dissolved in nitric acid, it leaves a residue, which is sulphate of lead. *Molybdate* of *Lead* crystallizes in square octahedrons, or very short prisms of four or six sides, of a dull yellow color, and translucent, which contain sixty-four per cent. of oxyde of lead. Sulphate of lead, in its general appearance, resembles the carbonate of lead, but is rather softer and less brilliant, and may always be distinguished from it by not effervescing with any acid. *Copreous* sulphate of lead has been found in a few places,—it is colored by the copper it contains.

*Galena*—(Sulphuret of Lead,) may generally be recognised by its crystallizations, and its very perfect cleavages, parallel to the faces of a cube, as well as by its pure lead-gray color. The surface is often dull, but the fracture always brilliant, and it is so brittle that cleavages may be obtained by a very slight blow. The same crystalline structure prevails where the galena is massive, sometimes resembling that of statuary marble; more rarely it is granular, or compact, with scarcely any lustre. Its specific gravity is 7·5 to 7·7. It is easily fused, giving off sulphureous fumes, and affording a globule of lead. Some galena is combined with sulphuret of silver. When this is in sufficient quantity to render it worth the expense of separating the silver, it is called *argentiferous* galena. In order to know whether galena contains any silver, dissolve a little in nitric acid, and dip into the solution a piece of copper; the silver, if there be any, will be deposited as a white metallic film on the copper.

*Bournonite* is a sulphuret of lead, copper and antimony—the former amounting to forty per cent.

*Graphite* (Plumbago, or black lead.) The substance called black lead is well known to most persons; but few, perhaps, are aware, that when they make use of a black lead pencil, they draw with a substance which is nearly identical with the *diamond.* It appears to be carbon, differing from it, when pure, only in the state of aggregation of its particles; it often contains a small portion of earthy matter, or of iron; but as the latter amounts sometimes to only one or two per cent. both are now generally considered as accidental impurities. Graphite has always a metalic lustre, with sometimes a fine scaly structure, and soils the fingers when handled. Specific gravity 2 to 2·45. It occurs in several places in the United States, and no doubt a large quantity is deposited in the slaty formation traversing Pennsylvania. For the best pencils, it is used without any other preparation than that of sawing it into thin strips, which are afterwards enclosed in wood, or formed into thin rods for-ever-pointed pencils; but great numbers of pencils of inferior quality are made of a mixture composed of black lead dust, intermixed with clay, and sometimes with lamp black, or sulphuret of antimony.

Near the village of Conshehocken, the Schuylkill assumes a most beautiful appearance; the banks, on both sides, are lined with stately trees, and foliage bending to the water's edge, while the stream is as smooth and clear as one broad sheet of glass. On the

LANDSCAPE ON THE SCHUYLKILL.

one side we have the Norristown Railroad and the Schuylkill Canal, and on the other the Reading Railroad, over which are seen passing an almost endless procession of black coal-trains, and as they wind around the projecting knolls, and intervening valleys, a great rumbling noise is heard, amidst the shrill whistle of the locomotive;

5*

the canal-boats lazily creep along, while around, as far as the eye can see, glorious nature spreads out in rich waving harvest-fields, and rolling elevations, with here and there a cluster of houses nestled amid the luxuriant scene.

The village of Conshehocken, though of recent origin, is quite an interesting and important place. It is supported solely by several manufacturing establishments, which are carried on in its midst, and which have sprung up under the facilities afforded by the lines of improvement, no less than the rich and varied resources adjacent. Of these, there is an extensive rolling-mill for the production of sheet-iron, and its manufactures are scarcely excelled by any similar article imported, having a close resemblance to the celebrated sheet-iron of Russia. There is a very large marble-mill, which has already been alluded to, with several workshops of different kinds, and smaller extent; while a company of enterprising men has just been organized to embark in the manufacture of a new description of useful ware. They intend to manufacture, principally from the lava of the furnaces in the vicinity, a description of glass for mantels, tables, and various ornamental purposes; and as the productions can probably be afforded at low rates compared with marble, and will prove equally durable, if not equally beautiful, there is a reasonable probability that this will ultimately form a distinguishing feature of the place, if it does not originate an entirely new branch of trade.

Between Conshehocken and Norristown, there are several extensive anthracite furnaces, as well as manufactories of various kinds. The banks of the river, on both sides, present many beautiful residences, and elegant, well-cultivated farms, while the scenery generally is of that soft and genial character, relieved occasionally by a rocky declivity or gently sloping hill, that pleases the eye of the observer, while it impresses him with the spirit of activity everywhere displayed.

Norristown, the county-seat of Montgomery, seventeen miles distant from Philadelphia, is one of the most beautiful boroughs in the United States. It lies on the east bank of the Schuylkill, rising to a gradual elevation from the water's edge. The streets are well drained, the houses substantially built, (many of them unusually elegant) the citizens remarkably intelligent, the society excellent, the location healthy, the general aspect that of an industrious and enterprising population, and, in short, the whole minutiæ of the borough is such as to render it one of the most attractive with which we are ac-

NORRISTOWN.

quainted. It contains several very extensive cotton and woollen factories, iron foundries, rolling-mills, and machine shops, with numerous other industrial establishments, of more or less extent, nearly all of which are located on the river bank, and are supplied, like Manayunk, with a splendid water-power from the canal. The present population is probably between eight and ten thousand, and must greatly increase in the future under ordinary circumstances of favor. A brighter day than we have known for the last five years is, we think, about to dawn upon our workshops. We hope so, at any rate.

Norristown was formerly included in the township of Norriton, comprising a manor belonging to William Penn. The land on which the town now stands was subsequently owned by several parties, when it finally came into the hands of Wm. M. Smith, who, in 1784, laid it out into town lots. During the revolution it was occupied as a farm, and belonged to a certain John Bull, who, notwithstanding his name, was a thorough-going Whig, and the British, on their way to Philadelphia, paid him the passing compliment of burning down his barn. A short distance below the town, on the banks of the river, are still to be seen the traces of former entrenchments or breast-works, thrown up by Gen. Du Pontel, by order of Washington, at a time when he expected the British to cross the river at this place.

The first canal improvement undertaken in the United States, was commenced at Norristown, about the year 1792, of which the outline features are still to be traced. The project contemplated both a navigable water course, and a water works for Philadelphia. For this purpose, the canal was to be carried to Philadelphia on *one level*, without locks or outlets. After completing several miles of the heaviest part of the work, and spending over $400,000, the company became embarrassed, and were compelled to abandon the enterprize, many of the principal parties having themselves been involved in commercial and financial ruin. The company, however, was afterwards consolidated into the Union Canal and the Schuylkill Navigation, of both of which we shall presently have something to say.

Montgomery is one of the richest and most favorably situated counties in Pennsylvania. In all the elements of real wealth and true prosperity, it is unrivalled. The soil consists principally of limestone and red shale—the latter performing well under good tillage, with the addition of the lime close at hand. The soil is traversed by several fine streams, of which the Schuylkill, comprising the greater portion of its southwestern boundary line, is the principal; the others next in importance being the Perkiomen and its branches, the beautiful and romantic Wissahickon and its branches, the Pennepack, Tacony, Neshaminy, and others—all of which afford excellent water-power. There are at this time not less than thirty merchant, one hundred and twenty grist, seventy-six lumber, eight marble, twenty paper, thirty-five oil, twelve clover, and about the same number of powder mills, in Montgomery county; besides which, there are fifteen or more iron-works of various kinds, twenty-five large cotton factories, ten woollen factories, twelve fulling mills, and some thirty-five tanneries. This, for a county comparatively small in area, exhibits a productive capacity not easily surpassed.

The county is supplied with some of the best turnpike roads and stone bridges to be found anywhere in Pennsylvania. One of these bridges (that over the Perkiomen) cost over $60,000, and was built nearly fifty years ago. The people seem early to have realized the importance of good roads, and an immense amount of money was freely expended to make them of the best and most durable character. This principle should always be acted upon by the constituted authorities, for nothing contributes more to the lasting benefit, or enhances property to a greater extent, than good roads and bridges.

The first settlers of Montgomery county were principally Welsh, with a few Germans and English. The Germans gradually dispersed to the northern part of the county, where the language is still spoken to some extent. The Welsh, or their descendants, have long since abandoned their native language, in favor of the predominating English. The early settlers comprised some of the best men who ever landed on our shores—remarkable for their morality, industry, intelligence, and uniform respectability of deportment. The oldest place of worship now standing in Pennsylvania, was erected by them near the present town of Manayunk, in 1695. It was a Friends' meeting-house, and is still occupied by that respectable society.

We have thus, somewhat briefly, given an exposition of some of the leading features of this interesting and remarkable valley, from the Falls of the Schuylkill to Norristown. Both sides of the river, we have previously remarked, are traversed by railways, running parallel with the Schuylkill navigation. The extensive manufacturing establishments, receiving their driving-power from the river, are all located on its eastern side, which will account for its dense population and busy aspect, as compared with the opposite shore, traversed by the Reading railroad. It is this fortuitous circumstance which creates the sole trade of the Norristown railroad, at the same time that it materially aids the Reading railroad;—for while the one is in the exclusive enjoyment of the local miscellaneous trade, the other has an abundance of tonnage in supplying a large portion of the fuel consumed. We should judge that at least fifty thousand tons of coal are annually transported by the Reading railroad for the supply of the lime-kilns and ordinary consumers, between Norristown and the Falls. The two railroads, therefore, (as well as the canal) are auxiliary to each other's success.

For the reason referred to, we have little of interest to note, between the points designated, on the western side of the river. The scenery, in general, is attractive; but owing to the splendid works of man, it has more of an artificial stamp, than the usually broader and more impressive one of Nature. The tunnel above Manayunk—the Schuylkill Navigation, with its numerous locks, and dams, and bridges—the towns and villages, with their busy work-shops and towering factories—the numerous lime-kilns, furnaces, and mills—the turnpike roads, with their ponderous teams and carts; the railroads, with their snake-like trains; the electric telegraph, with its lofty poles stretching out, single file, in magnificent procession; the canal-boats, with their faithful, tugging horses, and sun-burnt crews—all evince the restless activity of man, and proclaim his glory to the passing observer.

Leaving Norristown, we cross the Schuylkill by a splendid wooden bridge (indicated in the engraving), eight hundred feet in length, and again join the Reading railroad, which henceforth traverses every town and village on our route. Here, indeed, is a sprightly little village before us, just embarking in the world. It is but yesterday since "it set up," and already we find it a considerable town, under the name, style and title of Bridgeport. The Reading Railroad is the guardian and patron of the little fellow, and under its friendly auspices it will grow and thrive, until it becomes a good-sized, old-fashioned borough.

After leaving Bridgeport, four miles beyond, we reach the village of Port Kennedy, of which we have already spoken in connection with the production of lime, constituting its sole trade. Two miles beyond this place is Valley Forge. Here every inch of ground is sacred to the cause of liberty and patriotic suffering. There is not a heart in America—there is not a lover of liberal institutions anywhere, that will not swell with mingled awe and admiration, as he contemplates the scenes and incidents with which this region is identified. Here was concentrated, in the darkest hour of the revolution, the sole reliance of freedom against oppression; here were centered our hopes and our fears—here were quartered, amid the snows and blasts of a severe winter, without clothing, and almost without food, sick, famished, barefooted, and dying, Washington and his army.

REDOUBT AT VALLEY FORGE.

Valley Forge derives its name from a forge which stood near the mouth of valley creek, some time previous to the revolution. The grounds occupied by the largest portion of the encampment comprised both sides of the hill, south-east of the stream. The name of this hill is Mount Pleasant, and of that on the other side of the stream, Mount Misery. These terms were bestowed by William

Penn, who, on one occasion, lost his way on the latter hill, and having regained it on the former, bestowed the names accordingly. Several extensive redoubts and breast-works were thrown up at sundry places, some of which, on the south-eastern side of the hill, are yet distinctly visible, and of which the engraving on page 58 conveys a correct idea. These works consist of large embankments of earth, arranged one after the other, along the slope of the hill, so that, in case of attack, the men could remain behind them, secure from the fire of the enemy. These breast-works, moreover, were surrounded with deep ditches, thus rendering the approach of the enemy hazardous amidst the fire of the soldiers within the redoubts. The redoubts now lie in the depths of the forest, but their outlines, as well as the former sites of the miserable huts of the soldiers, are still distinctly visible. The head-quarters of General Washington were in a

WASHINGTON'S QUARTERS AT VALLEY FORGE.

small stone house, which stands near the railroad, and from which a good view of it is afforded. A slight addition has recently been made to the back buildings, which originally consisted only of a small kitchen, erected by Washington himself. The room occupied by the General had a secret closet, in which he kept his official papers. In other respects the house is quite small, and without interest.

Washington moved with his army to this romantic spot soon after the battle of Germantown. He had previously been following the British in their movements along the Schuylkill, and finally attacked them at that place. It was after this engagement, therefore, that he took up his winter quarters in this place—a step which was dictated by the best motives of prudence and the public good.

"His soldiers," says Mr. Day, "were too ill-clothed to be exposed to the inclemency of that season under mere tents; it was therefore decided that a sufficient number of huts or cabins should be erected of logs filled in with mortar, in which the troops would find more

ENCAMPMENT AT VALLEY FORGE.

comfortable shelter. The army reached the valley about the 18th of December. They might have been tracked by the blood of their feet, in marching barefooted over the hard, frozen ground between Whitemarsh and Valley Forge. They immediately set about constructing their habitations, which were disposed in the order of a military camp, but had really the appearance of a regular town. Each hut was 16 feet by 14. One was assigned to twelve privates, and one to a smaller number of officers, according to their rank. Each General occupied a hut by himself. The whole encampment was surrounded on the land side by intrenchments, and several small redoubts were built at different points. A temporary bridge was thrown across the river, to facilitate communications with the surrounding country.

The army remained at this place until the ensuing summer, when the British evacuated Philadelphia.

This was the most gloomy epoch of the revolution. For many weeks the army, although sheltered from the wind, endured extreme sufferings from the want of provisions, blankets, and clothing. The Commissary's department, through neglect in Congress, had been badly managed, and on one occasion the supplies of beef were actually exhausted, and no one knew whence to-morrow's supply would come. Gen. Washington says, "For some days there has been little less than a famine in camp. A part of the army have been a week without any kind of flesh, and the rest, three or four days. Naked and starving as they are, we cannot enough admire the incomparable patience and fidelity of the soldiery, that they have not ere this been excited to mutiny and dispersion. Strong symptoms of discontent have, however, appeared in particular instances." Such was the scarcity of blankets and straw that men were often obliged to sit up all night to keep themselves warm by the fire, and many were too ill clothed to leave their huts! The want of wagons and horses, too, was severely felt for procuring supplies, and almost every species of camp transportation was performed by the men without a murmur, who yoked themselves to little carriages of their own making, or loaded their wood and provisions on their backs. The small-pox threatened those who had not been inoculated. Provisions continued to grow more and more scarce; the country had become exhausted by the constant and pressing demands of both armies, and no doubt many provisions were concealed from the Americans by the disaffected tories, who found a better market at Philadelphia, and better pay in British gold, than in continental money. Washington stated that there were in camp on the 23d December, not less than 2898 men unfit for duty by reason of their being *barefoot and otherwise naked*, besides many others detained in hospitals, and crowded into farmer's houses, for the same causes. Happily for America, there was in the character of Washington something which enabled him, notwithstanding the discordant materials of which his army was composed, to attach both his officers and soldiers so strongly to his person, that no distress could weaken their affection, nor impair the respect and veneration in which he was held by them. To this is to be attributed the preservation of a respectable military force under circumstances but too well calculated for its dissolution.

In the midst of these trying scenes, a strong combination was formed against Washington, in which several members of Congress, and a very few officers of the army were engaged. (See Reading.) Gen. Gates, exulting in his laurels recently gained at Saratoga, Gen. Lee, and Gen. Conway, (neither of them native Americans!) were at the head of this movement, and the strongest attempts were made to involve Gen. Lafayette into it also, but he openly and promptly avowed his attachment to Washington, and spurned the insidious efforts to supersede him in favor of Gates. The result of this base conspiracy is well known—it did not injure Washington, while it consigned the authors to the contempt of the public. Conway, the principal party in the affair, an Irishman by birth, was called to account, and finally died from the effects of a wound received in a duel with Gen. Cadwallader. Gates never could give a satisfactory explanation of his conduct, and the consequence is a blur on his reputation, which no previous or subsequent act of his has been able to obliterate.

It was during the encampment at Valley Forge, that the brave and kind-hearted Baron Steuben joined the American army—a position having been vacated by the subsequent resignation of Gen. Conway. Steuben, as is well known, was one of the most thorough military disciplinarians in Europe, and it was through his talents and instructions that our men acquired a facility and precision in military tactics which soon after enabled them to carry the Revolution to a glorious termination. Mr. Headley, in speaking of Steuben, says: "A more sorry introduction to our army, for one who had served in Europe, could not well be conceived. He had found our cities in possession of a powerful enemy, and when he came to look for the force that was to retake them, he saw only a few thousand famished, half-naked men, looking more like beggars than soldiers—cooped up in miserable log huts, dragging out the desolate winter amid the straw. As the doors of these hovels opened, he beheld men destitute of clothing, wrapping themselves up in blankets, and muttering complaints against Congress, which could treat them with such injustice and inhumanity. He was astonished, and declared that no European army could be kept together under such sufferings. All discipline was gone, and the troops were no better than a ragged horde, with scarcely the energy to struggle for self-preservation. There was hardly any cavalry, but slender artillery, while the guns and accoutrements—a large portion of them—were uufit for use. Our army had never before been in such a state, and a more unpropitious time for Steuben to enter on his work could not have been selected. Nothing daunted, however, and with all the sympathies of his noble nature roused in our behalf, he began, as soon as spring opened, to instruct both officers and men. His ignorance of our language crippled him at first very much; while the awkwardness of our militia, who, gathered as they were from every quarter, scarcely knew the manual exercise, irritated him beyond measure. They could not execute the simplest manœuvre correctly, and Steuben, who was a choleric man, though possessed of a soul full of generosity and the kindliest feelings of human nature, would swear and curse terribly at their mistakes, and when he had exhausted all the epithets of which he was master, would call on his aid-de-camp and ask him to curse in his stead! Still the soldiers loved him, for he was mindful of their sufferings, and often his manly form was seen stooping through the doors of their hovels, to minister to their wants and relieve their distresses.

It was his practice to rise at three o'clock in the morning, and dress his hair, smoke, and take a cup of coffee, and at sunrise be in the saddle. By that time also, if it was a pleasant day, he had the men marching to the field for their morning drill. First, he would place them in line, then pass along in front, carefully examining their guns and accoutrements, and inquiring into the conduct of the subordinate officers. The fruit of his labor soon appeared in the improved condition of his men, and Washington was very much impressed with the value of his services. Owing to his recommendation he was made Inspector General. This branch of the service now received the attention it deserved, and discipline, before irregular, or practised only under particular leaders, was introduced into every portion. All the arrangements, even to the minutest, were planned and perfected by Steuben, and the vast machinery of our army began to move in harmony and order. He had one company which he drilled to the highest point of discipline, as a model by which to instruct the others. The result of all this was seen in the very next campaign, at the battle of Monmouth. Washington there rallied his men when in full retreat, and brought them into action under the very blaze of the enemy's guns. They wheeled like veteran troops into their places, and then moved steadily on the foe.

For some time previous to his encampment at Valley Forge, Gen. Washington had his head-quarters at Whitemarsh, in Montgomery county, (a few miles east,) a view of which is here afforded. The

WASHINGTON'S HEAD-QUARTERS AT WHITEMARSH.

whole surrounding country is full of incidents connected with the movement of the army in this vicinity, and all have more or less interest to the American reader; but we agree with the poet, that—

The camp has had its day of song—
  The sword, the bayonet, and the plume,
Have crowded out of rhyme too long
  The plough, the anvil, and the loom.
Oh, not upon our tented fields
  Are Freedom's heroes *bred alone;*
The *training* of the *workshop yields*
  More heroes true than war has known.
Who drives the bolt, who shapes the steel,
  May, with a heart as valiant, smite,
As he who sees a foeman reel
  In blood before his blow of might!
The skill that conquers space and time,
  That graces life, that lightens toil,
May spring from courage more sublime
  Than that which makes a realm its spoil.

Valley Forge contains a cotton factory, with some other minor manufacturing establishments, and has had a considerable accession to the population during the last few years. It is surrounded with a rich and populous agricultural district, in which are located several furnaces and iron works. The copper formation before alluded to outcrops here, and there is, besides, a considerable quantity of iron ore in the adjacent hills. The observatory on the summit of the hill was erected by Charles H. Rogers, Esq., the liberal-minded proprietor of the land and the cotton factory. It commands a magnificent landscape scene. The beautiful valley of the Schuylkill, richly carpeted with greensward and soft foliage, and traversed by several streams whose bridges rise above the swelling harvest-fields, stretches out before the eye. Far off the blue Kittatinny range is seen, into whose hazy atmosphere the picture gradually fades. The Schuylkill river, at the foot of the hill, winds gracefully around a broad projecting alluvial flat, beautifully shaded with tall trees, and fringed with wild bushes, very nearly in the centre of which stands the princely country house of Dr. Wetherill, and nearer the river the country mansion of John Price Wetherill, Esq. The *spirit* of the scene is greatly enhanced by the noise of the coal trains passing over the railroad, and which is echoed to the surrounding hills—no less than the view afforded of the trains themselves, often embracing one hundred and thirty loaded cars, each containing between four and five tons of coal! If any one desires to be impressed with the idea of stamina--of real greatness—of enterprise—let him stand on a commanding eminence, and behold a coal-train, nearly half a mile in length, rumbling and tearing by with extraordinary speed! But stand in the observatory and drink in the whole glorious scene—rich, and varied, and beautiful beyond description. Could unhappiness dwell amidst such plenty—such luxuriance—such inspiring incidents? It ought not; yet man is weak—

> Had he been made, at nature's birth,
> Of only flame or only earth,
> Had he been formed a perfect whole
> Of purely *that*, or grossly *this*,
> Then sense would ne'er have clouded soul,
> Nor soul restrain the sense's bliss!
> Oh, happy, had his light been strong,
> Or had he never shared a light,
> Which shines enough to show he 's wrong,
> But *not* enough to lead him *right*.

Four miles above Valley Forge, and twenty-seven from Philadelphia, is the borough of PHŒNIXVILLE, situated in the valley of French creek, at its junction with the Schuylkill. Phœnixville is a very pleasant borough, containing a population of some thirty-five hundred—of whom probably eight hundred are engaged in its industrial establishments. Probably the first nail-works in this part of the country were erected here, where the creek affords a fine head of water. After having passed through the hands of three or four different parties, the works, upwards of twenty-five years ago, came into those of Messrs. Reeves & Whittaker.

The present style of the firm at this place, is Reeves, Buck & Co.—Joseph Whittaker having retired a few years ago. His son, Dr. Joseph Whittaker, retains an interest however, and is one of the managers of the works. Joseph Whittaker lives in the stately mansion directly opposite the rail-road depôt, on the opposite side of the river. We believe he has partially retired from the more active pursuits of the trade—merely "keeping as many irons in the fire" as is consistent with his old-fashioned notions of leisure. He has some works, of small extent, in view of his residence, and a furnace or two near Easton; while two of his sons have an establishment at Havre-de-Grace, Md. The Phœnix Company (Reeves, Buck & Co.) own the iron establishments at Bridgeton, and the nail-works at Cumberland, N. J., besides those at Phœnixville;—while Mr. Reeves is the senior partner of the firm of Reeves, Abbot & Co., proprietors of the splendid railroad mill and iron works at Safe Harbor, in Lancaster county. The mill at that place, in connection with the one located here, produced all the iron used in constructing the Central Railroad; and it is not the least interesting feature of that road, that its rails are the most substantial and reliable of any similar route in the United States.

PHŒNIXVILLE IRON WORKS.

The works at Phœnixville embrace several extensive anthracite furnaces, machine-shops, rolling-mills, nail and cotton factories, etc., among which is the splendid establishment for the production of railroad iron. A visit to these extensive iron-works cannot fail to prove highly interesting — especially the railroad mill, where some two hundred men are employed. We shall describe the whole process of iron manufacture in connection with the trade of the Juniata, and beg leave to refer the reader to our book on the Central Railroad route for information on this subject.

A railroad from this place to Harrisburg, *via* Ephrata and Cornwall, and traversing the valley of French creek, is now being surveyed. The road will connect with the Reading railroad, and the Norristown Railroad below. That the enterprise will prove successful, there can be little doubt; as, in addition to the local trade of the route, it will probably become the favorite thoroughfare of travel to the West, and thus strip the State railroad of one of its most important resources. However, the business of the interior is increasing with such rapidity, that there will soon be enough for both railroads. This route will be the shortest, as well as the most attractive for travellers, and for that reason, will be preferred. It never can do much business in the transportation of coal, because the route cannot afford sufficient gravitation to carry the extraordinary loads so peculiar to the Reading railroad. This feature of the Reading railroad renders it, in respect to the transportation of coal, the most wonderful improvement of the age.

There are several very good schools and academies in the vicinity of Phœnixville, and it is worthy of remark that while Chester and some of the adjoining counties are celebrated for the number and excellence of their seminaries of learning, a large portion of their support is derived from the Southern States. These counties are nearly all under the influence of the peculiar social and religious tenets of the Quakers, and though their political sentiments are sometimes contaminated with sectionalism—the ghastly monster that is now gnawing the vitals of our Nationality—yet, in their social and moral deportment, there is everything to admire. Intelligent and educated themselves, their benevolence of character, rigid discipline, and simplicity of manner, added to their known frugality, industry, and peaceful habits—give them peculiar qualities for the

> Delightful task! to rear the tender thought,
> To teach the young idea how to shoot,
> To pour the fresh instruction in the mind,
> To breathe the enlivening spirit, and to fix
> The generous purpose in the glowing breast.
> Oh speak the joy! ye, whom the sudden tear
> Surprises often, while ye look around,
> And nothing strikes your eye but sights of bliss,
> All various nature pressing on the heart;
> An elegant sufficiency, content,
> Retirement, rural quiet, friendship, books,
> Ease and alternate labor, useful life,
> Progressive virtue, and approving Heaven!

The Chester or Yellow Springs are situated but a few miles from Phœnixville, and are approached by mail stages. This watering-place formerly enjoyed a high celebrity, and is still visited to some extent; but numerous similar establishments, springing up in every part of the country, have no doubt materially diminished its ancient attractions.

But it is time to leave this busy and pleasant village—pleasant to us with many recollections of the past—dear, as the residence of one of our most esteemed friends, "whose life is gentle," and, like lord Brutus, "the elements so mixed in him, that all Nature might stand up, and say, with a universal voice, this is a *man!*" But there are others—one of them a distinguished Poet and Traveller, who, even now, is traversing the broad desert plains, amidst the scorching climes of Asia—prominently associated with our "recollections" of Phœnixville. It was here that Bayard Taylor, while editor of the village paper, laid a portion of the broad and substantial foundation which is to support his present and his future fame. The beautiful valley stream, we are sure, will always retain a snug place in his memory; sporting on its clear, calm surface, with a cluster of admiring friends, the bright evenings were made musical. Rowing "by the light of the moon,"

Our oars kept time, and our voices kept tune!

After a considerable voyage, during which the poet would entertain us with incidents of his unpublished "travel's history,"—interspersed with the jokes, criticisms, and gossips of others of the adventurous party—we would reach the "head of navigation" and land upon the green sloping banks, which are sprinkled with gay wild flowers, and shaded with tall majestic trees. Here the perfume of the well-tilled harvest fields, borne along in the cool evening breeze, saluted the grateful senses; and then, with one accord, all would plunge into the stream,

Whose crystal depth
A sandy bottom shows,

and lave its pure bright waters until, late in the evening, and fatigued with the labors of the expedition, we sought

Tired nature's sweet restorer—
Balmy sleep.

Thus flew the happy, merry summer evenings when Taylor was a village editor, and when the fair prospect of a future glorious career was budding, and gradually opening out before him. Success to thee, poet!—thou more than poet—soaring eagle!—hail!

Proceeding on our tour, we pass through a tunnel, a short distance above Phœnixville, which is over 2000 feet in length. It is cut through a solid dark-red sandstone rock, and is probably one of the

TUNNEL AND BRIDGE ABOVE PHŒNIXVILLE.

heaviest sections of railroading ever executed in the United States, as, in fact, the entire road may be regarded as one of the most extraordinary, in many respects, in the world. Emerging from the tunnel, and crossing the splendid and substantial stone-arched bridge, the scenery is entirely changed. Here the eager eye may take in a glorious landscape. The Schuylkill, winding around the projecting hill through which we have passed, describes a half-circle in a distance of little more than a mile. As far as the eye can see, a broad and luxuriant valley, lying between gently sloping hills, stretches out, through which wanders the river. The scene is rich in its development of agricultural fertility, and the green fields sparkle with the neat and comfortable habitations of the farmer.

From the moist meadow to the withered hill,
Led by the breeze, the vivid verdure runs,
And swells, and deepens, to the cherished eye.
The hawthorn whitens; and the spicy groves
Put forth their buds, unfolding, by degrees,
Till the whole leafy forest stands displayed,
In full luxuriance, to the sighing gales.

LANDSCAPE.

Passing the stations of Royer's Ford, 32m. and Limerick, 34m. we reach the borough of Pottstown, forty miles from Philadelphia. It is very pleasantly situated, in a rich undulating country, on the right bank of the Schuylkill. The houses, which are generally plain but comfortable, are built principally upon one broad street, lying above the railroad, and lined with numerous gardens and shade-trees. The scenery of the country is very fine, but has nothing of the boldness mingled with it which characterizes some other spots along this river. The valley is here equally as fertile as it is below, and the Manatawny creek, crossed by a romantic looking old stone bridge, and emptying into the Schuylkill, furnishes the driving-power of several extensive flour and saw-mills. The Schuylkill navigation passes along on the opposite side of the river.

Pottstown derives its name from John Potts, who held a large tract of land in this quarter, including that upon which lies the town. West of it, beyond the Manatawny, is a stately but unique stone mansion, commanding a view of the valley, which was erected by him before the revolution. It was at that time the admiration of the people, and they came from a great distance around to look at it! Mr. Potts was an enterprising speculator in iron-works, and had an establishment in each of the adjoining counties of Chester and Berks.

He was a descendant of Thomas Potts, who early settled at Burlington, N. J., and was the father of Isaac Potts, who erected the ironworks at Valley Forge, from which that place derives its name. His son, Isaac, was at that time sole owner of the land where Pottsville now stands, but sold it long before it was known to contain coal. This tract afterwards came into the hands of a German named Potts, some of whose descendants still reside there, and we may probably allude to them again, in speaking of the coal formation, and the trade which it has originated. The population of Pottstown may be estimated at two thousand. There are several quite handsome churches, two large boarding schools, &c. The machine shop, and car factory, belonging to the company, are quite prominent and imposing buildings.

The extensive copper formation already spoken of, has several outcrops near this borough, where mining operations were prosecuted several years ago. The ore, however, was not rich enough to justify the continuance of the enterprize. We are not sure whether the ore proved deficient in quality, or whether the difficulties of mining it were too great and expensive to pay. Operations have, however, been discontinued for the present.

After leaving Pottstown, we soon enter the county of Berks—a rich and populous county, originally settled by Germans, and still more or less under the influence of its primitive characteristics. The general aspect and quality of the soil is rich, and its fertility is maintained, in the absence of scientific principles elsewhere called to aid, solely by *hard labor.* The first lesson (and often the only lesson) a Berks county farmer teaches his children, is upon the subject of labor, accompanied always with practical illustrations. The philosophy of the "shovel and de hoe," the plough, the harrow, and the team, is thoroughly expounded. It is no uncommon sight to see the father in the field plowing, with a little boy, scarcely able to walk, sitting on the horse, with a whip; while it is equally as common to see boys of fourteen guiding the plough, and turning over as pretty and graceful a furrow as could be desired. The old principles of cultivation are thus inculcated and handed down from father to son; and education—scholastic, social, or moral—has thus far had very little influence upon them. Like their fathers, they neither read nor travel—but believe in religion, democracy, and General Jackson.

An anecdote is recorded of two Berks county farmers, which exhibits the awkward simplicity of their business transactions. The individuals were neighbors, and frequently borrowed small sums of money from each other, which was promptly paid back at the specified time. They lived thus for many years, and both prospered by their indefatigable industry. At length one of them was compelled to provide himself with a new and larger barn, and as his "available" means did not quite suffice, he concluded to call on his old neighbor for the balance. His request was promptly complied with, and, after the money had been paid over, it was prudently suggested by the borrower himself that a promissory note should be *received*, "so as dat he might know dat de money must be baid." The note was drawn, whether *payable* to the bearer or not, we do not know—but it was mutually concluded that, as he had *received* the money, and was to return it at the specified time, he was the proper person to take charge of the note, which he did!—thus reversing the usual order of things. Time flew round, and, promptly at the time specified, the borrowed money was restored, and with it the note, so "dat de lender might know as dat the money haf been baid!"

Douglassville, 44½ m. and Birdsboro', 49 m. are two unimportant stations on the road.

Between Pottstown and Reading, there are several very pretty landscape scenes, which the observant traveller will not fail to notice. When within a mile or two of the latter place, the railroad winds along upon the side of a high precipitous hill, and penetrating a stratum of hard rock, irregular fragments of which are left standing by the side of the road, in bold and craggy peaks. Below the railroad, in almost perpendicular descent, flows the Schuylkill, which gracefully winds round a projecting mound of land on the opposite side, and reflects, in its clear and unruffled surface, the dark moss-covered rocks and wild bushes overhanging its banks. Winding swiftly around this mountain spur, we emerge into a wide valley or basin, hemmed in with high and sloping mountains, at the foot of which the city of Reading is situated. The city is a beautiful and healthy place, and has long been the retreat of strangers and travellers during the summer months. During the revolution, while the city of Philadelphia was constantly intimidated with sudden incursions from the enemy, Reading was the principal place of resort and refuge. Here some of the most distinguished citizens of the commonwealth temporarily established themselves. The effect of their

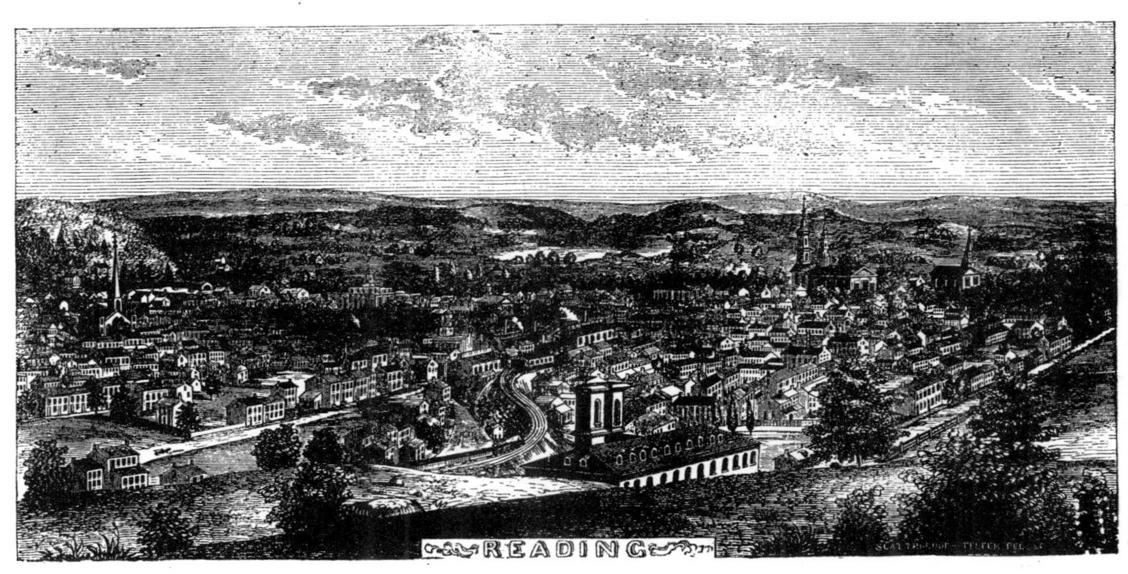

READING

RAILROAD CURVATURE NEAR READING.

presence and social intercourse with the citizens, was subsequently felt upon the society and general tone of the place, which is now, indeed, noted for its substantial, liberal, and comprehensive spirit, no less than the social good feeling, hospitality, and intelligence of its citizens, individually; for while the people of the county cannot generally be complimented for their intelligence, it must not be inferred that Reading is included with them. Nothing would be more unjust—for while it is true that the people are somewhat influenced by the tone of the country sentiment, it is equally true that the latter is also very much directed by the city; so that, considering their mutual dependency, and the *equilibrium* existing between them, it is owing principally to the popular sentiment of Reading that the people of the county have progressed, as far as they have, in education and the usages of modern society.

Reading possesses, to a very remarkable extent, all the requisites for great industrial enterprise. The agricultural resources of the county—of which it is the judicial seat—are truly enormous. The soil is drained by numerous streams of large volume, which, following the narrow alluvial valleys intervening between the mountain ranges, afford water-power of unlimited extent, and applicable to every description of manufacture. Iron ore, of various qualities, abounds throughout the whole county, and there are several rich deposits in the vicinity of Reading. The calcareous conglomerate, previously alluded to as appearing in the red shale deposits along the Mine Ridge and Blue Mountain ranges, in this county lies near the Schuylkill, in the vicinity of Reading. It is known as the "Potomac marble," and when not too hard to polish, must be considered as very valuable. Copper ore also occurs at several points, but generally in such small quantity, and so mixed with iron, as to render the expediency of working it rather doubtful. But what is most important to this city, and which has given it, within the last few years, an impulse of great industrial vigor, is the coal trade, from whose beds it is distant thirty-six miles. Added to this, is its accessibility, by canal, to the Susquehanna, and by both railroad and canal to Philadelphia and Pottsville, giving it a commanding interior position, which must ultimately be used to its great and permanent benefit.

*The Union Canal,* which unites with the Schuylkill Navigation at Reading, was the first canal route ever surveyed in this country, and a brief notice of some of the persons and circumstances associated with it, will probably not be without interest. George W. Smith, Esq., in an article first published in Hazard's Register, says that William Penn, in his proposals for a second settlement in the province of Pennsylvania, promulgated in 1690, alludes to the practicability of effecting a communication by water between the Susquehanna and the Schuylkill. Canals and turnpikes were unknown at this period, even in Great Britain. Numerous interesting letters of distinguished citizens are extant, which prove that the Union is indebted to this State for the first introduction of canals and turnpikes to public attention. Their views were regarded at that early period, (1750 to 1760) with but little interest in England, and excited the attention of but few in the colonies. At the present day it is difficult to determine to whom we are chiefly indebted for introducing the subject to public attention. If our information be correct, we may attribute to David Rittenhouse (the astronomer), and Dr. Wm. Smith, provost of the University of Pennsylvania, the credit of being the first laborers in this hitherto untrodden field. Afterwards Robert Morris, the financier of the Revolution, and still later Robert Fulton, the engineer, and inventor of steamboats, of whom Pennsylvania is justly proud, lent their powerful assistance. The writings of Turner,

Comac, Wm. J. Duane, and Samuel Breck, Esqrs., and subsequently of Gerard Rallston, Richard Peters, Jr., Matthew Carey, Samuel Mifflin, Wm. Lehman, John Sergeant, and others, are too well known to require enumeration. In 1762, David Rittenhouse, (and Dr. Smith, above mentioned, associated with him,) surveyed and levelled a route for a canal to connect the Susquehanna with the Schuylkill, by means of the Swatara and Tulpehocken creeks—the former emptying into the Susquehanna at Middletown, twelve miles below Harrisburg, and the latter emptying into the Schuylkill near Reading. The Union canal, which has since accomplished this object, passes over a portion of the route thus surveyed—and this is the first region ever surveyed in the colonies for a canal. The views of the projectors of this work were, if the difficulties of that period be duly considered, far more gigantic and surprising than have been entertained by their successors in any part of the Union. They contemplated nothing less than a junction of the eastern and western waters of Lake Erie and of the Ohio with the Delaware, on a route extending 582 miles! The Alleghany mountain was deemed to offer an insuperable obstacle to a continuous navigation—and to overcome this a portage was accordingly recommended;—an expedient which we, at a very recent period, were compelled to adopt, but which now, in the full era of steam locomotion, will soon be proudly overcome by our iron horses.

Duly to appreciate the enterprise of that age, we ought to consider that the great valley of the Ohio and Mississippi was almost one boundless forest, uninhabited but by the beasts of the forest, and the Indians. Attainable monied capital was then almost unknown in the colonies; the very term "engineering" was equally unknown in the vocabulary of those days. No canal was then in existence in England. Public opinion, even then, had yet to learn that canals were not visionary undertakings. The sneers of many were to be encountered; nevertheless, under all these trying discouragements, the earliest advocates for inland navigation commenced their efforts in Pennsylvania. In 1769 they induced the American Philosophical Society to order a survey for a canal to connect the Chesapeake bay with the Delaware—a work long since in successful operation. The provincial legislature, about the same period, authorized a survey of a route, extending five hundred and eighty-two miles, to Pittsburg and Erie. This survey was performed, and a report made strongly recommending the execution of the project. The adoption of the plan was only postponed in consequence of the Revolution. After the termination of that struggle, several works were commenced in North Carolina, Virginia, and Maryland. The canal through the Dismal Swamp, connecting the Chesapeake bay and Albemarle Sound, with the works on the Potomac, James, and Rappahannock rivers, were commenced and partially finished between the years 1786 and 1791. The great project of Pennsylvania was allowed to slumber until the 29th of September, 1791, about a century after William Penn's first prophetic intimation, when the Legislature incorporated a company to connect the Susquehanna and Schuylkill by a canal and slack-water navigation. Robert Morris, David Rittenhouse, Wm.

Smith and others, were named as commissioners. The intention of connecting the eastern and northwestern parts of the State is distinctly expressed in this, and a subsequent act of the 10th of April, 1792. By the terms of this last act a company was incorporated to effect a junction of the Delaware with the Schuylkill river, by a canal extending from Norristown to Philadelphia—a distance of seventeen miles, which has already been alluded to. The Schuylkill river, from the former city to Reading, was to be *temporarily* improved, and thus form, with the works of the Susquehanna and Schuylkill Company, an uninterrupted water communication with the interior of the State; with the intention of extending the chain to Lake Erie and the Ohio river. Experience soon convinced the two companies that a greater length of canal was requisite, in consequence of the difficulties of improving the channels of the rivers; hence the company last mentioned determined (in compliance with the suggestions of Mr. Weston, a British engineer, whom they had imported,) to extend *their* canal from river to river, a distance of seventy miles. In conjunction with the former company, they nearly completed fifteen miles of the most difficult parts of the two works, comprising much rock excavation, heavy embankments, deep cuttings, and several locks, which were constructed with bricks. In consequence of commercial difficulties, (in which, it is known, some of the chief stockholders were shortly after involved—including the patriot, Morris, who was, in fact, on one occasion, imprisoned for debt!) both companies were compelled to suspend their operations, after having expended upwards of $450,000. The suspension of these works, and, some time after, that of the Chesapeake and Delaware Canal, had a very disastrous effect on every similar enterprise which was projected for many years after.

Frequent abortive attempts were made, from the year 1795, to resume operations; and notwithstanding the subscription of $300,000 stock, subsequently tendered by the State, these companies continued in a languishing condition. In 1811 the two bodies were united, and re-organized as the Union Canal Company, which has ever since been the style of the company. They were specially authorized to extend their canal from Philadelphia to Lake Erie, with the privilege of making such further extension, in any other part of the State, as they might deem expedient. In 1819 and 1821 the State granted further aid by a guarantee of interest, and a monopoly of the lottery privilege. The additional subscriptions obtained in consequence of this legislative enactment, enabled the managers to resume operations in 1821. The line was re-located, the dimensions of the canal changed, and the whole work finished in about six years from this period; after thirty-seven years had elapsed from the date of the first work, and sixty-five from the date of the first survey. It is ninety miles in length, including the branch extending to the coal region at Pine Grove, in Schuylkill county. That portion between Pine Grove and Middletown, was enlarged in 1851, and is now equal to the capacity of boats running on the State canals; but the other section can pass boats of twenty-five tons only. The summit of this canal, about six miles in length, passes over a limestone deposit, and in consequence of the fissures abounding in this rock, a large portion of the water was lost. A number of experiments were tried to overcome this loss, such as lining it with clay, planking, &c. The difficulty was, we believe, entirely remedied on the occasion of its enlargement, last year, and the work may hereafter be regarded as complete in all its arrangements and in all the details of its construction, while the coal trade alone will probably soon make it a paying concern.

*The Schuylkill Navigation*, already alluded to in connection with the Union Canal, was incorporated as a separate and distinct concern in 1814, without mining and trading privileges, and hence it has ever been the interest of the company to invite tonnage from all sources, and in every quarter. It was originally designed for the products of the forest, the field, and the mine—all of which abounded in the counties drained by the river and its numerous tributary streams. The forests, especially, were at that period remarkable for the quality of their timber, and the height and symmetrical proportions of the trees; and, among intelligent and sagacious men, little doubt was entertained of the future importance of the coal trade, then without any existence whatever.

The Schuylkill Navigation is one hundred and eight miles in length, extending from Philadelphia to Port Carbon, in Schuylkill county. It was erected at a cost of nearly three millions of dollars. It was sufficiently complete, in 1818, to allow the descent of several boats, and tolls to the amount of two hundred and thirty dollars comprised the receipts for the season. From this year to 1825, no account was kept of the different articles for which tolls were received, and we are unable, therefore, to determine the amount of tonnage on coal descending the valley during this period. The Navigation, however, owing to the imperfection of the structure, was not in a favorable condition for the prosecution of business during any portion of this period. This arose from the obvious inexperience of the people of that day in canal-building: and obstructions of every description were of course to be expected. Of these, the most frequent were breaks in the banks of the canal, which would not only retard the progress of boats, and render the business extremely hazardous and uncertain, but subjected the Company to heavy expenses for repairs. The revenue to the stockholders was of course very limited; and at no season, we believe, previous to 1830, was it sufficient to yield a dividend of over one-half per cent. —while quite as often a *loss* would be experienced at the close of the business season.

A few years afterwards, however, (1830,) when the coal trade began to assume some importance, the stock of the Navigation yielded very handsome dividends, and continued annually to do so, until it encountered a formidable rival in the railroad, which was extended to Pottsville in 1842. From this period, the coal trade became immensely important, and the canal company determined, in 1846, to deepen the channel, and otherwise enlarge and improve their works. The canal was sufficiently enlarged to float boats of 180 tons burthen, while the number of locks was reduced from 109 to 71—eleven of which are guard locks without lift, of which the gates generally stand open, and are, in fact, closed only during freshets. The average time of passing a lock with a boat is about four minutes, at which rate all the locks on the canal could be passed in about five hours; or, making a reasonable allowance, six hours would give ample time to overcome the total descent of 620 feet—and if, at every lock, a descending boat should meet an ascending one, the whole time lost in effecting the cross passage does not exceed twelve hours. This is an immense improvement over the old navigation.

Above the Blue Mountain nearly all the canals are almost equal in width to the slack-water pools formed by the dams. Below the Blue Mountain, the water line of the canal, which is never less than sixty feet, widens frequently to one hundred feet and more. Taking these things in connection with the fact, that about half the length of the navigation consists of wide slackwater pools, and it will be observed that in point of width everything practically desirable has been attained.

Several attempts have been made to introduce steam in the navigation of the Schuylkill,—and though apparently attended with some success, have not led to any practical end, as yet. The only steamboats now plying on its waters, are those between Fairmount and Manayunk. If coal could be used for fuel—(of which, by the way, there can be no doubt,) and the machinery made sufficiently light to correspond with the tonnage of the boat, there would, indeed, seem to be no practical reason why steam should not supersede horses. The splashing of the water against the banks of the canal, occasioned by the evolutions of the paddle-wheel, presents the most potent objection;—yet this is but a trifle, and might readily be overcome, if sufficient attention were bestowed upon the subject. We look forward to the day, when Prof. Page's brilliant experiments in electro-magnetism will find practical application in the ordinary pursuits of life. His electric engine already possesses eight-horse power; and, inasmuch as the *entire machinery* consists of but a single wheel, or iron circle, this would be the exact thing to introduce for the propulsion of *canal boats*. The whole weight of an electric engine of sufficient capacity to propel *five canal boats*, together with the *fuel*, for twenty days, would not, probably, exceed three hundred pounds!

The entire length of the Navigation, as previously remarked, is 108 miles—its lockage 620 feet—the burden of its boats 180 tons—the size of its locks, 110 by 18 feet—the width of its canals, never less than 60 feet—and the least depth of water upon the mitre sills 5½, and in the clear levels 6 feet.

The five leading railroads, and their laterals, to the Schuylkill Navigation and the Reading Railroad, are the Mine Hill and Schuylkill Haven, terminating at Schuylkill Haven; the Mount Carbon, terminating at Mount Carbon; the Mill Creek, terminating at Port Carbon, and the Schuylkill Valley, terminating at Mount Carbon.

At Schuylkill Haven a very fine dock, nine hundred feet long, sixty feet wide, and six feet deep, with its rail seventeen feet high above water, shute and landings on both sides, has been constructed by Mr. Dundas. This dock alone is capable of shipping, in an active season's work, at least two hundred and fifty thousand tons of coal, and is leased by the Navigation Company.

At Port Carbon, the Navigation Company have constructed an extensive series of landings. A part of these landings below the Mill Creek Railroad bridge, consists of a dock, about nine hundred feet long, sixty feet wide, and six feet deep, with its rail elevated eighteen feet above water, with shutes and landings on both sides. There is room at this landing for thirty boats of one hundred and eighty tons burden to load at once, and it is capable of shipping five hundred thousand tons of coal per annum.

In the pool of dam No. 1, the company have erected six new landings, with their rails elevated sixteen feet above the water, and so arranged that six large boats may load at once, without interruption. In addition to these, and also in the upper dam, the Navigation Company have leased and fitted up the long dock, which accommodates six large boats at the same time. Thus, the Company have a variety of fine landings to ship coal coming from the Schuylkill Valley and Mill Creek Railroads, and capable together of shipping near seven hundred thousand tons in a season's work.

In addition to the foregoing, the Company have constructed a dock and landings at Mount Carbon, similar to the Firth Dock at Port Carbon, and of about the same capacity. We shall probably again refer to these landings when speaking of the coal trade.

To guard against the danger of a deficiency of water, to which the Navigation is exposed in dry seasons, they have erected several large dams upon tributary streams at the head of navigation, from which to draw supplies in cases of deficiency. The dam at Silver creek covers nearly sixty acres, and contains about forty million cubic feet of water, which is estimated to be capable, of itself, of floating about one hundred and twenty thousand tons of coal annually!

The Reading Railroad, of which we now propose to give a brief description, was chartered on the 4th of April, 1833, and surveys were made the same year, and forty-one miles placed under contract and construction a year afterwards. The charter authorized subscriptions to the amount of twenty thousand shares, of fifty dollars each, being a capital of one million dollars, with the right to double it, if found necessary. It provided for an annual meeting on the second Monday in January, and the right of stockholders to cast one vote for every share, not exceeding two; one vote for every two shares not exceeding ten; and one vote for every five, for any amount above ten, that may belong to them in their own right or as trustees. Proxies to be dated within six months, and only to be used for purposes

L

expressly stated. No blank proxy to be good, and no third person to be substituted. The government of the road is vested in a President and six Managers, who are authorized to make By-laws and all needful regulations, subject to the approval of the stockholders at their annual meeting. The President and Managers have full power to manage and operate the road. Special meetings may be called, but no business can be done without a majority in interest of the stock is represented. No dividend to be declared except from the *net profits*, so that the capital shall remain unimpaired. The charter is perpetual.

It was originally designed for its present purpose, an outlet or avenue to market for the Schuylkill Coal Region; but its first charter extended only to the city of Reading, fifty-nine miles from its terminus on the Delaware River, near Philadelphia, as the right of constructing a railroad between Reading and Port Clinton, twenty miles, had already been granted another corporation, the Little Schuylkill Railroad Company, extending from Tamaqua to Port Clinton, twenty miles. From insufficient means, this company was unable to extend their road, and yielded their right and charter to the Reading Railroad Company, who, with a further extension of their charter, beyond Port Clinton to Pottsville, went into an active prosecution of the whole work, from Pottsville to the Delaware, ninety-four miles, under one charter, now known as the Reading Railroad.

Every Pennsylvanian is familiar with the great embarrassments to the business of the country, checking commercial enterprise, disastrous to every branch of industry, and fatal to public and private credit, during the period from 1833 to 1842. Notwithstanding all these difficulties, the friends of this road pushed steadily on with its construction, taxing their energies, their means and their credit to the utmost, to insure its speedy completion; and on the first day of 1842, the first locomotive and train passed over the whole line between Pottsville and Philadelphia.

The event was celebrated with military display, and an immense procession of (seventy-five) passenger cars, twelve hundred and twenty-five feet in length, containing two thousand one hundred and fifty persons, three bands of music, banners, &c., all drawn by a single engine! In the rear was a train of fifty-two burden cars, loaded with one hundred and eighty tons of coal, part of which was mined the same morning four hundred and twelve feet below the water level.

The whole was under the charge of Mr. Robinson, chief engineer, and Mr. G. A. Nichols, superintendent. The entire capital invested up to this time, including all its vast real estate, locomotives, workshops, wharves, etc., amount to *over sixteen millions of dollars.*

From that date to the present, its business, its revenue and its credit have increased, in a degree scarcely paralleled by any similar improvement, until its tonnage and its receipts are measured, as at present, by millions.

Two continuous tracks of railway extend the whole distance of ninety-four miles, from Pottsville to the Delaware river, at Port Richmond—situated three miles above the heart of the city, and one hundred and four from the sea, while a branch road extends from the Falls of the Schuylkill *via* Fairmount, to Broad street, in the city. This portion of the road formerly belonged to the State—but upon the completion of the road to avoid the inclined plane, the canal commissioners sold this section to the Reading Railroad company, who, with characteristic enterprise, put it into immediate repair, and laid down upon it a strong and substantial rail. They also materially strengthened and otherwise improved the railroad bridge across the Schuylkill, so that, instead of awaiting the slow process of being hauled over with horses, the passenger trains are drawn over by locomotives without delay or hindrance. This branch of the road is used altogether for the coal and miscellaneous trade of the city, including passengers.

The rail used on this road is of the H pattern, with both top edges alike, and weighs forty-five and one-eighth, fifty-two and one-half and sixty pounds to the yard; the lightest having been first, and the heaviest last used. A few tons of other rails, purchased before a further supply of the pattern adopted for the road could be obtained in England, and varying from fifty-one to fifty-seven pounds per yard, are also in use.

The track is laid in the most simple manner, the lower web or base of the rail being notched into white oak cross sills, seven by eight inches in thickness, and these laid on broken stone, fourteen inches deep, and well rammed. This method is found admirably calculated for the enormous tonnage of the road, being rapidly and economically repaired and replaced, securing a thorough drainage, and preserving its line and level true, at all seasons of the year.

The grades of this road are the chief elements of its success in revolutionizing public opinion, on the subject of the carriage of heavy burdens by railway. From the most important branch Coal-feeder of the road, at Schuylkill Haven, to the Falls of Schuylkill, a distance of eighty-four miles, the grades all descend in the direction of the loaded trains, or are level, with no more abrupt descent than nineteen feet per mile. At the Falls, an assistant locomotive engine of great power pushes the train, without the latter stopping, or any delay, up a grade of forty-two and one-half feet per mile, for one mile and a quarter, thus placing it on a *descending grade*, within four miles of Richmond, whither it is readily conveyed by the same engine which started from Pottsville, never leaving the train.

The bridges on this line are of great variety in plan, and material of construction; stone, iron and wood being used. The most perfect and beautiful structure on the road, if not in the State, is a stone bridge across the Schuylkill near Phœnixville, built of cut stone throughout, with four circular arches, of seventy-two feet span, and sixteen and one-half feet rise each, at a cost with ice-breakers, of $47,000. (See engraving—page 69.) There are seventy-five other stone bridges and culverts, varying from six to fifty feet span; all of circular arcs, spanning water courses, branches of the Schuylkill and roads. There are seven bridges from twenty-five to thirty-eight feet span each, built of iron, trussed after the Howe plan, with wrought iron top and bottom cords, wrought iron vertical ties, and cast iron diagonal braces. These bridges are stiff and light, and present a very neat and handsome appearance. As, however, the flooring is of wood, and therefore liable to decay and accident, they have only been used where the width and depth rendered stone bridges impracticable; the latter being always used in replacing wooden structures, wherever it is practicable. There are twenty long wooden bridges, varying from forty-one to one hundred and sixty feet span, built on various principles, chiefly of lattice work, assisted by heavy arch pieces. Of this latter description, the bridge over the Schuylkill at the Falls is a fine specimen. It is six hundred and thirty-six feet long, consisting of four spans of one hundred and thirty-four, two of one hundred and fifty-two, and one of one hundred and sixty feet over the river. There is one bridge built on Burr's plan, with double arch pieces of one hundred and forty-nine feet span; and one on Howe's plan, one hundred and fifty-six feet span, also assisted by arch pieces. Besides the above, there are about twenty wooden bridges of short spans, from fourteen to thirty feet, built of King post, Queen post, Bowe's truss, and joists. There are also several small iron and wooden bridges.

There are four tunnels on the road. The longest of these is near Phœnixville, one thousand nine hundred and thirty-four feet cut through solid rock, worked from five shafts and two end breasts; deepest shaft one hundred and forty feet; size of tunnels, nineteen feet wide, by seventeen and one-quarter high; total cost, $153,000. Another tunnel at Port Clinton, is one thousand six hundred feet long, worked from the two ends only; material, loose and solid rock mixed; one thousand three hundred feet are arched; depth below the surface of the ground, one hundred and nineteen feet; total cost $138,000. The Manayunk tunnel is nine hundred and sixty feet long, through very hard solid rock, worked from two ends; depth below surface, ninety-five feet; total cost $10,000. Another tunnel under the grade of the Norristown Railroad, and through an embankment of the latter, is one hundred and seventy-two feet long, formed of a brick arch, with cut stone façades.

The depôts on this road are all substantially built, but with a view to use, rather than ornament. At Schuylkill Haven, four miles from Pottsville, is erected a spacious engine house, round, with a semicircular dome roof, one hundred and twenty feet diameter, and ninety-six feet high; with a forty feet turning platform in the centre, and tracks radiating therefrom, capable of housing sixteen second class engines and tenders. The principal depôts for making up the coal trains are at Mount Carbon, Palo Alto, (situate on the Schuylkill, about one mile, in an angle, from Pottsville and Mount Carbon;) Schuylkill Haven, and Port Clinton. At all of these places, there is extensive side-railway to arrange the cars in trains, as they arrive from the numerous branch roads. Sometimes upward of one hundred and fifty *loaded cars* are attached to a single locomotive, which, at five tons to each car, gives an aggregate tonnage of seven hundred and fifty tons! No other road in the world can do this!

At Reading are located the most extensive and efficient workshops and railroad buildings of every description to be found in the country. The company's property covers, altogether, besides the railway tracks, some thirty-six acres, the greater part of which is in use for the various occupations required to keep this vast thoroughfare in life and active motion. These shops embrace various departments, in

8

which every description of mechanical work required for the machinery of the road, can be supplied. A description of the dimensions of the several buildings is probably unnecessary—the reader will be good enough to take our assurance that they are large, very large, enormously large, and, in point of interest and extent, are second to no iron establishment in the United States. About four hundred hands (including men and boys) are employed in the establishment, which embraces an iron foundry and machine shop, brass foundry and machine shop, carpenter's shops, furnaces, smiths, and various other subordinate shops. In short, the establishment builds and repairs all the running-machinery of the road, as locomotives, cars, tenders, smoke pipes, etc., for which purpose all its waste *scrap iron* is consumed, being remelted and puddled, and thereby a great saving is effected, probably equivalent to some *fifty thousand dollars* per annum, besides the accommodation and perfect *adaptation* of the machinery to the road, which it affords. We do not know the items of cost of this establishment; but it must be regarded as one of the company's most valuable features, and it is now in complete and successful organization. To arrange the vast *details* of this road required many years of patient and persevering toil; and no words can express too strong a compliment upon the business talents of those persons under whose auspices it has finally attained its present admirable working condition.

For many years the company have been extremely anxious to introduce anthracite coal, instead of wood, as fuel for their locomotives. In point of economy, over *one hundred thousand* dollars would annually be saved, could coal be successfully substituted. Various and numerous experiments have been made, and latterly with success. Several engines, calculated to use coal, are now being constructed at their own workshops at Reading, under the direction of Mr. Mulholland. They will be completed and put on the road in a few weeks hence. They are of great capacity, and built with a view, also, to swiftness. Wood is getting scarce along the line of the road, and the introduction of coal, which can be had on the beds at a mere trifle, will prove highly advantageous to the interests of the company. The difficulty hitherto in the way of using anthracite, we may add, was the intense concentrated heat it would create, materially injuring the works of the fire-box, as well as the boiler. There never was much difficulty in burning the coal—but, under its destructive effects, there was no advantage in using it, and *all coals* are very nearly similar in this respect. They emit a heat which eats into the iron of the boiler, and, in time, renders it unfit for use. Thus a boiler, heated with anthracite, will last, say six months; one heated with bituminous coal will last nine months, and heated with wood, twelve, fifteen, or eighteen months. Now, all the money saved in the cheaper cost of coal over wood, is lost by the injury entailed on the locomotive, for the cost of a new boiler may be stated to be some two or three hundred dollars, besides the loss of time required to repair. But the difficulty can, will, and must be overcome. We know it can, and we will aver that it *will*, for the Reading Railroad Company have undertaken to do it, and with them there never has "been such word as fail." (No allusion to you, 'squire Trail!)

A merchandize depot, recently completed at Reading, is one hundred and twenty-four by eighty-four feet, to accommodate that rapidly increasing branch of business. About a mile below the Reading depot, where the railroad is nearest the river, most efficient water-works are constructed, consisting of a reservoir on the Neversink hill side, fifty-one feet above the rails, holding seven hundred thousand gallons of water, supplied by a force pump worked by a small steam-engine. Attached to this station are also two separate tracks, with coal chutes beneath, three hundred and four hundred and fifty feet long each, for the use of the town; two wood and water stations; a small portable steam-engine for sawing wood, a refreshment house for crews of engines stopping to wood or water; a brass foundry, passenger car-house, passenger rooms, offices, &c., &c. All the machinery of the main shops and foundry is driven by a very handsomely finished stationary engine, with double cranks, of thirty-five horse power, built entirely on the works.

At Pottstown station, eighteen miles below Reading, extensive and efficient shops have also been erected, chiefly for work connected with the bridges and track of the road, and new work of various descriptions. The principal shops here are one hundred and fifty-one by eighty-one, one hundred and eighty-one by forty-one, and eighty-one by forty-four feet. The first shop is covered with a neat and light roof, built of an arched Howe truss, forming a segment of a circle, seventy-eight and a half feet span by sixteen feet rise.

At Richmond, the lower terminus of the road, at tide water on the river Delaware, are constructed the most extensive and commodious wharves, in all probability, in the world, for the reception and shipping, not only of the present, but of the future vast coal tonnage of the railway; forty-nine acres are occupied with the company's wharves and works, extending along twenty-two hundred and seventy-two feet of river front, and accessible to vessels of six or seven hundred tons. The shipping arrangements consist of some twenty wharves or piers, extending from three hundred and forty-two to eleven hundred and thirty-two feet into the river, all built in the most substantial manner, and furnished with chutes at convenient distances, by which the coal flows into the vessel lying alongside, DIRECTLY FROM THE OPENED BOTTOM OF THE COAL CAR FROM WHICH IT LEFT THE MINE. See engraving, page 33. As some coal is piled or stacked in winter, or at times when its shipment is not required, the elevation of the tracks by trestlings, above the solid surface or flooring of the piers, affords sufficient room for stowing upwards of two hundred and fifty thousand tons of coal. Capacious docks extend inshore, between each pair of wharves, thus making the whole river front available for shipping purposes; over one hundred vessels can be loading at the *same moment*, and few places present busier or more interesting scenes, than the wharves of the Reading Railroad at Richmond. A brig of one hundred and fifty-five tons has been loaded with that number of tons of coal in less *than three hours time*, at these wharves. The whole length of the lateral railways extending over the wharves at Richmond will probably exceed ten miles, and affording a shipping capacity for upwards of *three millions of tons!* and it will probably not be many years before this amount, extraordinary as it may seem, (as, indeed, it really is,) will be annually transported over this great thoroughfare. The company has laid the *foundation* for a trade as broad as the future destiny of the coal trade itself.

A very convenient and neat engine-house is erected at this station; it is of a semi-circular shape, with a forty feet turning platform outside; from which tracks radiate into the house, giving a capacity for twenty engines, and their tenders, of the largest class. The building is three hundred and two feet long on the centre line, by fifty-nine feet wide. It is built in the simple Gothic style, the front supported by cast iron clustered pillows, from the tops of which spring pointed arches, and the whole capped with turretted capping. Immediately adjoining are built spacious machine and work shops, for repairs of engines and cars, all under one roof, two hundred and twenty-one by sixty-three feet. A visit to this chief outlet of the Pennsylvania coal trade will give the best idea of its magnitude, and of the various branches of industry connected with it.

The extraordinary business of this road requires, of course, a large amount of running machinery. The latter consists of about one hundred locomotive engines and tenders, including six or seven in constant use on the lateral railroads in the coal region; about five thousand iron and twelve hundred wooden coal cars, six hundred cars for merchandise, and some thirty elegant passenger cars.

The engines vary from ten to twenty-four tons weight; two very powerful engines, of twenty-seven tons weight each, are used exclusively on the Falls grade, before mentioned. The iron cars weigh over twenty-four tons when empty, and carry five tons of coal. The *average* load of each engine, during the busy months of the year, is very nearly five hundred tons of coal, (of twenty-two hundred and forty pounds.)

The total length of lateral railroads, connecting with the Reading Railroad, under other charters and corporations, but all contributing to its business, using its cars, and returning them loaded with coal and merchandize, is over one hundred miles. Some of these railroads are constructed in the most substantial manner, with the best superstructure at present used in the country.

Of these, it connects with the Mount Carbon Railroad, and the Mount Carbon and Port Carbon Railroad, at Mount Carbon one mile below Pottsville, and with the Mine Hill Railroad and its numerous radiating branches, at Schuylkill Haven, (this road is about being extended to connect with the Shamokin Railroad, thus affording a connection with the Susquehanna, and passing through the great Mohanoy coal region:—it will thus bring an incalculable amount of additional tonnage and passengers to the Reading Railroad) also at Port Clinton, with the Little Schuylkill Railroad extending to Tamaqua, and thence into several lateral branches to numerous coal districts adjacent. The roads have each many miles of branches, penetrating all the coal districts of this unparalleled region, and the greater portion of their tonnage is, and always will be, transferred to the Reading Railroad; for so firmly has it established itself into the *local arrangements of the lateral railway trade* of Schuylkill county, that it can always *command* a large portion of the trade.

Such, "my jolly companion," is a brief exposition of some of the leading features of the Reading Railroad. Look at it—starting out, with one hundred miles of railway branches, from the most extensive deposit of anthracite coal in the world, lying some seven hundred feet above tide-water—look at the road, as it winds its way amidst the rich fields and sloping banks of the Schuylkill, and gradually sinks into the bosom of the most beautiful and populous city on the American continent! One hundred miles in length, sloping gracefully from the coal-beds to the river Delaware—is not that a beautiful idea to contemplate? Nature has had a hand in it, and enterprising man has improved what she carefully prepared. She made the route, and raised the coal-beds to their present height with the express purpose, no doubt, of rendering them available to our wants. For this, all thanks!

> I'm not romantic, but, upon my word,
> There are some moments when one can't help feeling
> As if his heart's chords were so strongly stirred
> By things around him, that 'tis vain concealing
> A little music in his soul still lingers,
> Whene'er the keys are touched by Nature's fingers.

The ground upon which Reading is situated originally belonged to Thomas and Richard Penn, who disposed of the lots, subject to an annual ground-rent. This rent through neglect, had been left unpaid after the Revolution, and when attempted to be collected, some years ago, the accumulated amount occasioned a great deal of surprise and excitement in the place. The rent was stoutly resisted, but a compromise was soon after effected between the town authorities and the claimants. The public buildings of Reading are amongst the handsomest in the State. The Court House, the Prison, and several of the Churches, are models of architectural skill, and reflect great praise upon the liberality and taste of the citizens. Reading, says Mr. Trego, was formerly celebrated for the manufacture of wool hats, and the business is still carried on extensively; but of latter years other branches of manufactures have so much increased as to have given this ancient trade but a secondary rank. Previous to 1836, hats, boots, shoes, and stone-ware were the principal manufactures; since that time establishments have been put in operation for rolling iron, making nails, casting in iron and brass, manufacturing locomotive and stationary steam-engines, rifle-barrels, augers, &c.; a steam saw and chopping-mill, and several shops for the manufacture of thrashing-machines, corn-shellers, ploughs, harrows, and other agricultural implements. Besides these manufactories, some of which are very extensive, and employ a large number of mechanics, a cotton factory is now in operation, embracing some three hundred looms, and employing about the same number of operatives. The mill was finished a year or two ago, and is built in the most substantial as well as ornamental style—with a stock capital of some two hundred thousand dollars.

Besides producing excellent ale and porter, Reading enjoys some celebrity in connection with the manufacture of wines. The vineyards are said to be quite extensive, and the wine is certainly "not hard to take." For certain kinds of wine, we can see no reason why the banks of the Schuylkill should not prove available—the grape attaining here all the pulpy sweetness that characterizes it in some of the most favored lands. The weather, however, is too wet ever to permit the grape to attain the *dryness* so necessary for the production of the higher grade of wines;—but, under ordinary circumstances of favor, wine can be produced at least equal, if not far superior, to the horrid adulterated stuff palmed off as wine, and which, heavily charged with *impure liquor*, makes it justly obnoxious to the friends of the "Maine Law."

The common language of Berks county, and some of those lying adjacent, is an impure German, so corrupted and mixed with the more popular English words, that it would scarcely be understood by a well educated German from the fatherland. In many parts of the county, where the inhabitants seldom leave their own neighborhood, English is neither spoken nor understood; but this language is rapidly gaining ground among those of the people who have business communications with others than their immediate neighbors. It will probably not be long before English and German will be equally used, except in some secluded portions of the county.

Among the natural curiosities in the county, may be mentioned Dragon's cave, in Richmond township, which is thus described in Trego's Geography by a gentleman resident in the vicinity. "The entrance to this cave is on the brow of a hill, on the edge of a cultivated field. Passing into it, the adventurer descends about fifty yards by a rough and narrow passage, and then turns to the left at an acute angle with the passage hitherto pursued. After proceeding about thirty yards farther, he enters the great chamber, about fifty feet long, twenty wide, and fifteen to twenty feet high, in a rock of limestone. Near the end of this chamber, opposite to the entrance, is the 'altar,' a large mass of stalagmite, which rings under the hammer, and is translucent. Formations of stalactite are found in other parts of the cave, though none so large as the mass just mentioned." Sinking Spring, near the Harrisburg turnpike, five miles from Reading, is a considerable curiosity to those who are not familiar with the circumstances frequently attending large springs in a limestone region. The water here rises and *sinks again in the same basin,* which is very deep; thence finding its way again, under ground, through fissures and hidden caverns in the limestone rock, probably once more to seek the light of day in some other place. A similar phenomenon is found in Sinking Spring valley, in Blair county, which is elsewhere noticed in this work.

We have already stated that, during the revolution, Reading was (as it is now during the summer,) a place of resort for the citizens of Philadelphia. It was here that the conspiracy (for so it should be termed) against Washington was supposed to have had its birth, while the popular sentiment was by no means enthusiastic in favor of the commander-in-chief, owing, probably, to the exposed position of the frontier settlements to the ravages of the Indians, and who, in their bold incursions, rendered Reading itself sometimes obnoxious to their attacks. Alexander Graydon, who was, at that time, on parole, (having been captured by the British, near New York,) gives the following in his memoirs: "The ensuing winter, at Reading, was gay and agreeable, notwithstanding that the enemy was in possession of the metropolis. The society was sufficiently large and select, and a sense of common suffering, in being driven from their homes, had the effect of more closely uniting its members. Disasters of this kind, if duly weighed, are not grievously to be deplored. The variety and bustle they bring along with them, give a spring to the mind, and when illumined by hope, as was now the case, they are, when present, not painful, and when past, they are among the incidents most pleasing in retrospecting. Besides the families established in this place, it was seldom without a number of visitors—gentlemen of the army, and others;—hence the dissipation of cards, sleighing parties, balls, &c., was freely indulged. Gen. Mifflin at this era was at home—a chief out of war, complaining, though not ill; considerably malcontent, and, apparently, not in high favor at head-quarters. According to him, the ear of the commander-in-chief was exclusively possessed by Greene, who was represented to be neither the most wise, the most brave, nor most patriotic of counsellors. In short, the campaign in this quarter was stigmatized as a series of blunders, and the incapacity of those who had conducted it unsparingly reprobated. The better fortune of the Northern army was ascribed to the superior talents of its leader, and it began to be whispered that Gates was the man who should, of right, have the station so incompetently sustained by Washington.

"There was, to all appearance, a cabal forming for his deposition, in which it is not improbable that Gates, Mifflin, and Conway were already engaged; and in which the congenial spirit of Lee, on his exchange, immediately took a share. The well-known apostrophe of Conway to America, importing that 'heaven had passed a decree in her favor, or her ruin must long before have ensued, from the imbecility of her military councils,' was, at this time, familiar at Reading; and I heard him myself, when he was afterwards on a visit to that place, express himself to the effect, 'that no man was more of a gentleman than Gen. Washington, or appeared to more advantage at his table, or in the usual intercourse of life, but as to his talents for the command of an army, (with a French shrug,) they were miserable indeed!' Observations of this kind, continually repeated, could not fail to make an impression within the sphere of their circulation; and it may be said that the popularity of the commander-in-chief was a good deal impaired at Reading. As to myself, however, I can confidently aver that I never was proselyted, or gave in to the opinion, for a moment, that any man in America was worthy to supplant the exalted character that presided in her army. I might have been disposed, perhaps, to believe that such talents as were possessed by Lee, could they be brought to act subordinately, might often be useful to him; but I ever thought it would be a fatal error to put any other in his place. Nor was I the only one who forbore to become a partizan of Gates. Several others thought they saw symptoms of selfishness in the business, nor could the great eclat of the Northern campaign convince them that its hero was superior to Washington. The duel which afterwards took place between Gen. Conway and Gen. Cadwallader, though immediately proceeding from an unfavorable opinion expressed by the latter of the conduct of the former at Germantown, had, perhaps, a deeper origin, and some reference to this intrigue; not that Gen. Cadwallader was induced from the intrigue to speak unfavorably of Conway's behaviour at Germantown. That of itself was a sufficient ground of censure. Conway, it seems, during the action was found in a farm-house, by Gen. Reed and Gen. Cadwallader. Upon their inquiring the cause, he replied, in great agitation, that his horse was wounded in the neck. Being urged to get another horse, and at any rate to join his brigade, which was engaged, he declined it, repeating that his horse was wounded in the neck. Upon Conway's applying to Congress, some time after, to be made a Major-General, and earnestly urging his suit, Cadwallader made known this conduct of his at Germantown, and it was for so doing that Conway gave the challenge, the issue of which was his being dangerously wounded in the face from the pistol of Gen. Cadwallader. He recovered, however, and some time after went to France. While laboring under the effects of this wound, (which was at first supposed to be mortal,) he wrote a letter to Gen. Washington, apologizing for his previous conduct towards him, and expressing the highest admiration of his military career;—as I had the means of knowing that Gen. Cadwallader, suspecting Mifflin had instigated Conway to fight him, was extremely earnest to obtain data from a gentleman who lived in Reading, whereon to ground a serious explanation with Mifflin. So much for the manœuvering which my location at one of its principal seats brought me acquainted with, and which its authors were soon after desirous of burying in oblivion."

Conrad Weiser, a celebrated Indian agent and interpreter, spent the latter part of his life in Reading, where he kept a trading house. He was born in Germany, but came to this country in early life, and settled about the year 1714. He lived much among the Six Nations of New York. He was a great favorite among them, was naturalized by them, and became perfectly familiar with their language. Desiring to visit Pennsylvania, the Indians brought him down the Susquehanna to Harris' ferry, (now Harrisburg, the capitol of the State,) and thence he came across to the Tulpehocken, and thence to Philadelphia, where he met Wm. Penn for the first time. He became a confidential interpreter and special messenger for the province among the Indians, and was present at many of the most important treaties between the proprietary government and the Indians. In 1737 he was commissioned by the Governor of Virginia to visit the Grand Council at Onondaga. He started very unexpectedly, in the month of February, to perform this journey, of five hundred miles, through a wilderness, where there was neither road nor path, and at a season when no game could be met with for food. His only companions were a Dutchman and three Indians. In 1744 he was despatched in like manner to Shamokin (now called Sunbury) "on account of the unhappy death of John Armstrong, the Indian trader." On both these journeys he has specially noted interesting observations relating to a sincere and general belief among the Indians, in the interposition of an overruling Providence, and their habit of acknowledging with gratitude all such interpositions in their favor. Mr. Weiser had an Indian agency and trading house at Reading. In 1755, during alarms on the frontier, he was appointed colonel of a regiment of volunteers from Berks county. The Indians always entertained a high respect for his character, and for years after his death were in the habit of making visits of affectionate remembrance to his grave. Col. W. was the grandfather, on the maternal side, of the late Hon. Henry A. Muhlenburg, formerly Minister to Austria, and during his life one of the most distinguished citizens of Reading, where his family still reside.

The country from Reading to Hamburg is more hilly than that which we have already passed, but still maintains a high degree of cultivation. The rolling aspect of the soil, clothed in the richest verdure, affords here and there a splendid landscape; but the scenery is, for the most part, monotonous, until we arrive at Hamburg, where we take leave, for a time, of the pleasant harvest-fields and scenes of agricultural industry, and penetrate the region of mountains. Here the Kittating or Blue Mountain range crosses our course, and, as far as the eye can see, traverses the country in bold and majestic ridges, sometimes sloping gradually into the valley below, and again rising in towering grandeur to the overhanging clouds.

Hamburg, 75m. is situated on the left bank of the Schuylkill, near the Blue Mountain, and about a mile from the railroad. It embraces a population of about one thousand, and, being situated in Berks county, is composed mostly of Germans. The surrounding country is a rich agricultural district, and the village is at least very pleasantly situated. The trade of the place is unimportant. But let us hasten on, for—

> Our heart's in the mountain—our heart is not here,
> Our heart's in the mountain a-chasing the deer;
> A-hunting the deer and pursuing the roe—
> Oh, our heart's in the mountains wherever we go!

TUNNEL.

The Kittating is a formidable barrier to our progress, but the railroad has a way to overcome it—or to pass through it. Plainly, the road *pierces* (no allusion to you, General, or any other democrat!) the mountain, and the first thing we see, on emerging from it, is Port Clinton, seventy-eight miles from Philadelphia, and about six hundred feet above the Delaware river. Here we have a tolerable specimen of the scenery the traveller may expect for some time to come—for he is now in the midst of those bold parallel layers of mountain, broken and distorted into irregular fragments, which constitute the outlines of the great Apalachian system, and which, under various local names, traverses several States, and divides the lakes and rivers, east and west of it, into separate systems. The scenery here is bold, wild and picturesque, while the whole country looks like a vast

"Ocean into tempest tossed."

At some places, the mountain sides are steep, rising from eight to twelve hundred feet almost perpendicularly, at the foot of which flows the Schuylkill, or some of its tributary streams. The red shale, which support the outlayers of conglomerate rock, decomposes under exposure to the atmosphere, and the effects of rain, snow and frost, and the debrïs, borne off by the streams winding round the mountains, leave the conglomerates, and more durable rocks, reposing in awful cliffs and precipices, frequently overlooking the valleys below. Sometimes the mountains slope gradually from their base to the summit, and the harder rocks are strewn over its surface in the wildest confusion, in pieces of all sizes and shapes. The smallest of these stones are carried down the mountain sides by heavy rains, and the noise which the descending mass makes, as the stones are pushed along by the impetuous torrent, is both exciting and novel. It is thus that the narrow valleys have been gradually formed, which will be more minutely illustrated in our geological treatise, which we shall very soon commence. What can be more interesting to the eye of the traveller—to the man of care and business, "doom'd, for a certain time," to the daily rounds of city-life—than the change of scene which these bold, rolling mountains afford? Where is the invalid, accustomed to the dull monotonous scenes of level plains, or breathing the low and impure atmosphere of the populous city, who would not be invigorated, mentally and physically, in the midst of this primeval terrestrial ocean?

Thrice happy he! who, on the sunless side
Of a romantic mountain, forest-crowned,
Beneath the whole collected shade reclines;
Or in the gelid caverns, woodbine wrought,
And fresh bedewed with ever-spouting streams,
Sits cooly calm; while all the world without,
Unsatisfied, and sick, tosses in noon.
Welcome, ye shades! ye bowery thickets hail!
Ye lofty pines! ye venerable oaks!
Ye ashes wild, resounding o'er the steep!
Delicious is your shelter to the soul,

As to the hunted hart the sallying spring,
Or stream full-flowing, that his swelling sides
Laves, as he flows along the herbaged brink.
Cool, through the nerves, your pleasing comfort glides;
The heart beats glad; the fresh expanded eye
And ear resume their watch; the sinews knit;
*And life shoots swift* through all the lengthened limbs.

Before leaving this place, which is a point of divergence, it is proper that we should have an understanding with the reader. If the traveller desires to proceed to Wilkesbarre, or to Mauch Chunk, it would be advisable for him to leave the car, and place himself in the train for Tamaqua, twenty miles distant, where stages run directly to the place mentioned. For our part, we must proceed to Pottsville, fifteen miles distant, from which place, dear sir, we'll join you at Tamaqua, and then

Follow thee
With truth and loyalty.

We would cordially invite you and your carpet-bag to accompany us, but that there is no railway communication between the two places, and we have a horror for stages in warm weather. So, farewell!

If we do meet again, why we shall smile;
If not, why then this parting was well made.
* * * Come, ho! away! *All a-b-o-a-r-d!*

Well, leaving Port Clinton, we go puffing, and blowing, and thundering amid the wildest mountain scenery, but still keeping by the side of the Schuylkill, which gradually becomes smaller as we approach its head waters—(though we can't see that there is anything "un'nat'ral" in the circumstance.) We pass two unimportant post stations—Auburn and Orwigsburg—the former a promising candidate for village importance, and the latter a mere off-shoot of its unfortunate god-father, two miles distant—formerly the seat of justice of Schuylkill county. Eighty-nine miles from Philadelphia is SCHUYLKILL HAVEN, containing a population of nearly three thousand. It is the principal depot for the shipment of coal, both by canal and railway. Lying in a beautiful valley, it affords the only belt of tillable land to be found in the county. The valley is long but narrow, and is dotted with numerous pleasant farms, and surrounded with bold and romantic scenery, of which the annexed figure is an illustration.

LANDSCAPE.

The Mine Hill and Schuylkill Haven Railroad commences here, and following the valley for a short distance, throws out several radiating branches, connecting the main road with all the coal operations in the Mine Hill, and Swatara ranges, embracing the rich coal districts of Minersville, Tremont, Llewellyn, Branchdale, etc. The tonnage of the road is enormous, and like the Reading railroad to which it is tributary, it has a descending grade throughout its combined length. A train of passenger cars runs between Schuylkill Haven and Tremont, *via* Minersville. The route is a pleasant and attractive one—penetrating the richest coal districts of Schuylkill county. The company have recently obtained the right to extend their road (which is among the most profitable to the stockholders of any other in the United States, at the same time that it is one of the most substantial in its structure,) over the mountain, so as to connect with the Shamokin railroad at Sunbury—thus uniting the Schuylkill with the Susquehanna at that place. It is proposed, we believe, to ascend the mountain by inclined planes, constructed in the usual manner, or upon the plan of those at Mauch Chunk, hereafter described. This route will afford an outlet for the great and prolific Mahonoy coal region, and the road will probably prove as profitable, at no distant day, as the main line, with its numerous projecting branches, now is.

Three miles above Schuylkill Haven we reach MOUNT CARBON, which was formerly the terminus of the Reading Railroad. A large quantity of coal is also shipped from this place, from which several lateral railroads extend to the coal mines in the vicinity of Pottsville, Port Carbon, St. Clair, Tuscarora, and other mining districts. The handsome cottage on the slope of the hill on the opposite side of the river, is the residence of Mr. Walker, superintendent of this section

MANSION HOUSE NEAR POTTSVILLE.

of the railroad. The stone octagonal building in front of it, is his office. On the left, and near the railroad, is the Mansion Hotel, now conducted by Mr. Head, one of the most distinguished caterers on the American continent. His reputation, in connection with hotels, is so well established, and so preëminently superior to what is ordinarily associated with country inns, that no remark of ours could add one jot to its value. While proprietor of the Mansion House in Third street, Philadelphia, his guests—always few in number—comprised some of the most distinguished and opulent citizens which the country could boast. His wines were recognized as indisputably superior to those of any public or private gentleman in the city, while his *table d'hote* literally groaned beneath the sumptuous dishes spread out upon it.

This hotel has recently been materially enlarged and improved.

It is the only establishment, in this part of the country, specially adapted for the accommodation of summer visitors to the coal region —being large and airy, and sufficiently near Pottsville to render it readily accessible, and sufficiently distant to avoid its dust and business excitements. It has an extensive and beautiful park attached, with bowling house, and other arrangements for out-door amusements. The location, as may be supposed from a glance at it, is extremely cool and pleasant in the summer, as well as quiet and retired. The nights are particularly refreshing, and sleep is to be enjoyed, after the heat of the day, with a vigor perfectly unknown in the crowded city. Some time since, the family of Iterbide, formerly Emperor of Mexico, and the family of Mr. Tucker, the distinguished President of this road — (and the Emperor of American Railway Managers) made this hotel their annual summer quarters.

POTTSVILLE, nearly a mile above this hotel, is the great theatre of the anthracite coal trade. It is situated principally on the northern slope of Sharp Mountain, which constitutes the boundary of the coal formation. The present population is about eight thousand, included in which are some of the most active merchants, coal operators, and business men to be found anywhere in the State. The citizens are remarkably intelligent and enterprising, and there is probably no place in the commonwealth where the people combine a greater amount of practical intelligence with the accomplishments of travel and scholastic learning. The evidences of their industrial energy are scattered broadcast throughout the coal region—above as well as *below* ground. Schuylkill county presents a perfect net-work of railroads and canals, and there are probably upwards of *one hundred and fifty miles* of the former laid down below the surface of the earth. At nearly every turn in the road, the stranger will hear the loud puff of the colliery steam-engines, and the shrill whistle of the locomotive resounding through the narrow valleys and passes of mountains. Pottsville itself contains several large machine-shops, as well as a railroad and bar-iron rolling mill, recently erected. All the stationary steam-engines used in the coal regions are made here or in some of the adjacent villages. The heavy machinery used in the railroad mills at Phœnixville and other places, was produced here, and it is probably a sufficient compliment to her mechanics to say, that their productions are properly appreciated where they are subject to the severest test, which is in their own immediate locality.

POTTSVILLE.

Pottsville, like all the other towns in the coal region, is of recent origin. Previous to 1824 there was scarcely a dwelling on the spot where the town now stands. The excitement which followed the discovery of coal, brought to the place a swarm of adventurous spirits, which rendered it the focus of unprecedented speculations in coal lands and town lots. In the midst of this excitement, the town took a run-and-jump into existence. It never went through the slow and gradual movements of a baby-existence; but with one tremendous bound, found itself nestling at the foot of a high mountain, swarming with hungry speculators and eager adventurers of every description—young, old, and ugly—green, black, and brown—all huddled together, and "eager for the 'fray."

The late Joseph C. Neal, who was one of the motley mass, some years afterwards, wrote the following humorous description of the speculating scenes:

In the memorable year to which we allude, rumors of fortunes made at a blow, and competency secured by a turn of the fingers, come whispering down the Schuylkill and penetrating the city. The ball gathered strength by rolling, young and old were smitten with the desire to march upon the new Peru, rout the aborigines, and sate themselves with wealth. They had merely to go, and play the game boldly, to secure their utmost desire. Rumor declared that Pipkins was worth millions, made in a few months, although he had not a sixpence to begin with, or to keep grim want from dancing in his pocket. Fortune kept her court in the mountains of Schuylkill county, and all who paid their respects to her in person, found her as kind as their wildest hopes could imagine.

The Ridge-road was well travelled. Reading stared to see the lengthened columns of emigration, and her astonished inhabitants looked with wonder upon the groaning stage-coaches, the hundreds of horsemen, and the thousands of footmen, who streamed through that ancient and respectable borough, and as for *Ultima Thule*, Orwigsburg, it *has not recovered from its fright to this day!*

Eight miles further brought the army to the land of milk and honey, and then the sport began—the town was far from large enough to accommodate the new accessions; but they did not come for comfort—they did not come to stay. They were to be among the mountains, like Sinbad in the valley of diamonds, just long enough to transform themselves from the likeness of Peter the Moneyless into that of a *Millionaire;* and then they intended to wing their flight to the perfumed saloons of metropolitan wealth and fashion. What though they slept in layers on the sanded floors of Troutman's and Shoemaker's bar rooms, and learned to regard it as a favor that they were allowed the accommodation of a roof by paying roundly for it, a few months would pass, and then Aladdin, with the Genius of the Lamp, could not raise a palace or a banquet with more speed than they!

One branch of the adventurers betook themselves to land speculations, and another to the slower process of mining. With the first, mountains, rocks, and valleys changed hands with astonishing rapidity. That which was worth only hundreds in the morning, sold for thousands in the evening, and would command tens of thousands by sunrise, in paper money of that description known among the facetious as slow notes. Days and nights were consumed in surveys and chaffering. There was not a man who did not speak like a Crœsus, even your ragged rascal could talk of his hundreds of thousands.

The tracts of land, in passing through so many hands, became subdivided, and that brought on another act in the drama of speculation: the manufacture of towns, and the selling of town lots. Every speculator had his town laid out, and many of them had scores of towns. They were, to be sure, located in the pathless forests; but the future Broadways and Pall Malls were marked upon the trees; and it was anticipated that the time was not far distant when the deers, bears and wild-cats would be obliged to give place, and take the gutter side of the belles and beauxs of the new cities. How beautifully the towns yet unborn looked upon paper! the embryo squares, flaunting in pink and yellow, like a tulip show at Amsterdam; and the broad streets intersecting each other at right angles, in imitation of the common parent, Philadelphia. The skill of the artist was exerted to render them attractive; and the more German text, and the more pink and yellow, the more valuable became the town! The value of a lot, bedaubed with vermillion, was incalculable, and even a sky parlor location, one edge of which rested upon the side of a perpendicular mountain, the lot running back into the air a hundred feet or so from the level of the earth, by the aid of the paint box, was no despicable bargain: and the corners of Chesnut and Chatham streets, in the town of Caledonia, situated in the centre of an almost impervious laurel swamp, brought a high price in market, for it was illustrated by a patch of yellow ochre!

The bar-rooms were hung round with these brilliant fancy sketches; every man had a roll of incheate towns in the side-pocket of his fustian jacket. The most populous country in the world is not so thickly studded with settlements as the coal region was to be; but they remain, unluckily, in *statu quo anti bellum.*

At some points a few buildings were erected to give an appearance of realizing promises. There was one town with a fine name, which had a great barn of a frame hotel. The building was let for nothing; but after a trial of a few weeks, customers were so scarce at the Red Cow, that the tenant swore roundly he must have it on better terms, or he would give up the lease.

The other branch of our adventurers lent their attention to mining; and they could show you, by the aid of a pencil and piece of paper, the manner in which they must make fortunes, one and all, in a given space of time—expenses, so much; transportation, so much; will sell for so much; leaving a clear profit of 000,00! There was no mistake about the matter. To it they went, boring the mountains, swamping their money and themselves. The hills swarmed with them; they clustered like bees about a hive; but not a hope was realized. Calculations, like towns, are one thing on paper, and quite another when brought to the test.

At last the members of the expedition began to look haggard and careworn. The justices did a fine business; and Natty M., Blue Breeches, Pewter-Legs, and other worthies of the catchpole profession, toiled at their vocation with ceaseless activity. When the game could not be run down at view, it was taken by ambuscade. Several bold navigators discovered that the county had accommodations at Orwigsburg, (at that time the seat of justice, now located at Pottsville,) for gentlemen in trouble. Capiases, securities, and bail-pieces became as familiar as your garter. The play was over, and the farce of "*The Devil to Pay*" was the after-piece. There was but one step from the sublime to the ridiculous, and Pottsville saw it taken!

Gay gallants, who had but a few months before rolled up the turnpike, swelling with hope, and flushed with expectation, now betook themselves, in the gray of the morn, and then the haze of the evening, with bundle on back—the wardrobe of the Honorable Dick Dowles tied up in a little blue and white pocket handkerchief—to the tow-path, making, in court phrase, "mortal escapes;" and, in the end, a general rush was effected—the army was disbanded—*suavi qui peut?*"

The coal region, twenty-five years ago, stood in a position equally as tempting to the people of the surrounding States, and especially those of our own, as California recently did, and still does. It was a new and unexplored region, and, in the midst of the scenes which characterized it, every one thought to play a part, and receive the smiles of fortune. Many, of course, were disappointed:—but the more practical were enabled to sustain themselves, and with the aid of the improvements made in the moments of excitement and speculation, finally established themselves permanently in the successful pursuit of their business.

Pottsville is much frequented in the summer by strangers and travellers, but principally by those who, having investments in the improvements connected with the coal trade, or in the land itself, combine business with the pleasures of travel. The place, at this season, is therefore generally pretty well filled, and adds somewhat to its interest—though there is never a lack of gaiety and spirit in the society of the town. Indeed, from what we know of it, we should pronounce it inferior, in no essential, to that of any other community in the Union—characterized, as it is, by a high tone, governed by sound intelligence and fine social feelings.

The rides in the vicinity are magnificent—for while the roads are always in the midst of the wildest and most picturesque mountain scenery, they are also enlivened with the varied scenes of industry and activity peculiar to the region—little mining villages, colliery works, saw-mills, extensive forests, rocky promontories—now looking down from the tops of mountains, then from the narrow deeply-shaded valleys looking up; these, in continual and varied succession, are among the scenes to be enjoyed in a drive, in any direction, in the vicinity of Pottsville. The roads are generally very good; indeed, in many places they are unsurpassed. Nor are they, as might be supposed, very hilly; but winding around the mountains, they attain the summit without any steep ascents, and the trees and wild bushes always afford a shade, which, while it protects the road from the sun, and prevents the accumulation of dust, only renders it more inviting for the traveller.

There are, as we have already intimated, a variety of objects in the vicinity, which the stranger might visit with satisfaction. Amongst others of similar character, is Swatara Falls, situate about nine miles from Pottsville, over a beautiful summer-road. The falls, lying about a mile from the carriage-way, are accessible over a very stony path through the bushes; but the exercise required to approach them will be amply compensated by the view afforded. The outlines are sketched in the engraving from the recollection of the writer, who visited them some four years since, and it is very probable the picture is deficient in some of the minor points. It represents the stream with a full flow of water, which, however, is not peculiar to it during the dry weather of the summer. A wilder scene than is here presented we have never found, nor a cooler spot during the heat of the summer, and after the fatigue of reaching it from the road. It is often selected, with good taste we think, as the scene of Pic Nic parties, with which we have associated it in the engraving.

Inasmuch as the general character of the country will be frequently referred to in our article on the coal formation, it would be a useless repetition to extend our remarks at this place. We will, therefore, continue the journey, and meet our old friend (who has thought well enough of our good intentions to entertain him thus far,) at Tamaqua. We take the cars near Pottsville to Tuscarora, via the Schuylkill valley railroad, and thence by stage to Tamaqua. The country, as we pass along, is grand—wild—sublime, and so forth, and all that; at Tamaqua the gorge in the mountain, through which the wild little Schuylkill, like a lost child, runs on to meet her anxious maternal

SWATARA FALLS.

parent at Port Clinton—is truly magnificent. Sharp mountain rises to a lofty height in the rear of the town, and its summit affords a splendid view of the whole interior coal basin, which is here remarkably rich in its development. There are several characteristic mountain scenes along the railroad, between here and Port Clinton—did *you* observe them? Ah, it is well; it saves us the trouble of a description.

Tamaqua is beautifully situated, and is a thriving and important place. The little Schuylkill coal and railroad company own nearly all the adjacent coal lands, embracing upwards of five thousand acres, and it is through their auspices that the town has grown into its present prosperous condition. It is of comparatively recent origin, like all the towns and villages of the coal region; but its progress in population and business has been much more rapid, at the same time that it has been substantial and durable. During the depression of business which has characterized most of the mining districts, for the last few years, Tamaqua maintained a degree of comparative prosperity; and such is its peculiar position amongst the most important elements of industrial progress, that the place, at no distant day, must become one of great commercial interest. Tamaqua lies on the railroad route connecting Philadelphia with Lake Erie; and from the importance of the road, and the natural advantages favoring its construction, no doubt can be entertained of its ultimate prosecution. A large portion of the route, in fact, is already finished—a still larger portion graded, and but a comparatively short distance yet to be graded and equipped, to complete the whole, which would thus furnish a shorter route, from Philadelphia to Lake Erie, than is afforded from New York to Dunkirk, via the Erie railroad. Six miles from Tamaqua is Summit Hill, a small mining village, the spot where anthracite coal was originally discovered, and the site of the great open quarry of the Lehigh coal and Navigation Company. This quarry has been abandoned in favor of the usual process of mining by drifts. The coal strata, on arriving at this point, converge towards each other, and appear to have been overtilted, thus forming an almost solid area of coal of great thickness. Lying near the surface, it was for many years mined in open quarry. A railroad extends from this place to Mauch Chunk, over which the coal mined in the vicinity is carried. The road is nine miles in length, and has a gradual descent from the summit of the mountain to Mauch Chunk, where the coal is transferred to canal-boats. The entire descent is, we believe, over six hundred feet. The cars thus descend by their own weight, under the charge of a conductor; and the passage down affords one of the most unique and exciting trips which the imagination could picture. The scenery—oh! the wild, the bold, the terrible mountain scenery as you pass along—swifter than the winged messengers of the air! You look around and below with a feeling half fear and half inward unearthly satisfaction! Heavens! *could* anything be more pleasing—more delightful; and then, winding round a curvature, where the cars *might* run off the track, and precipitate their enthusiastic and excited contents several hundred feet into the little agricultural valley below—you tremble and shrug, and wonder *if* anything could be more *dangerous!* A-ha! Look at those lazy old conglomerates there, reposing in awful cliffs and massive columns on the mountain's side; and here, let us "calmly" survey the harvests fields, the fresh-mown hay, the little white cottages looking like children's playthings, scattered over the valley "away down below." How beautiful, in this tremendous ocean of untamed and unchiseled mountains, the little narrow patches of the farmer appear! They look like long pieces of richly figured carpet, while the stately pines give them a border of the darkest and softest green. Rising one after the other, like an army of soldiers, to the mountain top, their tall spear-shaped plumes pierce the region of clouds, while beneath they bury their quivering shadows in the solemn depths of solitude; for

> The sound of the church-going bell
> These valleys and rocks ne'er heard,—
> Ne'er sighed at the sound of a knell,
> Or smiled when a Sabbath appeared.

At the terminus of the railroad flows the Lehigh river, a stream of no great volume, except in times of long continued rain. At such seasons the banks are overflowed, and some of the villages and property on its banks occasionally suffer material injury. The banks are deep and much worn in consequence of the velocity of the stream, and the large amount of debris collected in the mountains are carried down in the current. The Lehigh empties into the Delaware river at Easton, a distance of thirty-six miles from Mauch Chunk. It is made

navigable to Easton, and also to White Haven, twenty-six miles above, for coal boats of one hundred tons. Beyond White Haven to Stoddartsville, the river has been improved for descending lumber, which forms a large trade on this river, besides that of coal. The Lehigh Coal and Navigation Company, under whose auspices these stupendous improvements were made, own upwards of ten thousand acres of the coal land embraced in this region, while their works afford an outlet for the adjacent coal districts of Beaver Meadow, Spring Mountain, Hazleton, Buck Mountain, White Haven, etc. Projects for the navigation of the Lehigh were set on foot as early as 1792; but it was only after the discovery of coal, and when its importance began to be righly appreciated, that its entire completion was effected. A vast amount of capital has been sunk unnecessarily, which has involved the present company to a serious extent; but the growing importance of its trade must and will ultimately place the works in a paying condition.

To effect the transfer of coal from the cars to the canal boats, extensive steam and other works are employed. In the first place, there is an inclined plane, running from the railroad, (which terminates at a point about one hundred and fifty feet above the level of the river), to the banks of the canal-basin, where tressel-works are erected, projecting over the river. Over this plane the loaded cars descend. The rapidity of their descent is in a measure checked by the weight of the empty cars ascending, which, being fastened at the other end of the rope, and moving on a parallel railway on the same plane, necessarily mount as rapidly as the loaded cars descend. But the partial counterpoise is still insufficient to moderate properly the speed of the descending cars. This object is effectually gained by an iron band which clasps the drum, (to which the rope is attached,) and which, compressed by a lever, controls its motion. Accidents have rarely occurred in this descent, but the cars have sometimes deviated or broken loose. They are now guarded against by a very simple, yet ingenious contrivance. The railway is double until the most rapid part of the descent is passed, when both ways curve and unite into one. Should a car break loose, therefore, its momentum will be so great as to prevent its following the curve, and as soon as it reaches the spot, it is thrown off the track, overturned, and lodged on a clay bank formed for this purpose below. Farther down, a bulwark is constructed, over-arching the railway, to intercept the loose coal as it flies from the cars. When the car arrives at the foot of the inclined plane, it pitches into a downward curve in the railway, and a projecting bar, which secures the lower end of the car, and which, for this purpose, is hung in a horizontal axis, knocks it open, and the coal slides down a steep funnel or shute, into the canal boat, which, receding from the shore by the impulse thus given it, occasions the coal to spread evenly over its bottom.

In addition to this inclined plane, however, there are shutes connecting directly with the railroad and the banks of the canal. These shutes are probably upwards of two hundred feet in length, lined with an iron flooring. The coal is thrown in from the car above, and slides down to the boats in the canal, thus saving the trouble and expense of hoisting the cars up and down. These works are all indicated in the engraving annexed.

After the cars are unloaded, they are returned to the summit, in precisely the same manner as they came down—that is, by gravitation. To effect this, two inclined planes are used, one of which, on the top of Mount Pisgah, is indicated in the engraving. A stationary steam-engine is placed at the head of each plane, by means of which the empty cars are drawn up. After reaching the top of Mount Pisgah, they descend by gravitation a distance of six miles, when the other plane is reached. Raised over this, they descend again, in like manner, until they get their supplies of coal, when they are returned on the road already described. An imaginary railway circle is thus described, over which the cars proceed with the swiftness, almost, of lightning, without any motive-power whatever. Mount Pisgah plane is twenty-two hundred and fifty feet in length, overcoming a perpendicular height of nearly seven hundred feet. This is probably the greatest elevation overcome by any other single inclined plane in the world. Previous to the completion of this new road, the empty cars were drawn back by mules, that always accompanied the loaded trains in their descent, having had cars expressly appropriated to their accommodation. Upwards of six hundred mules were thus employed, occasioning, as may well be supposed, a heavy item of expense, including that necessary for the support of their drivers. It is, in our opinion, a serious obstacle to the success of this company, that so much machinery, and that of an expensive and complicated character, has to be maintained to carry on their business.

Mauch Chunk is a remarkable village, and no engraving could hope

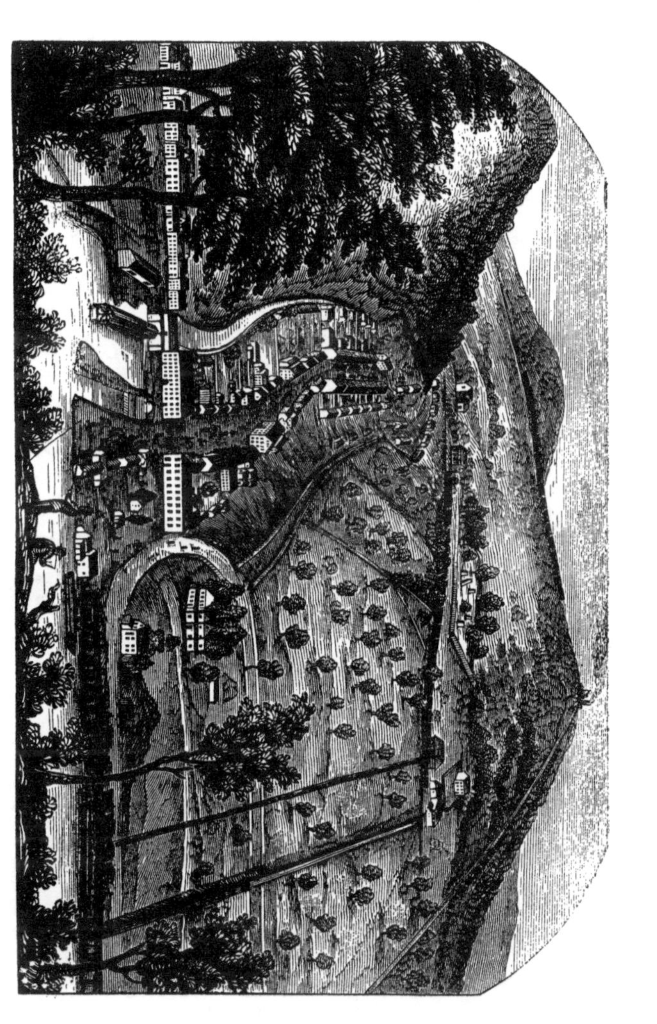

to portray its peculiar features—nor, indeed, could any considerable portion of it, small as it is, be exhibited in a single view, since the town creeps in amongst the narrow valleys of the numerous mountains, in which it is situated, for nestling places. It is a bird's-nest of a place—hemmed in by high and steep mountains on all sides—some gracefully curving around it, while others terminate abruptly in its midst, and seemingly frown down upon it. The houses, which are generally very neat and creditable structures, are built on the sides of the mountains—in some instances the fronts presenting three and four stories, the rear *one*. There are two principal hotels, which are much frequented during the summer months by travellers and tourists, but more especially by enterprizing capitalists and manufacturers interested in the vast resources of the country, and who, being usually accompanied by their wives and daughters, make their visit one of business as well as pleasure—mingling profit with the pursuit of novelty and entertainment—wild scenery, wholesome air, and so forth.

We have now—oh, reader!—we have now finished Part First of this, our "Travels!"

# PART II.

## The Anthracite Coal Formation.

"I will teach you to pierce the bowels of the earth, and bring out from the caverns of mountains metals which will give strength to our hands, and subject all nature to our use and pleasure."—*Dr. Johnson.*

COAL is indisputably a *vegetable production*, and occupies a position low down, among the earliest deposits of the primeval earth. For a long time it was hard to recognize its vegetable origin, because the fact opened the door to some of the grandest and most wonderful phenomena ever grasped by philosophy—originating theories and hypotheses, as it did, which conflicted not only with every previous opinion, but seemingly struck a heavy blow at the truths of revealed law. Instead of the earth being but a few thousand years old, the coal deposits prove it to be of great and incalculable antiquity—numbering its years not by centuries, but rather by thousands and millions far anterior to the flood. Nor is its wonderful antiquity inconsistent, as was at first supposed, with the doctrines of divine revelation.* Science and Religion are identical in their true mission,

* Prof. Silliman, during a recent course of lectures on Geology, before the Smithsonian Institute at Washington, stated, decidedly, that there is no conflict between geology and the scripture history. The case is widely different from that of astronomy, more than two hundred years ago, which was condemned as heretical, because the scriptures described the *appearances* of the heavens only, which were all that in general mankind could be acquainted with.

But in the case of geology, there is not even a literal discrepancy. On the contrary, all the geological formations correspond in the order of time, and, as far as they are described in the scriptures, with the nature of the deposits, and cannot fail to harmonize when properly understood. The beneficent doctrines of the great Mediator were promulgated at a period when the world was unprepared either for the startling truths or the practical benefits of science—hence it was left for the Bible to *gradually*

especially in the succession of created beings endowed with life, and man in both systems crowns the whole.

The only change required is extension of time, so as to afford enough to allow the events to happen by natural laws, established by the Creator, and expressive of his will, which is thus distinctly recorded in the earth itself. "The beginning" is not limited in time, and may extend as far back as the case may require; thus providing for all the early formations.

The periods called days are not necessarily such as we now denote by that word. There could be no regulation and division of time, as we now have it, until the sun was set to rule the day. Morning and evening may be, before that time, figurative expressions, denoting merely beginning and ending, as we say the morning and evening of life. The word day is used, in this short narrative, in all the senses in which it is ever employed in language, and significantly in the recapitulation or summary; in the beginning of the second chapter, day is used for the whole period of the creation, and in the same sense in various other parts of the scriptures.

The periods required for all the amazing series of events recorded in the earth are necessarily long; and if time was measured by natural days in the fifth and sixth periods, during the creation and sepulture of innumerable races of marine and terrestrial animals, there must have been a repetition of very many of those days to make out a long epoch, which might as well be regarded at once as a period of sufficient length for the work.

The Sabbath stands by itself, after the work is finished, a moral institution, having no necessary connection with the preceding physical events. By it man is every week reminded of his Maker and his destination, and although neither morning nor evening are in the Genesis named in connection with the Sabbath, it has no doubt always been of the same length as now, and does not belong to the geological epochs.

If this view is not acceptable, it is still indispensable, that in some way the time should be found, and no person fully acquainted with the structure of the earth can doubt that the time was very long, and no other person can be admitted as qualified to judge in the case. There is no reason to believe that man has been in the world more than six thousand years and the antiquity of the planet refers to ages before man was created. The allusion in the commandments, and in other parts of the scriptures, to the six days would of course be made in conformity with the language adopted in the narrative, which, being for the mass of mankind, was necessarily a popular history, although of divine origin; and the historian adopted a division of time that was in general use, *prepare* and guide fallen man to a higher and nobler destiny, rather than to confound and corrupt him with a premature and *unnatural* perception of the mysteries of the universe, and the great social and physical laws that impel him onward. This is the mission of the

although as to half the time, at least, it was inconsistent with astronomical laws. Extension of the time to such a length as to cover the events by the operation of physical laws removes every difficulty, interferes with no doctrine of religion, and prepares us to exclaim with our divine poets—

These are thy glorious works, Parent of Good:
Almighty! thine this universal frame,
Thus wondrous fair: Thyself how wondrous, then,
Unspeakable, who sits above the heavens,
To us invisible, or dimly seen in these thy lower works.
Yet these declare thy goodness beyond thought,
And power divine.—*Milton.*

Thou giv'st its lustre to an insect's wing,
And wheel'st thy throne upon the rolling worlds.
From Thee is all that cheers the life of man,
His high endeavour and his glad success—
His power to suffer and his will to serve;
But oh! Thou bounteous Giver of all good,
Thou art of all thy gifts thyself the crown.
Give what thou canst; without Thee we are poor—
And with Thee rich, take what thou wilt away.—*Cowper.*

Bayard Taylor who, at the last advices received from him, was in Egypt, gives the following in connection with his visit to the great temple of Abou-Simbel. "The sculptures on the walls of the grand hall are, after those of Medeenet Abou, and on the exterior wall of Karnak, the most interesting I have seen in Egypt. On the end wall, on either side of the entrance, is a colossal bas-relief, representing Remeses slaying a group of captive kings, whom he holds by the hair of their heads. There are ten or twelve in each group, and the features, though they are not coloured, exhibit the same distinction of race as I had previously remarked in Belzoni's tomb, at Thebes. There is the Negro, the Persian, the Jew, and one other form of countenance which I could not make out—all imploring with uplifted hands the mercy of the conqueror. On the southern wall, the distinction between the Negro and the Egyptian is made still more obvious by the coloring of the figures. In fact, I see no reason whatever to doubt that the peculiar characteristics of the different races of men were as strongly marked in the days of Remeses as at present. This is an interesting fact in discussing the question of the unity of origin of the races. I

Bible, and wherever its pages have been freely unfolded, there has science followed, and proclaimed her unyielding laws.

Referring the primary origin of the earth itself to the nebular theory of Herschell, it is supposed to be filled with everlasting fire—the result, probably, of its internal chemical organization, or the original incandescence of the planet. However this may be, the existence of universal heat within it, is amply demonstrated by the variation of the temperature of the atmosphere as we proceed downwards—a descent of a few thousand feet bringing us into a region entirely too warm to sustain life. It is further demonstrated by volcanic eruptions, which have through all time, and at various places, vomited out streams of burning lava and scoriæ, overflowing vast regions of country, as well as filling up the bottoms of the sea; while in more recent times they have buried entire cities—men, women, and children; servant and master; resident and stranger; the princely palace—the capitol—the column, and the arch—all buried in one common grave! To such eruptions, as the inevitable consequence of inextinguishable fire, should also be added the phenomenon of earthquakes, which are no less disastrous and terrible in their effects, and no less frequent in their visits. These elements of destruction have probably been awakened at irregular periods, and when fully aroused, have operated generally throughout the globe, rather than in isolated districts; hence gradually arose the vast mountain chains that now traverse the face of the globe, from pole to pole, throwing back, in their ascent, the waters of the sea, and dividing them by impenetrable barriers. Thus was slowly produced a material refrigeration of the climate—for it must be understood that, previous to these epochs, the climate of the earth must have been universal, or nearly so, and that, at least, it was much warmer than it now is, even in the torrid zone. This is evident from the fact that coal is distributed in all quarters of the globe—in cold as well as in warm regions.

have as yet, though deeply interested in the subject, not looked into it sufficiently to take either side; but, admitting the different races of men to have had originally one origin, the date of the first appearance of Man on the earth, must have been nearer fifty thousand than five thousand years ago. If climate, customs, and the like, have been the only agents in producing that variety of race, which we find so strongly marked nearly four thousand years ago, surely those agents must have been at work for a vastly longer period than that usually accepted as the age of Man. We are older than we know; but our beginning, like our end, is darkness and mystery.

Whatever may have been the local characteristics of the primeval earth, it is certain that the climate was much warmer, and much more humid than it now is, or has been in modern times. The character of the vegetation abundantly establishes this. Of the large number of plants comprising the coal-bearing period, there are few which bear any analogy to existing species, and these are the exclusive productions of the torrid zone. Fossil Botany, it is true, is a comparatively recent science—yet enough has been elicited through its aid to afford a good idea of the vegetation of the ancient earth. The vegetation of the coal formation, according to Dr. Lindley, consisted of ferns in vast abundance; of large coniferous trees, of species resembling *lycopodiaceæ*, but of most gigantic dimensions; of vast quantities of a tribe apparently analagous to *cacteæ* or *euphorbiaceæ*, but perhaps not identical with them; of palms and other monocotyledons; and, finally, of numerous plants, the exact nature of which is doubtful. Of the entire number of species detected in this formation, two-thirds are ferns. The fossils are divided by Botanists into the following genera, determined by the character of their fronds; pachypteris, sphenopteris, cyclopteris, glossopteris, neuropteris, odontopteris, anomopteris, tæaniopteris, pecopteris,louchopteris,clathropteris, schizopteris, otopteris, caulopteris and sigillaria, etc., the two latter occurring only as stems, and the last being considered by many as a dicotyledonous plant. Of these, figure 1 exhibits a specimen of the neuropteris, or

FIG. 1.—NEUROPTERIS.

nerve fern, which are plentifully distributed in the coal. Figure 2 is a specimen of the odontopteris, or tooth-fern, not so numerous as the former, but still characteristic of this formation. The next, Anomopteris, are seldom met with, but nevertheless flourished in this era.

FIG. 2. ODONTOPTERIS.

FIG. 3.—ANOMOPTERIS.

The Pecopteris, figure 4, is by far the most numerous of all varieties of the fern, having upwards of sixty different species in the coal. The common brake, or fern, exhibits a type of the family of which the figure will serve as a specimen; but the aborescent ferns, which now grow only in the vicinity of the equator, present the closest analogy to those of the carboniferous period, which were lofty trees, far surpassing in height and magnificence, even their tropical congeners of the present day. From their number and variety, they afford some of the most interesting fossil remains which the vegetable kingdom has produced. Their leaves are generally elegant, and display great variety of form and diversity of venation; from these characters the generic and specific distinctions of the family are obtained. They are often preserved in great perfection, and even the organs of fructification are occasionally observable at the back of the leaf. Several fine specimens of the fern may be seen at the Pennsylvania Hall, in Pottsville, while nearly every coal operator has more or less of various kinds of vegetable fossil, which they exhibit for the gratification of friends. Martin Weaver, Esq., has shown us several of the finest impressions we have yet seen, and he had, at one time, if he has not now, a considerable collection.

Fig. 4.—Pecopteris.

Fig. 5 exhibits a variety of fruit, of the family of *Chara*, and comprised in the same class as the foregoing. The fruit is oval, and consists of five valves, twisted spirally, with a small opening at each extremity. The figure on the left, marked 1, exhibits the nut within

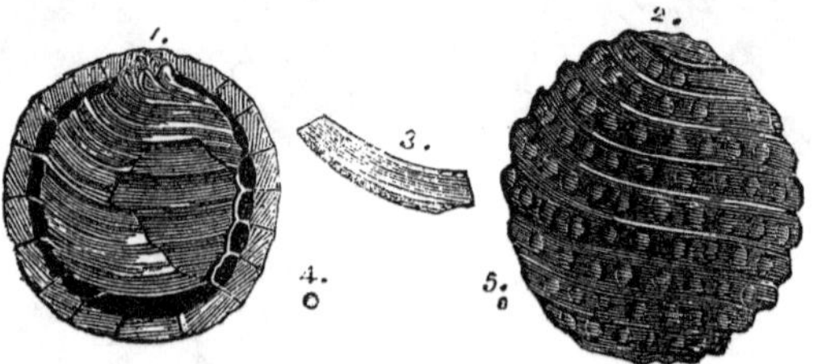


FIG. 5.—CHARA.

the pericarp; 2 shows the pericarp; and 3 a portion of the spiral valve, magnified, while 4 and 5 are the natural size of 1 and 2, magnified in the engraving. Of the family of the *club-moss* or *lycopodiaceæ*, there are numerous specimens, the most common of which are the lycopadites, lepidodendron, lepidostrobus, and stigmaria, a specimen of which we append—fig. 6. The stem of the Stigmaria was

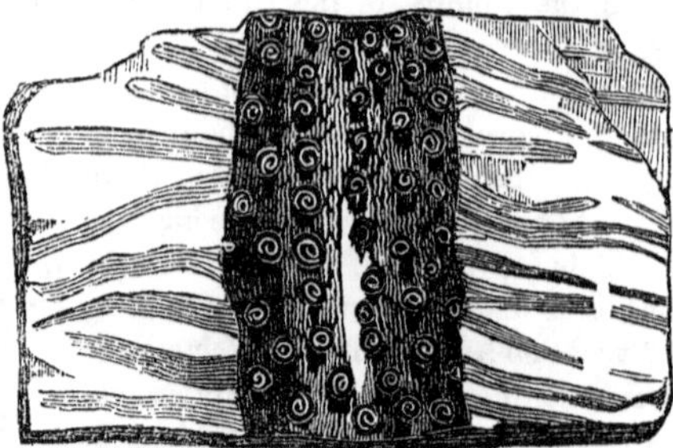

FIG. 6.—STIGMARIA.

originally succulent, marked externally with roundish tubercles, surrounded by a groove, and arranged in a direction more or less spiral —having a distinct axis, communicating with the tubercles by woody processes. Fig. 7 exhibits a specimen of the Pterophyllum, or wing-leaf, of the family of *Cycadeæ*, seldom met with in the coal, and of which the leaves only are known. Fig. 8, however, of the same family, called *Asterophyllites*, is one of the most numerous dicotyle-

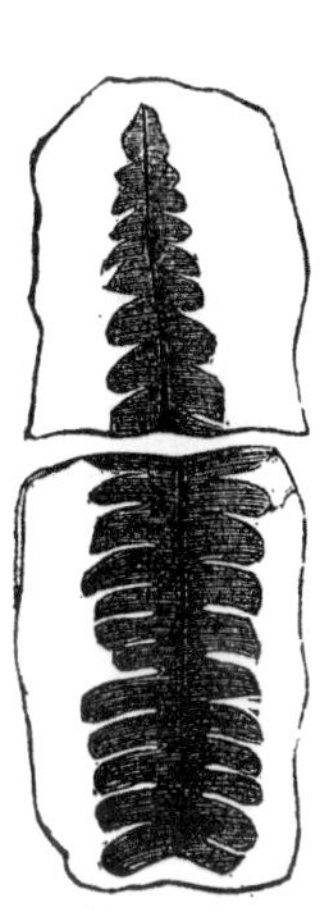

FIG. 7.—PTEROPHYLLUM.

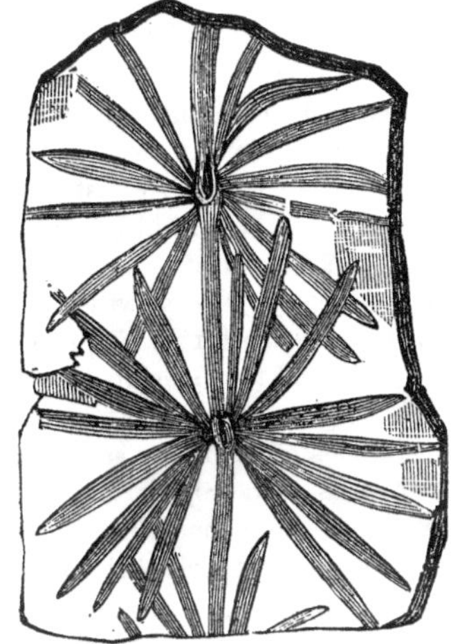

FIG. 8.—ASTERAPHYLLITES.

donous plants found in the coal, but unlike the other, the *stems* only are known. There is a variety of others which it is probably unnecessary to specify—the examples already afforded being, we think, quite sufficient to convey an idea of the several families comprising the coal vegetation. Of the numerous families composing the class of monocotyledonous plants, there are comparatively few to be found in the coal formation. The class of dicotyledonous, however, presents a greater variety and number, most of which belong to the family

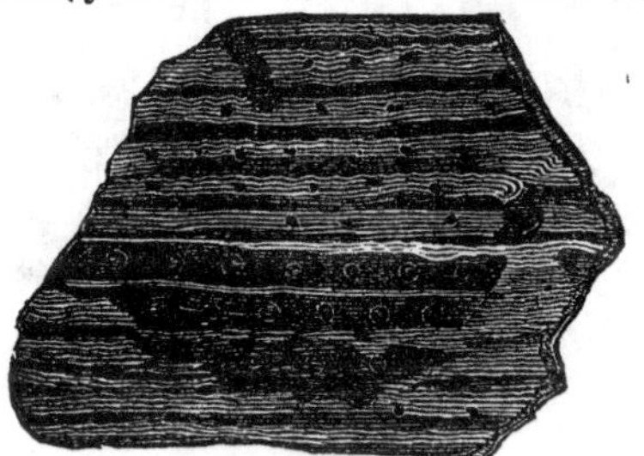

FIG. 9.—SIGILLARIA.

of *Sigillaria,* of which fig. 9 affords a specimen. The Sigillaria is one of the most important plants of the coal, and probably furnished a very large amount of its vegetable matter. The stem is conical, and deeply furrowed, with scars between the furrows in rows, but not arranged in a distinctly spiral manner. There are some forty species in the coal formation.

The most common of the coal-plants may be classified as follows: first, ferns and Sigillaria. Second, lepidodendron, a doubtful genus, variously associated by botanists. Thirdly, calamites. Fourthly, coniferous plants; and fifthly, stigmaria, which is probably an extinct family. To ascertain more satisfactorily the nature and circumstances attending the growth of the vegetable matter, it is necessary to institute some further inquiries. Thus, by comparisons with *existing species*, and the elements constituting their growth, we obtain considerable light on the subject, and are able to form conclusions which could not otherwise be safely arrived at. We shall append a few illustrations. The Sigillaria, so numerous in the coal, have generally been classed as monocotyledonous plants; but late observers contend that they properly belong to the dicotyledonous division. The irregular and longitudinal furrows of the surface of the stems —their swelling out at the base, angle of dip or downward direction of the roots, are characters constantly observable in the dicotyledonous, but never in monocotyledonous plants. Besides, these trees have a separable bark; and slices of it, prepared for microscopic investigation, have exhibited traces of medullary rays, which are universally recognized as proofs of dicotyledonous structure. While they are

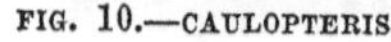

FIG. 10.—CAULOPTERIS.

FIG. 11.—EXISTING TREE-FERN.

thus regarded as dicotyledonous, or exogenous and compact trees, Dr. Lindley has divided from them another genus, termed *caulopteris*, which he considers true stems of tree-ferns. These are hollow, but the markings which they exhibit present so close a resemblance to existing tree-ferns as to leave no doubt of their identity with those plants. They are, however, comparatively rare in the coal; while of the true Sigillaria, over forty species have been discovered. We append a figure of the fossil stem, caulopteris, 10, and a figure of an existing tree-fern, 11, in juxtaposition for comparison.

The family of Lepidodendra have, by some writers, been supposed to belong to that of the club-mosses; while the larger species were regarded as forming a transition to the coniferous plants. The living species of their supposed analogues, fig. 12, abound in tropical climates;—they generally creep on the ground, some grow erect, but none exceed three feet in height; whereas, fossil specimens have been found over thirty feet high, while fragments have been discovered

FIG. 12.—CLUB-MOSS.

FIG. 13.—LEPIDODENDRA.

indicating a much larger size, figs. 13 and 14. Figure 15 exhibits a specimen of a numerous family, called *Crassula Tetragona,* probably allied to the above species, which are found at the Cape of Good Hope. They occur in the driest situations, where not a blade of grass nor a particle of moss can grow, on naked rocks, old walls, or hot sandy plains, alternately exposed to the heaviest dews of night, and the most intense rays of the noon-day sun. Soil is to them a something to keep them stationary, rather than a source of nutriment,

FIG. 14.—LEPIDODIA.

FIG. 15.—CRASSULA TETRAGONA.

which, in these plants, is conveyed in myriads of small cuticular pores, to the cellular tissue which lies beneath them.

The Calamites are not analogous to any existing species, as already noted, though they resemble some plants in structure, but differ widely in their proportions—the fossil indicating large trees, while the existing species which they resemble are but two or three feet high, and of corresponding diameter. Of the coniferæ of the coal, it has been observed that they bear a strong resemblance to existing *pines*—slices of the wood, when examined by the microscope, showing that the ducts or glands peculiar to this family of trees, are arranged in a similar manner, that is, alternately in double and triple rows, fig. 16.

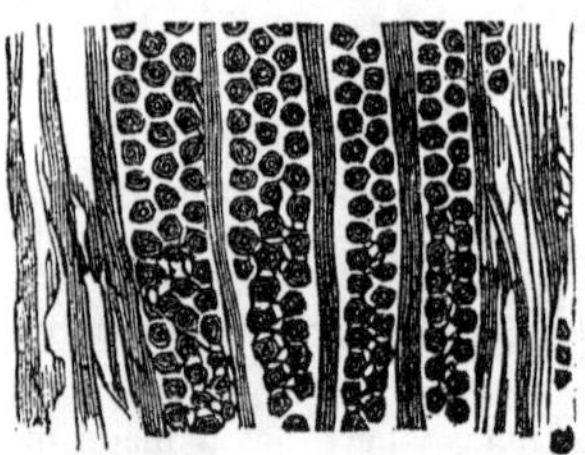

FIG. 16.—CONIFERÆ.

The stigmaria is generally supposed to have been a large succulent water-plant—the stem, in its compressed fossil state, varying from two to six inches in diameter, and has numerous processes, which proceed vertically, horizontally and obliquely, and traverse the beds in every direction. These processes have been traced to a distance of eight or ten feet from the stem, and had a horizontal range of twenty feet. From the extraordinary number of these plants, it is concluded that they have furnished the material for the great bulk of our coal beds.

From the general character of the vegetation, and the absence of the great mountain ranges which now conspicuously mark the earth's surface, it is probable that water covered a far greater area of country than it subsequently did, while, at the same time, its mineral qualities must have been essentially different from what they are now.

FIG. 17.—THE COAL VEGETATION.

The land, lying low and in broad marshes, must have resembled, in some respects, our great western prairies, so well known for their rank vegetation, which, added to the peculiar warmth and humidity of the climate, produced plants of extraordinary proportions—far exceeding our loftiest forest trees. The vegetable matter growing thus spontaneously under active stimulants, formed immense wild coverings, by which it was peculiarly adapted to receive the ascending charges of the elements constituting its growth. Fig. 17 exhibits an ideal view of the coal vegetation.

Regarding the manner of deposit, much difference of opinion exists among Geologists. For a long time an opinion prevailed (and is still entertained by some), that the vegetable matter was removed from the place of its growth by drift, and deposited in the bottom of the sea, or the estuaries of lakes and rivers, where it underwent a process of fermentation and pressure from the superimposed debris that accumulated upon it, and thus gradually changed into the state of coal. It is now, however, rendered probable that it grew on the identical beds in which we find it, and the supposition is supported from the fact, amongst others, that fossil trees have been found in the coal formation in an erect position, with portions of their trunks charred, and passing into the state of coal, which is, of course, inconsistent with the theory of their removal by drift.

Indeed, when we consider the enormous amount of vegetable matter entering into, and necessary to have produced even the smallest seam of coal, it is hard to conceive how it could have drifted from the place of growth—especially, too, as the floating mass would have been exposed to the liability of meeting and intermixing with various other substances, tending to impair the purity of the coal; whereas no such evidence is afforded. It is obvious, therefore, that the coal grew on the spots where it is now deposited, and the only remaining point to establish this view, is to account for the deposition of the *intermediate strata*. This, however, is not an easy task—for they comprise marine deposits of every description, in addition to those of sand, clay, and mud, which have produced the extensive sandstone rock that lies around the seams of coal. The only way it can be accounted for, is to suppose the submergence of the coal beds, time after time, and the deposition of the sea-shells and crustaceæ, that are now found over them, combined with the conglomerate detritris borne into the estuaries of lakes and rivers—after which the waters probably receded, and suffered another supply of vegetable matter to accumulate. That something like this process is at least probable, is sufficiently evident from the alternation of marine deposits and other matter with the coal beds, and their position high up upon the summits of mountains, hundreds of miles from the present flow of the sea. How else could they have been deposited there, in regular order and succession?—though it is nevertheless probable that extraordinary floods, internal convulsious and outbreaks in the earth's crust, as well as the general changes of land into sea, and sea into land, at that period, (and even now constantly going on,) contributed much as co-operative agents. We append a single figure, 18, showing the horizontal position of strata, which will also serve to illustrate the alternation of coal veins with other deposits in the same basins.

FIGURE 18.

The vegetable material, therefore, having been thus secured, a chemical process subsequently ensued, as before stated, by which the mass was turned into coal. The fermentation produced by the pressure of the overlaying strata, and the impossibility of the immediate escape of its gaseous elements, heated it sufficiently to produce a body of pitchy or bituminous matter, and the coal is consequently bituminous, or only partially so, in proportion as these gases were subsequently let out by the cracks, and fissures, and disruptions, going on in the surrounding strata. For we find that when the strata are undisturbed, bituminous or fat coals predominate; whereas, where the strata are inverted, and torn and disruptured, anthracite, or coals which have lost the greater portion of this pitchy matter, prevail. Thus, after long-continued and constantly increasing pressure, the vegetable matter becomes one compact body of coal; and now, after the lapse of countless centuries, during which the process of mineralization has still continued, we find it embracing every shade and variety of quality, according to the original ingredients constituting the growth, age, and local circumstances governing its deposition.

In connection with this branch of the subject, we present the following extracts from the opinions of Sir Charles Lyell, of England, who visited this region in 1841. In reference to the origin of coal, whatever dispute there may have been on the subject, he thinks was settled when a portion of the New Castle coal, some years ago, was submitted to a *microscopic examination*. After cutting off a slice so thin that it should transmit light, it was found that many parts of the pure and solid coal, in which geologists had no suspicion that they should be able to deduct any vegetable structures, not only were the annular rings of the growth of several kinds of trees beautifully distinct, but even the medullary rays, and what is still more remarkable, in some cases, even the spiral vessels could be discovered. But besides these proofs, from observing a vegetable structure in the coal itself, there has been found in the shales accompanying it, fern leaves and branches, as well as other plants, and when we find the trunks of trees and bark converted into this same kind of coal as we find in the great solid beds, no one will dispute the strong evidence in favor of the vegetable origin of this coal. If we find a circumference of bark surrounding a cylindrical mass of sand, we know that it has been a hollow tree filled up with sand, nor can there be any doubt that the coal is formed of vegetable matter. No less than three hundred species of plants have been well determined by botanists; some of whom have devoted a great part of their lives to this study. From this it is to be inferred that the carboniferous formation of Europe and America is made up of comparatively recent plants. He thus alludes to three or four of the most peculiar facts which lead to this conclusion.

In the first place, the boughs and leaves of ferns are the most frequently and strikingly met in America as well as Europe. So perfectly have they been preserved that there can be no doubt that they are really ferns; and in some cases even their infloresence has been preserved at the back of the leaves. Where we have not the flowers and prints remaining we have found it possible to distinguish the different species of fossil and ancient ferns by attending to the veining of the leaves. At least one hundred species are determined in this way. The most numerous of those vegetable veinings are those which have been called *Sigillaria or tree ferns.* Their stems are found to be fluted vertically, and in the flutings are little stars, as it were, each of which indicates the place where the leaf was attached; and it is evident, as M. Brongniart has shown, that although the bark of these trees is so well marked that forty-two species have been described, yet there is never found any leaf attached; while we have in the same beds leaves in abundance which have no trunks. The natural inference is, that they must have belonged to the aborescent ferns; as, for instance, the section *Cauloptoris* is admitted by all to have belonged to this species. The fact is also important because the tree-ferns, and especially the *Cauloptoris*, are now known to be exclusively the inhabitants of a *warm and humid climate*—much more hot and moist than in those parts of the globe where coal now abounds. For we find coal, not only in England and Nova Scotia, but as far north as Melville's Island and Baffin's Bay, in a climate where the growth of such fern plants is dwarfish and stinted. It is evident that when these vegetables existed there must have been a warmer, and probably a more equable climate than is now found even in warmer latitudes.

The climate in Northern latitudes was then much warmer and more moist than it is now in any part of the globe. The same thing is made evident by a comparison of their fossil *Sigillaria* with those which now attain their greatest size in the islands of the Pacific. He had found several plants, as the *Asterophyllites*, in the Apalachian Chain, and which are also found in Nova Scotia and Europe, which cannot certainly be referred to any living families. These all, however, bespeak a terrestrial vegetation, though occasionally found mixed with marine shells and corals.

Another class of fossils common in coal shales is the *Lepidodendra*, somewhat allied in form to the modern *Licopodiums*, or white mosses. Though the mosses of the present day are never more than shrubs, even in the warmest regions, yet, at the carboniferous period they attained an enormous development, being forty, fifty, or even seventy feet high.

There have been two theories to explain how these plants could have been carried into the sea, estuaries, or lakes, and drawn beneath the water and accumulated in the strata, so as to form coal. One of them asserts that the plants must have been drifted and buried in the water, since we find them intercollated between different strata of shales; just as plants lie between the leaves of a botanist's *herbarium*, and are pressed together, so have these ferns been found flattened between the seams of shale. They have been carried from the place where they grew, drifted out to a certain distance, water-logged and sunk in the mud, and other strata deposited above them, so as to form this intercollation between the different leaves of clay.

But many believed, from seeing the roots, that the plants grew on the spot where we now find them. But when we come to observe that these roots terminate in different strata, it will seem evident that they were carried down, sunk and struck in the mud, as snags are now in the Mississippi. * * * * This may appear contradictory to what has been said with regard to a change of climate since the carboniferous era; but it is not necessarily so. The opinion of Werner, confirmed by the speculations of Brongniart, led him to believe, contrary to his early impressions, that by far the greater part of the coal had grown on the spot where it is found. Accumulating like peat in the land, the land must have been submerged again and again, to allow the strata of sand and mud to be superimposed as we now find them.

In excavating for coal at Belgray, near Glasgow, in 1835, many upright trees were found with their roots terminating in a bed of coal; and only seven years ago, in cutting a section of the Bolton Railroad in Lancashire, eight or ten trees were found in a vertical position; they were referrable to the *Lipidodendra* species, and allied *Licopidiums*, or club mosses. All were within forty or fifty feet of each other, and some of them were fifteen feet in circumference at the bottom. The roots spread in all directions, and reached beds of clay, and also spread out into the seams of coal. There is no doubt that these trees grew where they are found, and that the roots are in their original position. The seam of coal has possibly been formed of the leaves which fell from the trees. This is a singular fact: that just below the coal seam, and above the covering of the roots, was found more than a bushel of the *Lepidostrobus*—a fruit not unlike the elongated cone of the fir tree. It has always been imagined that the *Lepidostrabus* was the fruit of the *Lepidodendra*, but here they are found beneath other trees.

Under every seam of coal in Wales is found the fire-clay—a sandy, blue mud, abounding in the plants called *Stigmaria*. First is the seam of coal, then the fire-clay, then another seam of coal, and then the sandstone. In one open part of the Newcastle coal field, about thirty species of *Sigillariæ* were discovered; the trunks were two or three feet in diameter. They pierce through the sand in a vertical direction, and after going for some eleven feet perpendicularly, the upper part bends round horizontally, and extends laterally into the sand—and then they are so flattened by the superincumbent strata, that the opposite barks are forced within half an inch of each other. The flutings are beautifully preserved in the flattened horizontal stems. Here had been an ancient forest growing in a bed of clay—buried in some way with sand to a certain depth, and then the upper part was bent and broken off by the water current, and buried in layers of shale and mud. There are many cases of this kind in Wales, where the roots of the trees evidently preserve their original position. Mr. Logan, an excellent geologist, has examined no less than ninety of these seams of coal in Wales. They are so exceedingly thin that they are but of little value in an economical light—yet, they are just as important for geological purposes, as if they were thick strata. Under every one of the ninety, he has found fire-clay, a sandy mud, containing the plants called *Stigmaria*. It was discovered years ago that this fire-clay existed with the coal mine; but it was not known that it was the floor of every coal seam, and not the root, which contained this plant in a perfect state. The *Stigmaria* appears in the under-clay (to use the term employed by the miners,) a cylindrical stem, from every side of which extends leaves—not only from the opposite sides, but from every side, they appear like tubercles, fitting on as by a joint. They radiate in all directions in the mud, where they are not flattened like the ferns. Had they been, we might have had leaves in two directions, but not on every side. These plants resemble the *Euphorbiaceæ* in their structure, and in some respects are analogous to the caniferous or fir tribes. In their whole structure, they are distinct from all living genera or families of plants. In one instance, a dome-shaped mass was found with stems and leaves—some of the branches being twenty or thirty feet in length, and sometimes longer. It has been thought by Dr. Buckland and other geologists, that those plants either trailed along in the mud at the bottom of the swamps, or floated in lakes like the modern *Stratiotes*.

After Mr. Logan had arrived at this remarkable fact, Mr. Lyell became particularly desirous to know if the same fact was true in the United States. When he arrived here in August, 1841, he had no idea how far it was true, yet it was known the *Stigmaria* did occur; and his first opportunity to inquire into the fact was at Blossburg, in the Bituminous field, in the northern part of this state. His first inquiry of the geologist was, whether he found *Stigmaria* there. He received in answer an affirmative reply; and on being asked if the plant occurred in the *under-clay*, he said that they could soon settle the point. Whereupon he had one of the mines lighted up, and the only plant *they could find in the under-clay was this Stigmaria*. It existed in abundance—its leaves radiating in all directions, just as in Wales, more than four thousand miles distant. The same cretal appearance was preserved. In the roof of the coal seam were seen different species of ferns—*Sigillaria* and *Calamites*, just as in North Carolina and in Wales. Afterwards another opportunity occurred in the Pottsville region of anthracite coal. Professor Rodgers, the state geologist, who, though well acquainted with the strata of the district, was as anxious as Mr. Lyell to know if the rule would hold good, examined, first at Pottsville, and then at Mauch Chunk, and the same phenomenon was observed at both points. In the first coal mine they came to, the coal had all been quarried away (for the work was carried on in open day), and nothing but the cheeks of the mine remained. The beds, as they have been horizontal, are now not vertical, but have gone through an angle of little more than ninety degrees, and turned a little over, so that what is now the under side was originally the upper; therefore, the cheek on the left side was originally the floor of the mine. They now looked at the lower cheek, and the first thing they saw was the *Stigmaria*, very distinct; on the other side, but a little way off were Ferns, Sigillaria, Calamites, Asterophyllites, *but no Stigmaria.* So it was at Mauch Chunk, where they found one thirty feet long, with leaves radiating in all directions.

It has now been ascertained for many years that Professor Caton was quite correct in affirming the anthracite and bituminous coals to be of the same age. This is shown, not only by their relative position with regard to the red sandstone, but from the plants found in both being identical.

All the coal fields, therefore, may be regarded as one whole, and the question will occur, how did it happen that the great floor was let down so as to prevent the accumulation of coal, and yet plants of so different textures should be found in it. It has been suggested that these plants grew in the swamps; and it is possible to imagine that there may have been morasses fitted only for the growth of the species of plants called *Stigmaria;* and that, as this marsh filled up, this and the other plants became dry, and the leaves accumulated one layer above another, so as to form beds of coal of a different nature from those that preceded. We know it is a common thing for shallow ponds to fill up gradually with mud and aquatic plants, and at last peat and trees are formed upon them. A corresponding change is constantly going on in different parts of Europe—the same transition from bogs and marshes to a soil capable of supporting various great trees is taking place, and then the ground is submerged; for always, again and again, we must refer to this subsidence of the soil.

Those who have seen the morass called the Great Dismal in North Carolina and Virginia, may possibly have had an opportunity of crossing the northern extremity of it on a railway supported by piles, from Norfolk to Welden. There is no less than forty miles from North to South, and twenty from East to West, covered entirely with various forest trees, under which is a great quantity of moss; the vegetation is of every variety of size, from common creeping moss to tall cypresses one hundred and thirty feet high. The water surrounds the roots of these trees for many months in the year. And this is a most singular fact to one who has travelled only in Europe, that, as is the case in the United

States, trees should grow in the water, and yet not be killed. This Great Dismal was explored some years ago by Mr. Edmund Ruffin, author of the valuable Agricultural Journal. He first calls attention to the fact that a greater portion of the vast morass stands higher than the ground that surrounds it; it is a great spongy mass of peat, standing some seven or eight feet higher than its banks, as was ascertained by careful measurements when the railroad was cut through. It consists of vegetable matter, with a slight admixture of earthy substance, as in coal. The source of peat in Scotland is, that one layer of vegetation is not decomposed before another forms. So is it in Chili, Patagonia and Terra del Fuego. Thus, also, is it in different parts of Europe, in the Falkland Islands, as Darwin has shown. Thus, too, is it in the Great Dismal, where the plants and trees are different from those of the peat in New York. It is found, on cutting down the trees and draining the swamp, and letting in the sun, that the vegetation will not be supported as it was before, beneath the dark shade of the trees. In the middle is a fine lake, and the whole is inhabited by wild animals, and it is somewhat dangerous to dwell near it by reason of the bad atmosphere it creates. It is covered by most luxuriant vegetation. It is found in some places in England, that there is a species of *walking-mosses*, which are sometimes seized with a fancy to walk from their places; the moss swells up, bursts, and rolls off, sometimes burying cottages in its path. In some places this peat has been dug into and houses have been found several feet below the surface—curious antiquarian remains. In the same manner the Great Dismal may spread itself over the surrounding country.

Having thus dwelt somewhat minutely upon the coal formation, and the geological phenomena to which it is allied, we will, in conclusion, take a retrospect view of the strata of the earth, and the means which have, from time to time, modified and changed its configuration. The origin of our earth, as already intimated, must have been a mass kept in a state of fusion by heat, its surface becoming hard by being gradually cooled. The most ancient portion of the earth, therefore, is composed of granite, which appears in an unstratified mass, and bears every indication of an *igneous* origin. There are some kinds of granite, however, of comparatively recent origin, which so clearly resemble the ancient rock as to be sometimes difficult to distinguish one from the other. Gneiss is a rock very analagous to granite. It is stratified, however, and seems to have been formed under water. It alternates with mica-schist, which ordinarily accompanies granite and gneiss. Next we have argillaceous schist, which was also formed under water, and which is of a soft, slaty nature, and easily split.

These rocks, whose origin is co-eval with the creation of the earth, are frequently found at the tops of mountains, as well as at the lowest depths of valleys, which goes to prove that the earth has, at various periods, been subjected to the severest upheavals and internal convulsions. Among these rocks *no fossils* have ever been found, and it is thus certain that animal and vegetable life did not exist at this early period of the earth's history.

It is in the next, or second geological epoch, called the transition formation, that the first traces of the existence of vegetable and marine life on the surface of the globe, are found. Previous to this period, and perhaps as a prelude to the introduction of life, the former rocks had been disturbed, as above mentioned, for we do not find the strata of the transition formation in parallel layers over the primitive beds; but, on the contrary, they are deposited in the greatest apparent confusion.

Geologists have divided this formation into three divisions, which are called respectively the Cambrian, the Silurian, and the Devonian systems of rocks. The former are the oldest *sedimentary rocks* known, and are composed of schistose grauwackes, mica-schists, and gneiss. The Cambrian rocks contain organic remains of various brachipods, polyparia, coral animals, &c.

The Silurian system, which is next above the Cambrian, comprises an upper and lower stratum, and is very nearly similar to those rocks. They are exclusively of marine origin, and whole beds are composed of shells, corals, &c., and those peculiar crustacea termed *Trilobites*, and which, being rarely found in other situations, are characteristic only of the Silurian and Devonian strata.

After the revolutions which seem to have terminated the primitive epoch, the earth must have remained for a long time in a state of repose, as we find in the third geological period, denominated the secondary formation, the stratum called the *old red sandstone*, consisting of a mass of rocks and pebbles, cemented together, having been transported and accumulated through the action of water, and upon which rest the *carboniferous deposits*. This formation is composed principally of marine fossils, the varieties of which are very numerous. The mountain limestone, and metalliferous limestone, in which are found ores of lead, copper, zinc, &c., besides numerous descriptions of organic remains, belong to this formation. Next comes the *coal formation*, and, as previously stated, this is exclusively composed of vegetable matter, formed as aforesaid, and in which marine fossils are rarely found.

A violent convulsion seems to have terminated the coal period, which was succeeded by what is called the Saliferous formation—being the fourth geological epoch. In this are found the red conglomerate, new red sandstone, &c., very often deposited in layers from one to five hundred feet deep. Few organic remains are found in these beds; but it was at this time that the animals belonging to the class of reptiles were created.

In this epoch are embraced several formations, (mostly of local names,) which, not being essential to our present purpose, it is unnecessary to enumerate.

The fifth geological epoch, (in ascending order,) comprises what are called the Liassic, the Jurassic, and the Oolitic systems. Previous to this epoch, the earth was inhabited only by certain plants, and a few inferior animals and reptiles; but at the commencement of this formation, a new fauna was created, composed of animals and reptiles of strange form and gigantic size. Rocks of the Jurassic system, as also those of the Liassic, are not met with in this country, and we therefore avoid a further reference to them, as well as the fossils which they contain.

In the sixth geological epoch, also in the secondary formation, we have the lower or inferior cretaceous system, abounding, as the latter mentioned series, in marine and animal fossils. This formation contains limestone, with here and there deposits of gypsum, clays, sands, iron ores, &c. In England, under the name of *Wealden formation*, are deposited, in alternate layers, limestone, sand, and clay, all of which are frequently of great thickness. Above the Wealden formation is a group of deposits of green sand, in which are distributed particles of silicate of iron, which are also found in New Jersey. Higher up are again found limestone, sandstones, and chalk marls, the stratification of which is only indicated by layers of flint in the latter. Beds of the cretaceous group are found in New Jersey and other parts of the United States, but they rest on the oldest secondary rocks, without the intervention of the Oolite.

The next formation, (and the seventh geological epoch) is called the *Tertiary*. Between the commencement of this epoch, and the termination of the chalk strata, all traces of ancient or primitive remains are lost; the fossils which are found in the subsequent formations being but types of existing organic creatures.

The Tertiary formation is divided by geologists into the Eocene,

Miocene, and Pliocene; or the older, middle and newer Tertiary groups. The first named stratum is developed in the states of Virginia, North and South Carolina, Georgia, Alabama, &c. It consists principally of greenish sands, nearly identical with the cretaceous series, and of the same mineral qualities. Near Paris it embraces layers of limestones, marls, and siliceous matter;—while in London it forms stiff and again plastic clays, which are useful for manufacturing purposes. Above these layers occur various kinds of clays, limestones, marls, gypsums, &c., the latter of which are extensively used in France for the manufacture of Plaster of Paris. Above the gympsum we find a more modern group, composed of marls, sands and flints—the first a marine, and the other a fresh water deposit.

The Miocene beds prevail on the Continent of Europe, and in America along the shores of the Chesapeake Bay, and in some parts of Virginia. They abound in fossils, and consist mainly of shells, sands, sandstones, and conglomerates of gravel, which are hard enough for building stones. In some portions of the globe, the Miocene series present combustible materials—and remains of dicotyledonous plants abound in them in Switzerland, Germany and Italy.

The Pliocene beds of the United States are of comparatively recent origin. They are found in New York, Kentucky, and along the banks of the Potomac in Maryland. In Europe, *brown coal*, or lignite, is found in layers, which can be advantageously worked. The beds extend all over the old world, and their mineral properties vary in different points; at some places they exhibit evidences of far greater age than at other points. They consist mainly of marls, sands, and remains of marine, fresh water, and land animals.

In this formation are also embraced superficial deposits of drift, consisting of gravel, boulders, sand, clay, &c. There are two kinds of drift, one called the ancient or *diluvium* and the other the modern or *alluvium*. In the former, which covers over the Tertiary formation, are found fossils which date not very far back from the present period,—as the diluvial period, in a manner, unites the Tertiary with the recent past. In these deposits are found bones of extinct and recent genera of animals, and among them those of the *Magatherium*, the skeletons of which measure eighteen feet in length, and about nine feet in height. This animal is much larger than any subsequent one, and the thigh-bone is believed to be three times as great as that of any known elephant. In this formation are found remains of elephants, horses, rhinoceroses, &c., while it is to this period also that geologists refer the immense masses of debris which contain gold, platina, and the diamond, in Brazil, Africa, India, and California, as well as the veins of tin in England and Mexico. The formation known as the boulder or erratic block formation, also belongs to the diluvial period. All over the world these boulders have been deposited. In some places they are of huge proportions and weight, while ordinarily they consist of gravel stones, of more or less greatness. They are composed of various mineral material, and not unfrequently are pure and hard granite.

In the United States, many of the valleys are filled up to a great depth with the modern or alluvial deposits. They consist mostly of a heterogeneous mass of earthy matter, brought down from the higher lands by rains and freshets. Bones of the buffalo, the elephant, and other animals, are found in these beds; and skeletons of the celebrated *Mastodon* have been exhumed at different localities.

It is in the modern formation, comprising the eighth geological epoch, that the first traces of the human family have been discovered; and although it is possible that its origin may date farther back than can be supposed from the evidences furnished by the exposed land, yet geologists generally unite in the belief that no earlier records appear in that portion of the earth covered by the sea.

Immediately previous to the modern epoch the earth seems to have enjoyed a repose of long duration. With the exception of a few upheavals occurring during the latter portion of the diluvial period, there has been no catastrophe of any moment; and all the changes which have taken place "since the great flood" have been brought about by various causes—by those gradual and almost imperceptible agencies which, continuing from century to century, and from thousandth year to thousandth year, will sooner or later have brought the world to another grand epoch.

Having thus desultorily traced the order of strata, we may add that it is always *regular*. We can never find coal, for example, below the more *ancient* formations; though we often see ancient rocks overlaying modern formations, the result of recent eruptions and upheavals. Thus we perceive the value, in an economical view, of scientific knowledge. Thousands of dollars have been, and are still expended by the uninformed, in explorations after mineral treasure, which, did they but enjoy a limited knowledge of those paramount laws which pervade throughout all the Creator's works, could be saved; besides the labor, anxiety, and bitter disappointments which invariably attend ill-directed enterprises.

In casting our eye over the surface of the earth, we everywhere perceive evidences of a universal and continual change. The frosts of autumn, the snows of winter, the rains of spring, the electricity of the summer—each contribute to this purpose. The substance of mountains is daily diminishing; and rocks, those silent historians of the past, gradually crumble into atoms, and unperceived, are borne off to new resting-places in the deep green ocean. Here they enter into new combinations, and by earthquakes and volcanic action, as well as by the natural accumulation of the beds, again appear to the

FIG. 19.

light of day, throwing back the surrounding waters, and presenting new "isles in the watery waste." Finally, one little island effects a friendly union with another, and thus, age after age, century after century, the undeviating, the everlasting laws of the great God are performing the functions contemplated in the creation.

Although our limits will not allow a minute description of the varied strata of the earth's crust, yet it is necessary to a proper elucidation of what has already been said as well as what is to follow, to point out some of the changes of position, of fracture, denudation and disruption which they have undergone. Fig. 18, will probably serve to show the original horizontal appearance of strata, one layer lying upon another. Fig. 19 exhibits the usual appearance of stratified rocks, lying also in a horizontal position, the lines of stratification being distinctly marked, dividing the rocks into layers very nearly detached from each other. This is peculiar to all aqueous rocks, and may be noticed in quarries of limestone, and similar stratified rocks. From a horizontal position, owing to the disturbing causes previously enumerated, the strata have in many instances been changed to a vertical position, as shown in fig. 20.

FIG. 20.

In other cases, they are changed to an inverted position by the intrusion of igneous rocks from below, and actually thrown back, as

FIG. 21.

exhibited in fig. 21. Sometimes the strata are disjointed, and fig. 22 represents a similar instance to the above of change in the direction, probably produced by a like cause; for the beds which at *b*, strike in a southerly direction, on reaching *a*, are thrown into vertical and dis-

FIG. 22.

jointed masses at *b**. In other instances, the strata are *curved*, as is frequently the case with gneiss, especially in the Isle of Wight, a delineation of an instance of which forms the frontispiece of Dr. McCullough's Western Isles, from which fig. 23 is extracted. In many cases, especially in the anthracite coal districts, the strata have a waving or arched position, similar to that indicated in fig. 23, while in other cases they are frightfully contorted, as illustrated in fig. 24.

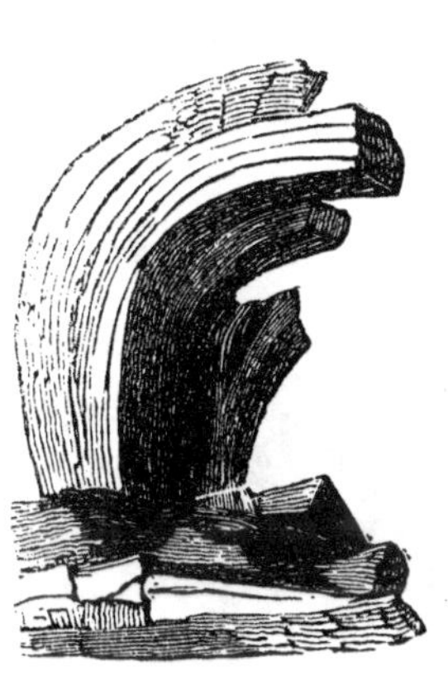

FIG. 23.

FIG. 24.

Such contortions were shown by Sir James Hall, by a simple experiment, to have resulted from lateral pressure, attended with some degree of resistance, both above and beneath. He took several pieces of cloth—some cotton, some linen—and having placed them horizontally on a table, *c*, fig. 25, covered them by a weight, *a*, placed also horizontally on the pieces of cloth. He then exposed

FIG. 25.

FIG. 26.

the *sides to pressure*, upon which the curved appearance indicated in fig. 26 was the result. It is thus that, by the chemical operations within the interior of the earth, the strata have been contorted and thrown into every imaginable shape and position, while the unstratified rocks have, at the same time, been heaved up, and thrown around in irregular shapes and quantities. The unstratified, which are the oldest of all rocks, differ from the sedimentary principally in having no lines or parallel markings; but present a shapeless and irregular mass of mineral matter, similar to fig. 27. But

FIG. 27.

while the granite, and other rocks of igneous origin, are unstratified, they still occur in veins, which are sometimes traversed by other veins newer than themselves. This is illustrated in fig. 28, where the new veins project over the old granite somewhat like the horns of a deer. These veins often penetrate the overlying deposits, and flow over the rocks which they displace, as exhibited in fig. 29.

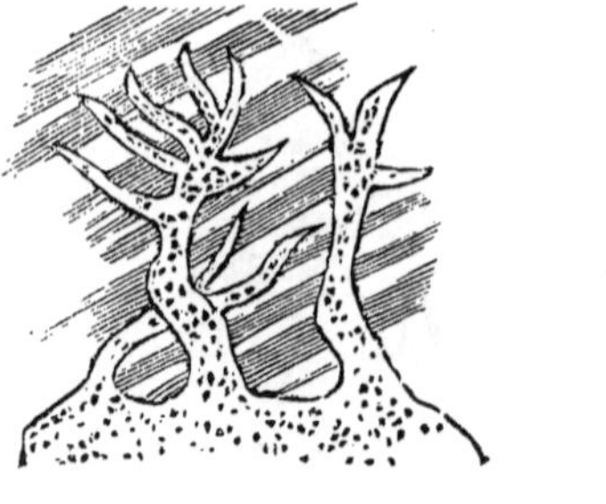

FIG. 28.

FIG. 29.

Sometimes they are so small that the markings of the cleavage are scarcely visible, yet they still resemble stratification, and might readily be mistaken for such. This deceptive appearance is often presented at the junction of granite with slate, and may readily be detected by observing the distinct mineral character of the two rocks. Some of the unstratified rocks, more especially basalt and greenstone, occasionally assume a columnar form, as indicated in fig. 30.

FIG. 30.

These columns are of various sizes, but have most generally from four to six sides. They vary, however, in length and shape, not unfrequently appearing in short blocks or prisms—sometimes standing vertically or inclined, and at others laying horizontally. In the celebrated Giant's Causeway, where they occur in a tabular mass, the columns are nearly in a vertical position, as illustrated in the engraving.

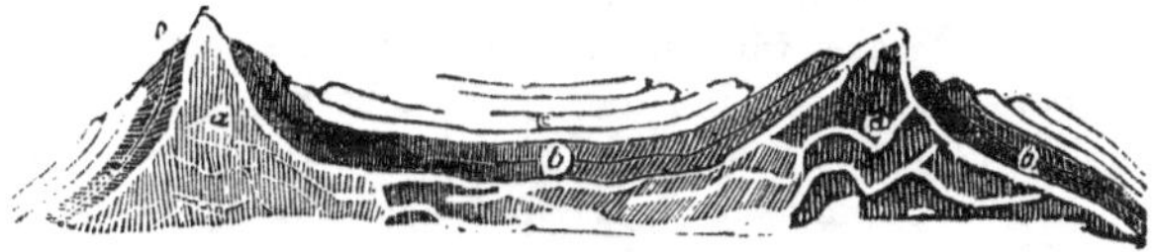

FIG. 31.

We have thus briefly described the unstratified rocks as constituting the frame-work or foundation of the whole superstructure of the globe. The accompanying figure, 31, exhibits the varied situations of the granite, (the oldest rock of the earth,) as forming the foundation upon which all the others repose, and the nucleus of the mountain, which, having been forced through the superincumbent rocks, has borne them upwards in its ascent; the strata in the vicinity of the mountain, *a*, being raised at an acute angle at *b*, and sinking to nearly a level position in the plains at *c*. The form and succession of these rocks, says Prof. Richardson, prevail all over the earth, with some local exceptions; so that its entire surface may be considered to form a series of basins, of which the largest, deepest, and thickest lie at the bottom, and are filled up by others, which become smaller, shallower and thinner as they approach the top—the deposits being uplifted and raised towards the edges of these basins, and become level, or nearly so, towards the centre.

The inclination of strata from a horizontal position is called their dip, the amount of the dip being the quantity of the angle which the line of inclination makes with that of the horizon, as in the accompanying figure, 32. If the angle made by the meeting of the lines

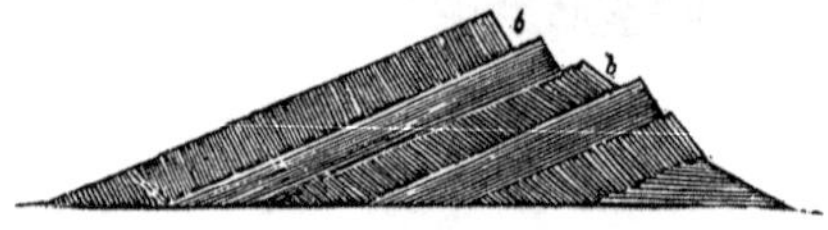

FIG. 32.

of the strata, *bb*, and the horizontal line, *a*, be equal to forty-five degrees towards the east, the strata are said to dip to that extent in that direction. Again, the terms *dip* and *strike* of strata will be further understood—(for these are terms in universal use in mining,) by the following illustration: The dip, as before observed, is the line which the strata makes with the horizon—the *strike* being a line at right angles to the dip. To illustrate; place a book on a table, with the edges of the leaves downwards, and the back of the book upwards, as in the accompanying figure, 33. If one side of the cover be removed a short distance, the cover so moved, *b*, will represent the line of *dip*, while the back of the volume, *a, a*, will exemplify the line of *strike*. If the cover of the book be extended only in a slight degree, the dip, of course, will be proportionally steep, and *vice versa*. Having thus ascertained the line of dip, we can determine the probable direction of the strike—for if the dip be towards the north or south, the strike will be east and west; and *vice versa*.

FIG. 33.

But the converse of this proposition by no means holds good; for though the line of dip gives the line of strike, the line of strike does not give the line of dip, since there are *two* lines of dip common to every line of strike; and strata having a line of strike running from north to south, may *dip* either to the east or west. In short, as we have moved one side of the cover of our book to the right, we can move the other to the left *b*, (fig. 34,) while the *back* of the volume, *a*, *a*, remains in the same position. The terms anticlinal and synclinal lines are frequently used in mining phraseology. The anticlinal line is, simply, that elevated central point from which the strata *diverge* in opposite directions. To illustrate this, we have only to extend both sides of our volume, as in fig. 35. The synclinal line is exactly the *reverse* of the above, being the point at which the strata *converge* towards each other. To illustrate this, we have merely to turn our book over, and open it only half-way, exactly in the middle, and the line between the two pages will present the synclinal line, or that point *towards* which the strata tend, as exhibited in fig. 36.

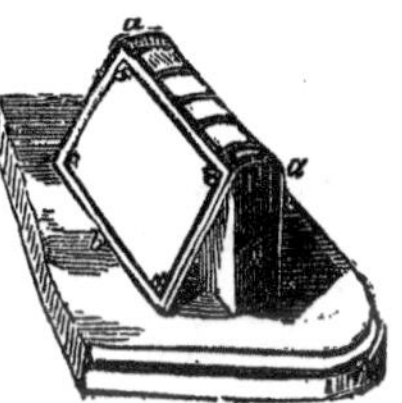

FIG. 34.

FIG. 35.

FIG. 36.

In speaking of strata, in mining phraseology, they are said to be *conformable* when their general planes are parallel, no matter what may be their dip, as in figure 37, where both the upper horizontal strata, *a*, and the lower inclined series *b*, are conformable to each other. When a series of upper strata, however, rest on a lower formation, without any conformity to the position of the latter, they bespeak a more modern series, showing that the *newest* of the underlaying group must have been deposited before the oldest of the latter. They thus occupy an unconformable position, as exhibited in the same fig. 37, wherein the upper horizontal beds, *a*, are unconformable to the lower inclined deposits, *b*. This simple illustration is very important, because it often shows the position of coal veins

FIG. 37.

lying in an unconformable position to the more modern overlaying surface.

Various writers have cautioned the observer against certain deceptive appearances of the strata in particular lines of coast, (which are no less frequent in our mountain regions,) where beds, apparently horizontal, in reality dip at a very considerable angle. The following fig. 38, exhibits a headland as seen from the south, in which the strata appear to the eye perfectly level. There appears to be no mistake about their horizontal position; but if the headland treads

FIG. 38.

off, at the point *p*, in fig. 39, to the northward, affording a view of the cliffs westward, it will be seen that the appearance from the south is defective, for the lines here show a considerable angle to the north, and gradually increasing in their dip, finally become *vertical* at *a*.

It has already been intimated that fossiliferous rocks follow an invariable *order* of succession, but that the arrangement, although never reversed, is sometimes imperfect; so that, while we never meet *b* going before *a*, or *c* preceding *b*, yet we occasionally miss not only a single letter, but a succession of letters, and find, in certain locali-

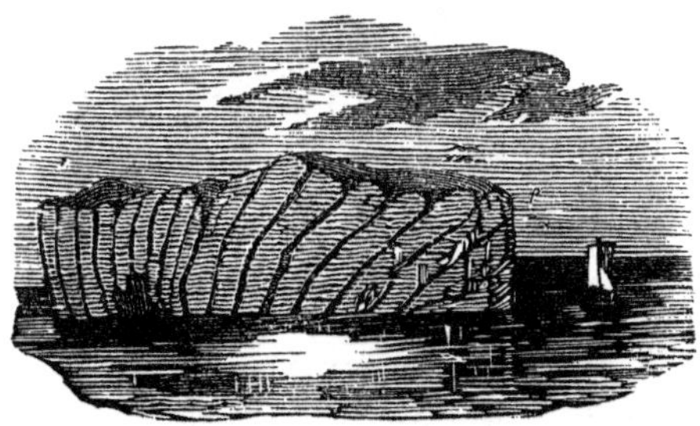

FIG. 39.

ties, that entire groups of strata are wanting, which occur in other places of like geological character. This effect may have resulted either from the missing beds never having been deposited in this spot, or from their having been denuded, and carried away by the abrading power of water, before the new strata were deposited. Similar causes may have occasioned either the partial deposition, or partial denudation of a single bed, and produced the thinning out of a particular stratum, as exhibited in fig. 40. The conformable or unconformable position of the strata affords a safe and satisfactory guide to many investigations of interest and great practical importance. From data thus furnished, we learn that the mountain-chains were not all of contemporaneous origin, but have been raised at different periods, and sometimes under different circumstances and agencies. Thus, if on the sides of one mountain, fig. 41, we find a series of strata, *a*, raised and covered unconformably by another group, *b*, it is obvious that the central chain must have been thrown up *after* the series *a* had been deposited, but *before* the formation of the beds *b*. But if, on the sides of another mountain, fig. 42, we find both the series *a* and *b* tilted, and covered unconformably by another series *c*, we have proofs *that this mountain chain* is of more modern date than that on the sides of which the same strata, *b*, are undisturbed.

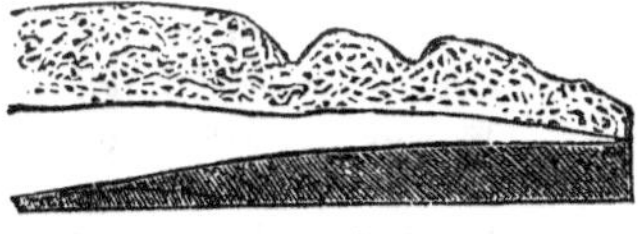

FIG. 40.

We have already remarked, that in all mineral regions, and espe-

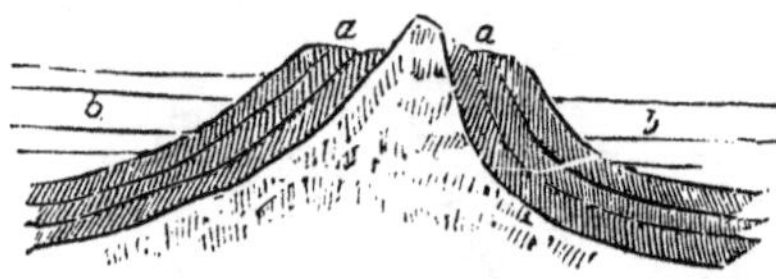


FIG. 41.

cially in that of coal, where the basins are generally more or less disturbed—as, from the very nature of the deposit, they must be—the dip and strike of the strata are matters of great practical moment. Prof. Richardson, in his geological work, supposes a case, where a land-owner, aware that coal exists on an adjacent estate, is desirous

FIG. 42.

of ascertaining whether it may also be found on his own, and whether an attempt to discover it might be instituted with probabilities of success. In this case, the dip is almost the sole reliance. If the dip of the strata in the vicinity be *towards* the land where the trial is to be made, it is highly probable that the coal may be found under

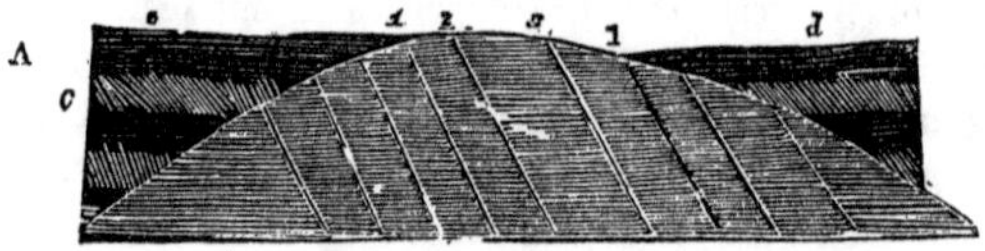


FIG. 43.

it; but if it is in a contrary direction, the search ought not to be undertaken (unless, on examination, the veins should prove to be broken, and have a backward pitch.) The lines outcroping at the surface, fig. 43, and numbered 1, 2, 3 and 4, represent coal veins, dipping towards *d*, on the right-hand side; the unconformable strata, *c c*, are beds of sandstone lying over the coal veins. Supposing coal vein No. 4 to rise to the surface at that point on the estate of A, adjoining the estate of B, which lies towards *d;* it is apparent that A would find only a point of the vein on his land, and that it would be useless to search in the direction of *b* for it, since the dip of the veins is sufficient to show that none exists there. But on the estate of B, though no coal came to the surface, still the dip of that which exists on the estate of A, would render it probable that coal could be readily found—the circumstances of its lying too deep for successful mining being considerations which would depend very much on the angle of dip, and the nature of its position in other respects. Strata are said to form outlayers when they constitute an isolated portion, detached from the principal mass of the same bed or region of which they once

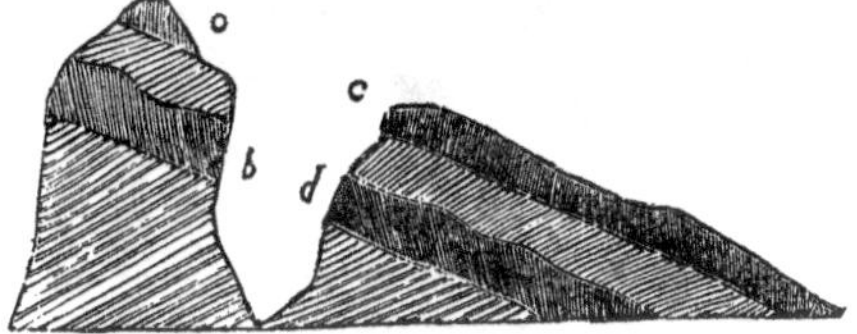


FIG. 44.

formed a part. (The anthracite coal regions are merely outlayers of the great Alleghany bituminous field, which originally comprised one immense body.) Thus, in fig. 44, the beds *a* and *b* form outlayers of the main strata, *c* and *d*—the *missing* portion having been removed by denundation, while their original identity is fully established by the accordance of the mineral deposit and position. Strata are also said to form an escarpement when terminating abruptly, as in the above figure, *a* and *b*.

The origin of valleys has most generally been referred to the agency of water; but there are other causes besides this. The surface, as

FIGURE 45.

well as the interior strata, are first dislocated by enormous fissures, caused by the upheaval of the region of which they form a part.

Fig. 45 represents such an upheaval, and shows the steep escarpements which follow as a natural consequence. It is in these fissures, therefore, that the formation of valleys commences, gradually enlarging until two or more unite. It is thus that most of the mountains east of the Alleghany have been formed, the water traversing them having carried off the material lying over them, and thus left steep and

FIGURE 46.

rugged ridges, with narrow intervening valleys. Valleys of undulation, however, are produced directly by two neighboring elevations, which, by lifting the strata on either side without occasioning fracture, leave the valley between. This is the most usual structure of the coal beds of Schuylkill County, fig. 46, and hence it follows that that county contains a much greater amount of mineral, in proportion to the superficial area, than any other district in the United States. The whole county is but a succession of wave-like elevations, with narrow intervening valleys, all of which are full of the valuable mineral for which the region is so remarkable. Valleys of erosion are formed by the action of water. Imagine a nearly level plain, and then, at one end of it let a stream of water issue forth: in a comparatively short time, with the assistance of snows and rains, and alternate dry seasons, it will scoop out a hollow similar to fig. 47, while, in course of time, it will form a deep valley, surrounded with high elevations, or table-lands. The Alleghanies have, for the most part, been scooped out in this manner, and the debris deposited in the table-flats sloping out from its loftier ranges.

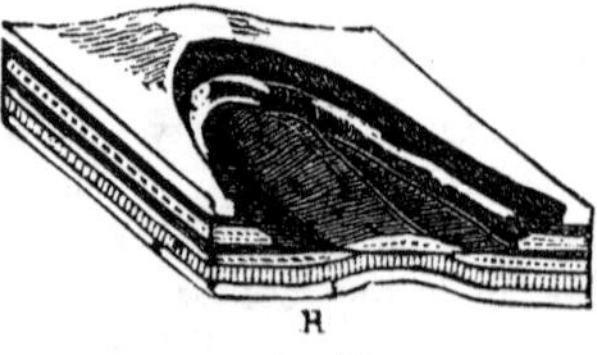


FIG. 47.

The word *fault* is one very extensively used in mining, and refers to the dislocations which interrupt the continuity of the strata. They are of various kinds and forms, and constitute a source of great expense and annoyance in mining coal. Fig. 48 represents an example, where the strata, which were once continuous, either by their subsidence on one side, or their elevation on the other, have been dislocated and displaced. Suppose that *b*, on the left, were a coal vein: on arriving at the *fault*, and penetrating it, the coal vein is lost, and a great expense necessarily ensues before it can be found. While faults are a source of great annoyance, generally speaking, they still afford some corresponding advantages, since they somewhat counteract the tendency of the coal veins—pitching, as they do in this region, at a steep angle, to plunge into inaccessible depths; and when the fissures are filled with solid rocks, as they most generally are, they form strong supports for the overlaying strata, as well as embankments to keep back the water from the mine. There is, we have stated, a

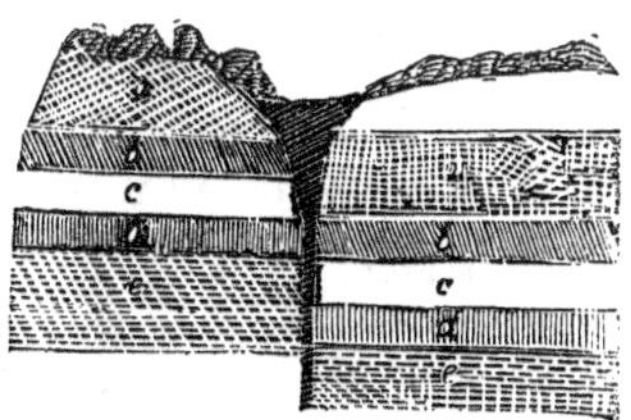

FIG. 48.

FIG. 49.

variety of faults, both of rock and clay, or soft earth. We present another idea in reference to them, in fig. 49, where the strata have been still more disturbed than in the preceding.

We have thus, somewhat briefly, pointed out some of the leading features constituting the vegetation and stratification of the coal formation; to dwell upon them at greater length than is necessary to sustain the tenor of our object, would be a waste of time, and would usurp too much of our space. For much of what has been said, we are indebted to the able geological works of Prof. Richardson, and acknowledge our obligations with a high sense of appreciation of his researches. With these examples, therefore, we must leave this branch of the subject, and proceed to the direct consideration of the great anthracite coal beds now before us.

## DISCOVERY OF ANTHRACITE COAL.

The discovery of coal in the Lehigh district is said to have been purely accidental. There had been legends of long standing, supposed to have emanated from the Indians, that coal abounded in this section of Pennsylvania; and among some of the credulous German farmers in Lehigh, Berks, Lancaster, &c., one is occasionally reminded of them, and grave intimations thrown out that coal is reposing in "certain places" beneath the luxuriant soil of those counties. Such traditionary reports prevailed for a long time among the early settlers of the territory now comprising the several counties of the anthracite regions, and if similar ones in the counties above named should ever be realized in the same happy manner, all will unite in admiration of the German stoicism with which they are still maintained by the "older inhabitants." The story of its discovery near Mauch Chunk, in the present county of Carbon, is doubtless already familiar to many. Nevertheless, it is so curious and romantic in itself, and is fraught with such miraculous results upon the physical and mental condition of mankind, that we cannot omit it here. The account was given by the late Dr. James, of Philadelphia, who, in the year 1804, in company with Anthony Morris, Esq. of the same city, visited some lands, held jointly by them, near Sharp Mountain.

"In the course of our pilgrimage, we reached the summit of Mauch Chunk mountain, the present site of anthracite coal. At the time there were only to be seen three or four small pits, which had the appearance of the commencement of rude wells, into one of which our guide, Philip Ginter, descended with great ease, and threw up some pieces of coal for our examination. After which, whilst we lingered on the spot, contemplating the wildness of the scene, honest Philip amused us with the following narrative of the original discovery of this most valuable of minerals, now promising, from its general diffusion, so much of wealth and comfort to a great portion of the United States.

"He said that when he first took up his residence in that district of country, he built himself a rough cabin in the forest, and supported his family by the proceeds of his rifle; being literally a hunter of the backwoods. The game he shot, including bear and deer, he carried to the nearest store, and exchanged for other necessaries of life. But at this particular time, to which he then alluded, he was without a supply of food for his family; and after being out all day with his gun in quest of it, he was returning, towards evening, over the Mauch Chunk mountain, entirely unsuccessful and disappointed; a drizzling rain beginning to fall, and night rapidly approaching, he bent his course homeward, considering himself one of the most *forsaken* of human beings. As he strode slowly over the ground, his foot stumbled against something, which, by the stroke, was driven before him; observing it to be black, to distinguish which there was just light enough remaining, he took it up, and as he had often listened to the traditions of the country of the existence of coal in the vicinity, it occurred to him that this might be a portion of that *stone-coal*, of which he had heard. He accordingly carefully took it with him to the cabin, and the next day carried it to Colonel Jacob Weiss, residing at what was then known by the name of Fort Allen—(erected under the auspices of Dr. Franklin.) The Colonel, who was alive to the subject, brought the specimen with him to Philadelphia, and submitted it to the inspection of John Nicholson and Michael Hillegas, Esqs., and also to Charles Cist, a printer, who ascertained its nature and qualities, and authorized the Colonel to pay Ginter for his discovery, upon his pointing out the precise spot where he found the coal. This was readily done by acceding to Ginter's proposal of getting, through the regular forms of the patent-office, the title for a small tract of land, which he supposed had never been taken up, comprising the mill-seat on which he afterwards built the mill which afforded us the lodging of the preceding night, and which he afterwards was unhappily deprived of by the claim of a prior survey."

Coal was known to exist in the vicinity of Pottsville more than seventy years ago, and searches for it had been made repeatedly—but the coal found was so different from any previously known, that it was deemed utterly valueless—more especially as no means could be devised to burn it. Searches for it were abandoned, at least for a time, when a blacksmith, by the name of Whetstone, luckily chanced upon some, and immediately undertook to use it in his shop. After experimenting with it for a short time, his efforts proved successful, and his triumph having been duly communicated, in the shape of

14

local gossip, to the citizens of the surrounding neighborhood, attention was very soon after directed to the expediency of instituting further inquiries as to the nature and extent of the deposit, and its applicability for other purposes. Among those who at a very early period did not hesitate to declare his belief in the existence of coal in this district, was the late Judge Cooper; and it was through the influence of such persons that searches were continued through circumstances and prejudices at once discouraging, and seemingly foolhardy. Among the first, if they were not the first, who undertook explorations for coal, were the Messrs. Potts. They made examinations at various points along the old Sunbury road, but in no instance did success attend them. The late William Morris, soon after the operations of Messrs. Potts were terminated, became proprietor of most of the lands lying at the head of the Schuylkill; and about the year 1800 he was fortunate enough to find coal, and in the same year took a considerable quantity to Philadelphia. It was in vain that he held forth its peculiar virtues, and vast future importance—all his efforts to convince the people of its adaptation to use proved abortive; and when, occasionally, an individual was found who could be induced, through the force of argument and eloquence, to coincide in the merits of "stone-coal," the well-known lines—

> A man convinced against his *will*,
> Is of the same opinion still—

would be involuntarily forced upon his mind; and finally he had no other alternative but to dispose of his lands, and abandon his projects as altogether fruitless.

We do not know that any farther notice had now been taken of this coal, for six or seven years afterwards. Peter Bastons made some discoveries of its deposit, while erecting the Forge in Schuylkill Valley; and a blacksmith, named David Berlin, continued to improve upon the suggestions of Whetstone, (who, by this time, had discontinued business, and perhaps left the vicinity) and imparted his successes freely to others of his craft. But few, however, could be prevailed upon to use it. Prejudice—prejudice was ever keen, and it seemed to keep men of ordinary spirit at a respectful distance. Men of iron nerve could only oppose themselves to the current.

In the latter part of the year 1810, a practical chemist, combining science with practice, made such an analysis of the coal of this region, as convinced him that there was inherent in the mass all the properties suitable for combustion. He therefore erected a furnace in a small vacant house on Front street, between Philadelphia and Kensington, to which he applied three strong bellows. By this means he obtained such an immense *white heat* from the coal, that platina itself could have been melted! From this experiment was derived such proofs of its qualities, as ultimately favored its general introduction into that city.

But although it might easily be inferred that such experiments could not fail to have secured for it immediate favor, yet such was by no means the fact. Intelligent men, it is true, calmly deliberated over the subject, but that was all—the time had not yet come to act. Two years after this, the late Col. George Shoemaker and Nicholas Allen discovered coal on a piece of land which they had but recently purchased—in times past called Centreville—situate about one mile from Pottsville. They raised several wagon-loads of it, but no purchaser could be found. Mr. Allen soon became disheartened, and disposed of his interest in the lands to his partner; who, having received some encouragement from certain citizens of Philadelphia, persevered in his operations. He got out a considerable quantity, and forwarded ten wagon-loads to Philadelphia, in quest of a market. Its arrival there was, as usual, greeted with the warmest *prejudice*, and there were few who appeared to evince any curiosity or interest in the subject. Nearly every one considered it a sort of *stone*, and, saving that it was a "peculiar stone"—a stone-coal—they would as soon have thought of making fire with any other kind of *stone!* Among all those who examined the coals, but few persons could be prevailed upon to purchase, and they only a small quantity, "to try it;" but alas! the trials were unsuccessful! The purchasers denounced Colonel Shoemaker as a vile imposter and an arrant cheat! Their denunciations went forth throughout the city, and Col. Shoemaker, to escape an arrest for swindling and imposture, with which he was threatened, drove thirty miles out of his way, in *a circuitous route, to avoid the officers of the law!* He returned home, heart-sick with his adventure. But, fortunately, among the few purchasers of his coal, were a firm of iron factors in Delaware county, who, having used it successfully, proclaimed the astounding fact in the newspapers of the day. The current of prejudice thereafter began to waver somewhat; and new experiments were made at iron works on the Schuylkill, with like success, the result of which was also announced by the press. From this time, anthracite began gradually to put down its enemies—and among the more intelligent people, its future value was predicted.

The first successful experiment to *generate steam* with anthracite coal, was made in 1825, at the iron works at Phœnixville. Previously, however, John Price Wetherill, of Philadelphia, made several efforts to accomplish this, at his lead works—but we have understood that he only partially succeeded.

We will now pass by three or four years, during which little worthy of note occurred, and behold the coal trade, in the first smiles of infancy, starting into active existence. As early as 1812 the forests in the neighborhood of Philadelphia, and the vicinity of many of the principal towns of the adjoining counties, began rapidly to disappear. Cord wood, and every description of building timber, were held at high prices—the former, during the winter months, frequently ranging between thirteen and sixteen dollars per cord. There were no improvements except turnpike roads, by which the magnificent timber of some of the more distant counties could be reached; and under these circumstances, and as population and business increased, attention was directed to the necessity of rendering navigable the Schuylkill river. It was originally designed for the products of the forest, the mine, and the field; all of which abounded in the counties drained by this stream, and its numerous tributaries. The forests, especially, were remarkable for the quality of the timber, and the height and symmetrical beauty of the trees; and among intelligent capitalists little doubt was now entertained as to the destiny which awaited the product of the mine—satisfied that it needed but a fair start to ensure its onward progress.

We have thus glanced at some of the leading incidents connected with the early history of the coal trade; it now remains to consider the position, dimensions, and structure of the coal basins themselves, which, ever since their discovery, have annually grown in value and importance, and, in their future bearing upon the economy of trade, are more important to the people of Pennsylvania than all the gold of California.

## GEOGRAPHICAL POSITION OF THE COAL REGIONS.

Locality.—The anthracite formation of Pennsylvania lies in the Counties of Schuylkill, Dauphin, Lebanon, Carbon, Northumberland, Columbia, and Luzerne, in the middle part of the Eastern portion of the State. It is watered by the Susquehanna, Schuylkill, and Lehigh rivers, and their numerous tributary branches.

Extent.—The anthracite formation of Pennsylvania may be divided into three grand divisions, or large coal regions; the first, or most southern division, being known as the South Anthracite Region; the second division called the Middle Anthracite Region, and the third grand division is known as the North Anthracite Region, or Wyoming Coal-field.

The three great anthracite regions may again be divided into coal districts, as follows, viz.: The coal districts contained in the south anthracite region, commencing at its eastern end, and continuing thence westward, are the Lehigh, Tamaqua, Tuscarora, Schuylkill Valley, Pottsville, Minersville, Swatara, and the Lykens' Valley, and Dauphin—the Lykens' Valley being the north fork, and the Dauphin the south fork of the western extension of the south anthracite region.

The middle anthracite region, commencing at the western end, and continuing thence eastward, has the Shamokin, Mahanoy, Girardsville, and Quaquake coal districts; together with the small detached coal basins contiguous to the Lehigh river, as the Beaver Meadow, Hazleton, Black Creek, Sandy Creek, and others of still smaller area.

The north anthracite region, commencing west and continuing thence north-eastward, has the Shickshinny, Wilkesbarre, Newport, Pittston, Lackawanna, and Carbondale coal districts.

The south anthracite region extends in length from its eastern point-like end, near the Lehigh, to its western terminus near the Susquehanna—a distance of about seventy-five miles. The greatest breadth, including the coal formation on Broad Mountain, is about six miles. This measurement is across the widest and central portion of the region, and will only hold good for a short distance. The average width of coal ground of the south anthracite region is not more than about two miles. This region, as has already been remarked, is spread like a canoe, being broadest at Pottsville, and gradually contracting at each end at the Susquehanna and the Lehigh. Thus, at Tamaqua, sixteen miles east of Pottsville, the basin is a little more than a mile in width, and the arrangement of the strata is exhibited in the following figure, 50, which we extract from the book of the late Richard C. Taylor.

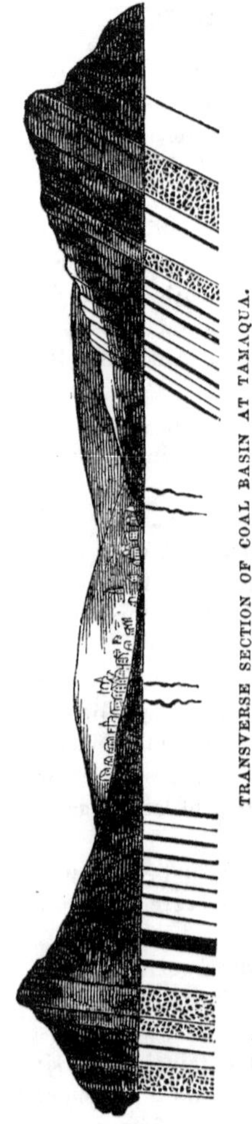

TRANSVERSE SECTION OF COAL BASIN AT TAMAQUA.

FIG. 50.

The middle anthracite region, with the detatched coal basins at its eastern part, on the Lehigh, extends in length to its point-like terminus at its western end, which point is about seven miles east from the river Susquehanna—a distance of about fifty miles. The middle region will average nearly as much *coal ground* as the first named region.

The north anthracite region extends from its north-eastern end, on the head waters of Lackawanna creek, to its western point at Skickshinny, on the north branch of the Susquehanna, a distance of upwards of sixty miles. This will not average so great an area of coal ground as either of the other two great regions.

Within the limits of the three great anthracite regions, are ridges and spaces composed of conglomerate, red shale, and sandstone strata, which lie between, and separate from each other the several basins of each of the three great divisions. In this stratification no coal exists. The value of the land which contains the coal is calculated by taking into consideration the number, thickness, character, and quality of the veins of mineral in each particular place, and from their adaptation for mining to advantage, and their accessibility to market.

## GEOLOGICAL CHARACTER OF THE ANTHRACITE COAL FORMATION.

The anthracite formation of Pennsylvania, as regards its geological character, especially in the south region, is very much distorted, and the coal veins disturbed, and irregular in their courses. In working the mines *faults*, both of a hard and soft nature, or, in other words, rock and slate, (or what is not inappropriately named *dirt faults*, some of which are of great magnitude) are frequently met with, which not only prove a great loss to the owners of the properties in which they occur, by diminishing the quantity of coal, but are often a serious inconvenience to the prosecution of the mine, and a great drawback upon the profits of the operators and lessees of the colliery, sometimes occasioning the abandonment of the work altogether.

In the middle anthracite region, taking as an index the mines in operation, the explorations already made, and the general kind appearance of the rocks, and great regularity of the surface, it is presumed that faults will not be found to exist to any great extent. Indeed, the whole geological character of the middle anthracite region—the general order and range of the stratification being so uniform and undisturbed—goes far to prove that *faults* of any magnitude will be rarely encountered. The mountains are very high, the coal veins, especially those of the bottom part of the series, are generally thick, and crop out high up the mountain sides; therefore, an inexhaustable amount of coal, of the very best quality, may be safely calculated upon as existing in this coal region.

In the north region the general character of the strata is undulating, and comparatively flat to what is found in the south or middle regions. The coal veins, which are those of the bottom of the formation, are generally of great thickness, and of good quality, but in quantity there is not that average amount per acre of coal as is found in the other great regions. This may be accounted for from the slightly undulating arrangement of the strata, and from the waters of the North Branch of the Susquehanna River, which flow through the central part of the coal valley, having changed its course from time to time, and swept or washed away much of the coal, leaving in places sand and gravel banks that cover considerable area of surface. The great Wyoming flats indicate the change which has taken place in the course of the river.

The basis of the anthracite formation of Pennsylvania is a conglomerate rock, consisting of white quartz pebbles of various sizes, imbedded in a strong siliceous cement; underneath the conglomerate is a thick mass of red shale and sandstone strata, which completely encircles, in a continuous mountain chain, the three great anthracite regions of the State.

The conglomerate, where the measures are perpendicular, forms high massive walls of rock on the summit of the mountains which bound the coal regions, and divide the coal basins; and it is of such a durable, undecomposing nature, that in some places where the strata is on edge, it rises a natural wall twenty to thirty feet in height above the level of the crest of the mountain, and not more than from two to three feet in thickness from the base up. In other places it lies *en masse* in immense blocks, covered with a variety of moss—giving it an imposing, extraordinary rough, and romantic appearance, as in the valley of the Swatara, Wolf Creek, etc.

As the coal measures—from their highly inclined angle of dip, which are in some places in the mountain that forms the southern boundary of the south anthracite coal region, overtilted—pass to a lesser angle of inclination, which gradually decreases in proceeding northward over the three great anthracite regions—the conglomerate becomes more thin and less abrupt in its character; and, indeed, its situation is at times only marked by the loose detached white pebble stones scattered over the surface of the ground, the cement which binds the parts together being in some situations of a more decomposing quality than it is at other places.

The red shale, by exposure to the air, and by the action of water, decomposes very freely, and is the great reason why the general character of the mountains which form the boundaries of the coal regions are so steep as they are found to be where streams of any size run along their base; while the conglomerate on their summits remains undisturbed a rock of ages, until the red shale, on which it reposes, crumbles away, and thus these immense rocks are hurled from their elevated natural position into the valleys below, and thus are immense boulders of the conglomerate carried away from their native beds to great distances.

The south anthracite region contains several elongated synclinal and anticlinal axis of stratification. The general order of the coal veins range parallel with the mountain chains that bound the sides of the troughs or basins, which is in an east and west direction—the general *dip* of the veins being north and south.

The first or south axis or trough of coal strata, of the south anthracite region, is bounded by Sharp Mountain on the south, and by a range of hills, parallel with Sharp Mountain, on the north. This axis is in shape like a *canoe*, its greatest width being about the town of Pottsville, which, in that place, is something over half a mile. The eastern terminus of this axis is a short distance south-east of Middleport. The western terminus is near the Susquehanna. Its continuation westward forms the southern fork of coal strata in Dauphin district. The extreme length of this axis is about fifty miles. At each terminus of this axis or trough of coal strata, the bottom veins end in a point, and are considerably elevated above the place of the same veins in the central part.

In the commencement of mining operations in Schuylkill county, and indeed down to the present time, it has been considered by many persons who profess a knowledge of these matters, that the range of Coal veins in Sharp Mountain, which are what is termed *overtilted* from the perpendicular, are not identical with those veins worked on the opposite side of this narrow trough or synclinal axis—*i. e.* they are not the uprising to the south of the coal veins worked in the range of hills on the north side of the trough, and which dip to the south, and the sections hitherto made and published tend to show that the veins on the north side of the axis are not connected with those of Sharp Mountain. It is true that the coal veins of both sides of this synclinal axis dip in the same direction to the south—those of Sharp Mountain on the south side the axis at an angle of about 80° to 85°, and those on the hills on the north side the axis, at an angle of 45° to 50°, and 60°,—yet there is ample evidence to prove the fact that the south and north ranges *connect with each other*, and will be found to basin beneath the surface in the valley.

In an excavation at Pottsville, made in the centre of the two ranges of coal strata of the first or south synclinal axis, is developed the curvature of the axis,—the stratification of rock overlaying the upper vein of coal is regularly continued and unbroken from one side of the range to the other, and at the extreme ends of this elongated trough, from the bottom veins of coal being highly elevated, and their dip thereby considerably decreased, they show the axis to be perfect throughout, and the south and north ranges identical and connected with each other. Thus we have at the extreme ends of the first synclinal axis, the bottom, and in the centre of it, the top of the stratification of which it is composed, in a perfect and regular basin-like and synclinal order—clearly connecting the coal veins which are found in Sharp Mountain, the south side of the axis, with those in the small range of hills, the north side of the axis.

A very important experiment has recently been made at St. Clair, in this basin, which goes to prove that the white ash veins of Mine Hill and Broad Mountain run into the Schuylkill basin, where they underlie the red ash coal. At the anticlinal axis, at the place above named, a shaft was lately sunk, which, after penetrating about a hundred and fifty yards, struck a vein of white ash coal, lying nearly horizontal, and thirty feet in thickness. The result of this discovery is, that the Schuylkill basin necessarily contains a much larger amount of coal, to the acre, than any other basin of either of the three great regions, and from its position at the head of navigation, renders the value of the land correspondingly great. The value of coal land, in this basin, is moreover greatly increased by the inclined position which the coal strata occupy, thus affording a greater *amount of coal to the acre*, than if they lay flat or horizontal. An acre of coal land, in Schuylkill county, estimated at twenty-five cents per ton, is worth from twenty to thirty thousand dollars, and it will not be long before such prices, high as they may now seem, will be freely realized. A single vein of coal in the Mine Hill region, for several years past, has returned an annual rental of over $62,000 to the proprietors, from four or five colliery works in operation upon it; and the same tract will probably continue to yield a like sum for many years to come.

A prejudice against the productive value of coal lands was early created by the scenes of speculation which formerly involved them, and from their outside connection with ill-conceived improvements. The day for speculation, however, is about disappearing; and people are now awaking to a sense of the true and *bona fide* value of coal lands, which must henceforth greatly increase with the increasing annual demands of the trade. We repeat, therefore, what we know to be the fact, that an acre of coal land, favorably situated in Schuylkill county is worth, on a fair average, at least three times the amount of money of a similar acre, situated in other districts where the strata are horizontal, the veins *flatened out*, the coal necessarily soft, and the facilities for mining correspondingly impaired. The value of Schuylkill county land is also greatly increased by the facilities for transportation to market, the numerous lateral roads penetrating every coal district, and the *natural adaptation* of the county for mining purposes. In respect to market, this coal region is the nearest, and for all time to come must rule the destiny of the trade. The real future value of its coal land is, in our opinion, far beyond estimate. The gold of California, centred in one huge stupendous lump, could not purchase a single basin of it, were we the fortunate owner!

In his description of the Sharp Mountain range of coal strata, our State Geologist and myself* do not agree, and it may not be out of place here to give his remarks thereon in full, with the reason why my opinion and his are at variance with each other, as to this particular part of the coal formation. Prof. Rogers, in his second report on the Geological exploration of Pennsylvania, p. 80, says: "By far the most conspicuous north and south disruption of the coal measures and their southern conglomerate barrier, is displayed in an enormous dislocation of the entire chain of the Sharp Mountain, about nine miles east of Pottsville, by which the whole mass of the mountain on the eastern side of the break, has been moved northward, through at least one-fourth of a mile, throwing, of course, all the coal seams far out of their regular position." From a careful examination of the place referred to by Professor Rogers, as above, I find that no evidence is shown that the coal measures of Sharp Mountain have been moved northward, or in any way displaced; but, on the contrary, a uniform regularity is maintained in this part of the coal region. The Sharp Mountain, it is true, is not continued eastward further than the place referred to, for the reason that the coal measures of the first synclinal axis of the south anthracite region having terminated there. The coal veins of this axis, as I before observed, are gradually elevated as they approach this point, one vein basining out after another, until the last or bottom vein of the axis runs up on the table land at the end of the mountain, bounded by the conglomerate. The red shale at the termination of the axis, from its soft decomposing nature, forms an abrupt declivity, occasioned by the streams which flow down its sides into the valley below—and this is

* Wm. F. Roberts, Esq., Geologist and Mining Engineer, to whom we are indebted for a portion of these remarks on the Geological structure of the coal formation.

the "conspicuous north and south disruption" of Mr. Rogers. Further north than the termination of the first axis, another mountain (not Sharp Mountain,) bounds the south side of the second axis of coal strata of the south anthracite region.

The second synclinal axis lies between the range of hills before named, and a range further north, which, in the vicinity of Pottsville, is called Peach Mountain. The coal veins of the Peach Mountain range are very much contorted in their disposition, having several undulations or axis of a minor synclinal and anticlinal character. In the more elevated land along the range of Peach Mountain, the curvatures of the coal veins are more duplicated than they are in the low parts of this mountain range. A better development of this peculiar coal formation may be seen in the lands north-east of Middleport, where the curvatures of the strata are more numerous and exposed by actual workings, than may be found in any other position along the entire range

The uprising of the coal veins at this place forms several synclinal and anticlinal axis—the lower veins curve over before they reach the surface, and the upper ones lie over them in an uniform way. In some places, where denudation has taken place, the continuity of the saddle, or anticlinal curve of the upper veins, is washed off, and the same veins form several north and south dips, which, previous to the nature of the formation having been clearly understood, were taken for so many different and distinct veins of coal. This misconstruction of the true geological character of the veins, and the reason why so many outcrops are exposed, not being considered, led to a great many errors in the estimation of the real value of the coal land in the Peach Mountain range, as regarded the quantity of mineral contained therein. In many other places, too, in the anthracite formation, the same causes have, and do even at the present time, lead to similar results, and is the reason why erroneous calculations are not unfrequently made.

The extreme length of the axis of Peach Mountain coal strata is about thirty-five miles. The eastern terminus of the synclinal axis is at the Old Summit coal mines on the Lehigh estate. This, the second axis, extends further east than the first axis a distance of about eleven miles. The western extreme point of the second synclinal axis is about twenty-five miles east from the western end of the south fork of coal strata in Dauphin district, and about fourteen miles east from the western end of the north fork of coal strata in Lykens'

Valley district. The point of termination of the second axis is where the two before-named forks begin to diverge in their westward prolongation from their course eastward. The terminus is seven miles north-west from Pinegrove. In the continuation of the axis of Peach Mountain coal strata, the undulations that are found in its central part do not continue through its entire length—as its eastern and western parts—for as the bottom coal veins of the axis become more elevated, the curvatures of the strata are diminished.

The third synclinal axis is between Peach Mountain and Mine Hill, and extends from the point-like terminus of the south anthracite region, near the Lehigh River, to a similar terminus, the end of the north fork in the Lykens' Valley district,—a distance of about fifteen miles. In this axis undulations and curvatures of the coal strata are found, but not of that frequent occurrence as in the Peach Mountain range. These undulations may be seen at Rhume Run, in the Lehigh district; on Silver Creek, in Schuylkill Valley district, (north of Pinegrove,) and in the Lykens' Valley district.

The fourth synclinal axis of coal strata is known as the Broad Mountain coal basin, which lies between Mine Hill and Broad Mountain; its eastern end is between the head waters of Wolf Creek and Silver Creek; its western end is west of "Woolaston's or Raulin's tavern." The length of this axis is about eleven miles.

The fifth synclinal axis of coal strata is on the summit of Broad Mountain; its eastern end is east of New Boston colliery; its western end, west of Raulin's Run. The length of this axis is about fourteen miles. The axis is narrow, and the coal is in places washed off—therefore, it is not so valuable in point of quantity of coal as it would be were the veins continuous through it.

The foregoing axis of coal strata constitute the south anthracite region—the first grand division of the anthracite formation of Pennsylvania.

The middle anthracite region contains, as well as the elongated synclinal and anticlinal axis of coal strata, several small and detached coal basins.

Between Mahanoy Mountain, the south boundary of the middle anthracite region, and the mountain ranging parallel thereto, and next north, known as Locust Mountain, are three synclinal and two anticlinal axis of coal strata. The valley containing these axis is about twenty-six miles in length. The eastern termination of the axis is about eleven miles east from Girardsville, near the head waters of the Mahanoy and source of the tributaries of the Little Schuylkill. The western termination is south of Shamokin. Both terminations of this axis break off in a similar manner to the eastern termination of the first axis, in the south anthracite region. Locust Mountain is the north boundary of the Mahanoy, and the south boundary of Shamokin coal valley. The north boundary of Shamokin coal valley is Big Mountain. In the Shamokin coal valley, taking its central part as a section, there are four synclinal and three anticlinal axis of coal strata, besides a roll of the outcrops of the lower veins of coal shown on the north slope of Locust Mountain. The first synclinal axis of the Shamokin coal valley is between Locust Mountain and Mount Carmel Ridge; the second between Mount Carmel Ridge and Mine Ridge; the third between Mine Ridge and Coal Run Ridge; the fourth between Coal Run Ridge and Big Mountain.

The anticlinal axis are Mount Carmel, Mine Ridge and Coal Run Ridge. The most complete and beautiful development of the coal strata of the anthracite formation of Pennsylvania, is the anticlinal axis of Mount Carmel Ridge, developed by the north branch of Shamokin Creek. The creek passes through the axis at a right angle to the run of the coal strata, about five hundred yards west from the centre turnpike at Mount Carmel—the arch of sandstone rock is cut down perpendicularly, forming a beautiful curve, and giving an admirable illustration of the regularity and perfection of this part of the coal field. The anticlinal axis of Mine Ridge is likewise cut by the same stream, and affords another example of the perfection of the coal strata of the Shamokin coal valley. Mine Ridge, from the Centre turnpike, gradually rises into a hill of great elevation eastward, where coal veins of great thickness and extraordinarily pure quality are opened—a strong evidence that this ridge or axis of coal strata, when thoroughly developed, will prove to contain mineral in quantity and quality inferior to few other places in the anthracite formation of Pennsylvania.

Big Mountain contains the bottom series of coal veins which crop out along its summit. These veins are the same as those developed in Locust Mountain, the thickest veins of the anthracite formation.

The Shamokin coal valley extends in length from its eastern terminus, on the head waters of Little Schuylkill and Quaquake creeks, to its western terminus within about seven miles from the Susquehanna—a distance of about forty miles.

The eastern terminus of the Shamokin coal valley has two forks of coal strata, similar to the forks of the western terminus of the south anthracite region, but much smaller in point of length and width.

North of these forks are the detached coal basins of Beaver Meadow, Dreck Creek, Hazleton, Black Creek, Little Black Creek, Sandy Creek, and Hell Kitchen, extending one after the other northward to the Nescopeck Mountain. The Nescopeck summit is conglomerate, the base of the coal formation; and from it to the Wyoming coal field, traces of the coal formation are found—a sufficient evidence that the three great divisions of the anthracite formation of Pennsylvania were, in former times, a connected and continuous formation of coal strata.

The Black Creek coal basin, laying about one mile and a half north from the Hazleton coal basin collieries, in Luzerne County, has for the last few years been attracting much attention, owing to the extensive explorations made by boring and shafting on the Black Creek coal estate, where a mammoth vein of coal has been struck and passed through, of the thickness of twenty-eight feet, and two other veins, one of six and the other of seven feet. These veins of coal are supposed to underlay the greater part of this estate, and show themselves to be of the very best quality. In fact, it is believed by many that when the Black Creek coal basin shall be fully developed, it will be found one of the richest basins in the Lehigh coal field. This basin, as far as the developments extend, would seem to afford to the miner easy facilities for taking out the coal, and its proximity to the Hazlelon Railroad, must in a short time render it of great consideration to those who are now the owners of the land. This basin is about fifteen miles in length, and has many advantages in procuring supplies, such as provisions, &c., being but a short distance from the beautiful farming district of Conyngham Valley, and only separated from it by Buck Mountain.

The south anthracite region contains white, red, and gray ash coal veins. The white ash are found in the Lehigh, the Broad Mountain, the Mine Hill, and underlie the Pottsville basin. The principal gray ash are in the Peach Mountain range, and the principal red ash coal are the south dipping veins of the first synclinal axis. The south fork in Dauphin district has, in its eastern end, a semi-bituminous coal, which gradually changes, going westward, into a pure bituminous. A similar graduated change from an anthracite to a bituminous coal is found in the coal formation of Wales, in Great Britain, and according to Professor Murchison, in several coal fields in Russia the coal veins which are bituminous at one part of the basin, become anthracite at the other. Lykens' Valley district yields coal of a semi-bituminous or free burning quality. Argillaceous iron ore, both nodular and in seams, is found varying with the coal veins in places through the coal region, and black band or carboniferous iron-stone is found in the Lykens' Valley district.

In the middle anthracite region is found white, gray, and red ash coal veins. In Big Mountain a superior vein of red ash coal, seven feet in thickness, which burns very freely and leaves no clinker, is opened with white ash coal veins above and below it. Red ash coal veins are found in other localities in this coal region. Argillaceous iron ore in the nodular form, and in regular strata, appears to be in abundance through this coal region; and bog ore exists in large beds in various places. Carboniferous iron-stone is likewise found in this region, and may ultimately become an article of great value for smelting purposes.

## HISTORY OF MINING OPERATIONS.

Schuylkill District.—Like every other branch of business, the mining of coal has undergone many striking changes and improvements, since its commencement. We have frequent cause for astonishment, while regarding the progress of improvement in every department of busy life; and although it would seem, standing upon the platform of the present, and taking in the whole perspective of the past, with its numerous shades, that we have really achieved the *ne plus ultra* of inventive genius;—still, as the world goes on, new enterprises are opened, new feelings are instilled, and new desires are to be filled:—so that the *field* for thought and scientific knowledge is continually enlarged, and the progress of invention must always be proportionably rapid.

When openings were first made for coal in the vicinity of Pottsville, the shafts were sunk to the depth of from twenty to thirty feet, and the coal hoisted in large vessels, by means of a common windlass. As soon as the water became troublesome, which was usually the case after penetrating beyond thirty feet, the shaft was abandoned, and another sunk, and the same simple process repeated.

This mode, however, was soon superseded by *drifts* (or openings above water-level, running in with a surface sufficiently inclined to drain off the water). These would be opened at the heads of veins upon the hillsides, and the coal brought out in wheel-barrows; but

FIGURE 51.

it was not until 1827 that railways were introduced into mines, and from that period until 1834 drifts were the only mode pursued for mining coal.

In the meantime, various experiments had been made for the use of shafts, the principal one of which was the substitution of horse-power and the gin, for the windlass, by which means the miners could clear the water from the shaft with greater facility, and penetrate somewhat farther down on the veins. But with this great improvement, as it was then regarded, they were enabled to run down on the vein for but a comparatively short distance, and the coal was, of course, inferior; for experience has since demonstrated that the crop of the coal is never equal to that taken out at lower depths, where the roof and floor have attained the regularity and hardness so necessary for effective labor and good coal.

At the period to which we have alluded, there was a total and perfect absence of every convenience which is now deemed necessary for mining operations. The country itself was, we were about to say, *uninviting;* but such we never could have esteemed it. There never *was* a more grand, picturesque region, beautiful at all seasons grand in all eyes, precious to the man of science, the capitalist, and to the whole world of business. But if it be wild and beautiful *now*, when jealous art has despoiled it somewhat of its wild aspect; stripped the mountains of their gaudy foliage, and levelled the venerable and sturdy forest trees to the earth, with here and there a few remaining, stripped of bark and branches, as if intended for monuments to their perished fellows; what must it *not* have been when the howls of wild beasts echoed in the solitary depths of the woods; in the deep ravines and mountain-passes until then unexplored by man? The country then, clothed in its rich spring garb, fragrant with its wild-flowers, musical with its numerous streams, majestic with its frowning crags and precipices, in its general range resembled the green ocean "into tempest tossed," and its primitive silence was the sleep of nature, when, like a miser, she had finished burying her treasures!

But what we wished to convey is, that the country at this period was destitute of those conveniences for sustaining life, and for carrying on a regular business, which are rightfully looked for by the laborer. The only mode of transporting coal from the mine, was by common wagons, over roads at all seasons bad, and through a country in which, from its mountainous character and wild state, the horse was enabled to accomplish but little, in comparison with what could be done in a level and more improved country.

But notwithstanding these difficulties, the work was still pursued, and that most assiduously. The prices commanded by coal afforded but a scanty pittance to the laborers employed, without insuring the least profit to the owner of the lands. Previously, the inhabitants of the country subsisted entirely by their skill in hunting. Every species of game was plenty, and the skins of bears, wolves, wild cats, foxes, &c., as well as the quarters of deers, and birds, were eagerly sought in the country and towns adjacent. The hunters, few in number, lived in rude cabins far from each other, and there was scarcely a path, in the rugged state of the country, by which the steps of the stranger could be directed. All the coal mined anterior to 1818, was mostly sold to blacksmiths in the surrounding country; for to haul it away for fuel, while wood was still plenty, could not be afforded nor justified by the economist.

Although the Schuylkill Navigation, as previously stated, had been completed in 1818, its facilities for transporting coal were not of such character as to warrant the mining of any considerable quantity. Having been thrown out of repair, time after time, by freshets, its use could by no means be relied upon, and thus, from 1818 to 1825, the trade, if it may be said to have had existence at all, was so extremely limited and uncertain in its general features and prospects, that little attention was bestowed upon it. The whole extent of the trade of the anthracite regions, from this period to 1824, did not exceed twenty thousand tons. In 1825, (the year following,) this amount was nearly doubled, of which the quantity sent down the Schuylkill was six thousand five hundred tons; that of the Lehigh twenty-eight thousand one hundred tons, and of the Susquehanna no account has been kept.

From this year, therefore, the existence of the Schuylkill trade may be dated—that of the Lehigh having commenced five years previously.

The introduction of railways into this region, which occurred in 1827, is, perhaps, one of the most important epochs in its history. The natural arrangement of the country is admirably adapted for grading and laying down railways, and it was on this account that their introduction was more welcome. The coal seams crop out by the sides of the mountains, and the valleys between them, usually affording small streams, allow sufficient descent to convey the loaded cars to the head of navigation. The distinguished credit of having been the first person who erected a railway in this region, is, we believe, assigned to the late Abraham Pott, who constructed one over half a mile in length, leading from his mines, east of Port Carbon, to the navigation at that place.

Their subsequent introduction into drifts, by which the cars were drawn in the mines by mules, gave a new impulse to the business, and greatly added to the capacity of each operator. In 1826 the amount shipped was nearly seventeen thousand tons, and in 1827 it was over thirty-one thousand tons. In 1828 it reached forty-seven thousand, in 1829 seventy-nine thousand, 1830 eighty-nine thousand, and in 1831 eighty-one thousand tons.

During this period coal was being generally used in stoves, in the more populous towns; and after the grate was introduced into them, which was accomplished more or less successfully between the years 1827 and 1831, the trade began to assume an imposing and gigantic attitude. For no sooner had the people become familiar with the peculiar properties of the coal, than its vast future importance in the arts and manufactures was readily acknowledged.

In 1826 and '27 large accessions had been made to the population and business of the region. The Schuylkill Navigation had been placed in excellent repair, and interruptions in its navigation were no longer experienced. This happy state of affairs continued until 1829, when a momentary pause was made in the trade, but it was a pause prophetic only of still greater triumphs, of busier scenes, and of more active life. It was at this period that scenes of excitement, speculation, and daring enterprise were enacted, which surprised and startled our good old Commonwealth from its Quaker propriety. Capitalists awoke, as if from a dream, and wondered that they had never before realized the importance of the anthracite trade! What appeared yesterday but as a fly, now assumed the gigantic proportions of an elephant! The capitalist who, but a few years previously, laughed at the *infatuation* of the daring pioneers of the coal trade, now coolly ransacked his papers, and cyphered out his available means, and whenever met on the street, his hand and pockets would be filled with plans of towns, of surveys of coal lands, and calculations and specifications of railways, canals, and divers other improvements until now unheard of! The land which yesterday would not have commanded the taxes levied upon it, was now looked upon as "dearer than Plutarch's mine, richer than gold." Sales were made to a large amount, and in an incredibly short space of time, it is estimated that upwards of *five millions of dollars* had been invested in lands in the Schuylkill coal field alone! Laborers and mechanics of all kinds, and from all quarters and nations, flocked to the coal region, and found ready and constant employment at the most exorbitant wages. Capitalists, arm-in-arm with confidential advisers, civil engineers, and grave scientific gentlemen, explored every recess, and solemnly contemplated the present and future value and importance of each particular spot. Houses could not be built fast enough, for where nought but bushes and rubbish were seen one day, a smiling village would be discovered on the morrow. Enterprising carpenters in Philadelphia, and elsewhere along the line of canal, prepared the timber and frame-work of houses, and then placing the *material* on board a canal boat, would hasten on to the enchanted spot to dedicate it to its future purposes. Thus *whole towns* were arriving in the returning canal boats, and as "they were forced to play the owl," a moonlight night was a god-send to the impatient proprietors, for with the dawning of the morning would be reflected the future glory of the new town, and the restless visages of scores of anxious lessees!

The days of speculation, however, were not terminated in '29; and a few words more remain to be said concerning them. Many persons who had purchased lands, moved thither with their families, designing to take up their permanent abode in the region, and pursue the mining business regularly, as they would farming, or any other calling. But, in a majority of cases, the lands were purchased in large tracts, by companies formed for the purpose, and these, as well as many tracts held by single individuals, were leased out to tenants. These joint-stock companies, or those composed of citizens of other States, obtained charters for the mining of coal from the Legislatures of their respective States, and thus evaded the statutes of *mortmain* in force here; and the lands owned by them were held by deeds of trust, and were thus used and occupied. But no sooner were companies chartered by the Legislature of this State, than a general law was passed escheating the lands of companies formed under charters not granted by this State, and held without its license and consent. This was done in 1833, when the trade had partially recovered from the speculations of the previous years.

It was under such circumstances as these that a vast amount of capital had been expended in the region, not only in the improvement of the lands, and the facilities for mining coal, but in the construction of railways and similar improvements, of the most stupendous character.

In contemplating these times, though we cannot but laugh at the ludicrous scenes they present, all will admit that they were the indirect and direct means of accomplishing incalculable benefit to the whole country. Nor was it possible, under the circumstances, to restrain the fever of speculation. The real value and resources of the lands were comparatively unknown, and in the hands of those who had no intention of "piercing the bowels of the earth, and bringing forth from the caverns of mountains treasures which shall give strength to our hands, and subject all nature to our use and pleasure," a fictitious value could not but be placed upon them. Calculations were cunningly made of the number of *square yards of coal* in an acre, and the quantity each acre was capable of yielding, without considering the labor and expense necessary to mine it, or without knowing in fact that it contained coal at all, and exhibiting such calculations, in glaring and *bona fide* figures, to the bewildered capitalists, land would sell for one hundred dollars an acre to-day; to-morrow for three hundred, and then for five hundred dollars. And when, at last, the tracts were cut up into small parcels, to suit the means of the purchaser, they would presently be esteemed as beautiful locations for towns, and straightway plans were laid out on paper, elegantly printed and colored; and, finally, the whole would wind up with a sale of "valuable town lots"—lying, perhaps, in the heart of a swamp, a forest, or upon the brow of a mountain. This *last* operation would frequently prove the "noblest Roman of them all;" for although the purchaser might have paid five hundred dollars per acre for the whole plot, he would realize the whole of that sum on a single "corner-lot," and if he could make five or six hundred lots, there would be no such thing as estimating his profits!

We shall dismiss this subject with a single remark. The speculating mania had involved hundreds of persons in utter ruin; and there were few persons of fortune who now ventured, voluntarily and alone, into the mining business. Companies were formed, not only for the purchase of the lands, but also for conducting mining operations upon them; and it was thus hoped, that by concentrating the lands and business of the region into the hands of a few, whose combined capital and influence could silence individual competition, the trade could be made obedient to their wild schemes. Coal had already been universally adopted; and by withholding supplies when they were absolutely needed, it was thought that it could be made to command from eight to twelve dollars a ton; and then, the price being thus established, another series of "calculations" of the value of each particular acre of coal land, and fresh ground for speculations, would be laid open. Advocates for coal companies were consequently not lacking, and many were chartered by the Legislature. But the practical experience of those interested in the trade soon awakened a powerful opposition to them, and this feeling has existed from very nearly the commencement of the trade to the present time. It was especially active from 1831 to 1839, during which the trade had thrice fallen off, in the gross amount of the annual product, from the years respectively preceding; and during the whole of which period, the influence of the public journals in the coal regions was directly arrayed against them. The country, through such aid, was happily saved from the calamities which threatened the trade, and which did much, during this period, to retard its annual growth.

## ANTHRACITE FOR SMELTING IRON.

Nothing worthy of special notice occurred in the progress of the anthracite trade, until 1838, 1839, and 1840. It was during this period, that the attention of intelligent and enterprising citizens was called to the practicability of using anthracite coal for the smelting of iron ore. Dr. Weisenheimer, of New York, had, in the latter part of 1838, and before similar results had been obtained, or at least promulgated in Europe, secured a patent for smelting iron with anthracite and hot blast; but Mr. Crane, having about this time succeeded in a series of experiments in Wales, having in view the same object, is understood to have purchased the claims of Dr. W., which were as follows: *First:* In the application of anthracite coal, exclusively or in part, for deoxidating and carbonating iron ore. *Second:* The application of anthracite coal, exclusively or in part, in combining iron in a metallic state, with a greater quantity of carbon; if bar-iron, for steel; if pig or cast-iron, for a superior quality, &c. *Third:* The smelting or reducing of iron-ore, so deoxidated and carbonated by the application of anthracite coal as aforesaid, into pig or cast-iron. *Fourth:* The refining or converting of iron ore, so deoxidated or carbonated by the application of anthracite coal, as aforesaid, into malleable or bar iron. *Fifth:* The application of anthracite coal as a fuel, in smelting or reducing iron ore raw or roasted, but not prepared by a separate process of deoxidation and carbonation, as above described, into pig or cast iron. *Sixth:* Though not claiming an exclusive right of the use of heated air for any kind of fuel, nevertheless he claimed the use of heated air, applied upon and in connection with the said principle and method discovered by him to smelt iron ore in blast furnaces, with anthracite coal, by applying a blast of air in such quantity, velocity and density, or under such pressure as the compactness or density, and the continuity of the anthracite coal requires, as above described, &c.

As soon as this transfer was effected, Mr. Crane obtained a patent in this country, which differed slightly from Dr. W.'s. But it was several months anterior to the dates of both these patents that a furnace had been blown in at Mauch Chunk, which used anthracite as fuel, and this enterprise was followed in a short time after by a more extensive and successful one at Pottsville. In consequence of this, and in view of the certainty of litigation, Mr. Crane never insisted upon an observance of his claims by priority of discovery, but, as we are informed, published a card, formally renouncing them.

Experiments for using anthracite coal in blast furnaces had been made at Mauch Chunk in 1830, by the Lehigh coal company; and up to the period of

Mr. Crane's method, vast sums of money had been expended, from time to time, in different parts of Europe, to effect the same object, but every attempt proved unsuccessful. The thing had been almost entirely abandoned as impracticable, when the great secret seems to have been imparted simultaneously in Europe and America—for while Mr. Crane was rejoicing over his triumphs in Wales, three enterprising gentlemen of Reading were repairing and blowing in their furnace at Mauch Chunk—and if not the very one previously abandoned, it was the ground, at least, which had sustained a former defeat!

From a letter, by Mr. Lawthrop, dated at Beaver Meadows, to Prof. Walter R. Johnson, of Philadelphia, we gather the following interesting particulars concerning this first application of anthracite coal for smelting purposes: The experiments, says Mr. L., were made by Messrs. Joseph Baughman, Julius Guiteau and Henry High, of Reading, in an old furnace which was temporarily fitted up for the purpose. They used about eight per cent. of anthracite, and the result was such as to surprise those who witnessed it, (for it was considered as an impossibility, even by iron masters,) and amply sufficient to encourage those engaged in it to go on. In order, therefore, to test the matter more thoroughly, they built a furnace on a small scale, near the Mauch Chunk weigh-lock, which was completed during the month of July, 1839. The dimensions, &c., were as follows: stack, 21½ feet high, 22 feet square at the base; boshes, 5½ feet across; hearth, 14 by 16 inches in the square, and 4 feet by 9 inches from the dam stone to the back. The blowing apparatus consisted of 2 cylinders, each 6 feet diameter; a receiver, same diameter, and about 2½ feet deep; stroke 11 inches—each piston making from 12 to 15 strokes per minute. An overshot water-wheel, diameter 14 feet; length of bucket, 3½ feet; number of buckets, 36; revolutions per minute, from 12 to 15.

The blast was applied August 27th, and the furnace kept in blast until September 10th, when they were obliged to stop in consequence of the apparatus for heating the blast proving to be too temporary. Several tons of iron were produced of Nos. 2 and 3 quality. Temperature of the blast did not exceed 200° Fahrenheit—the proportion of anthracite used not remembered.

A new and good apparatus for heating the blast was next procured, (at which time Mr. Lowthrop became personally interested in the works,) consisting of 200 feet in length, of cast iron pipes, 1½ inches thick; it was placed in a brick chamber, at the runnel head, and heated by a flame issuing thence.

The blast was again applied about the last of November, 1839, and the furnace worked remarkably well for five weeks, exclusively with anthracite coal; they were then obliged, for want of ore, to blow out on the 12th of January, 1839. During this experiment, says Mr. L., our doors were open to the public, and we were watched very closely both night and day—for men could hardly believe what they saw with their own eyes, so incredulous was the public in regard to the matter at that time. Some iron masters expressed themselves astonished, that a furnace *could work* whilst using unburnt, unwashed, frozen ore, such as was put into our furnace. The amount of iron produced was about 1½ tons per day, when working best, of Nos. 1, 2, and 3 quality—the temperature of the blast being still about 400° Fahrenheit.

The following season the hearth was enlarged to 19 by 21 inches, and 5 feet 3 inches from the dam stone to the back of hearth; and on July 26th, the furnace was again put in blast, and continued in blast until December, 1840, a few days after the dissolution of the firm, when it was blown out in good order. For about three months no other kind than anthracite was used, and the product was about 100 tons of iron, good Nos. 1, 2, and 3 quality. When working best, the furnace produced about two tons per day. Temperature of the blast was from 400 to 600° Fahrenheit.

The following ores were used: "pipe" ore, from Miller's mine, near Allentown; "brown hematite," commonly called *top mine*, or iron-face ore; "rock" ore, from Dickerson's mine, in New Jersey; and "Williams township" ore, in Northampton county. The last mentioned ore produced a very strong iron, and when it is considered that these experiments were conducted under circumstances wholly unfavorable, and that the furnace and machinery were thoroughly defective, the results obtained may be viewed as being in the highest degree satisfactory.

In December, 1840, this furnace was blown out, the work discontinued, and the firm dissolved. The furnace at Pottsville having at this time been in operation, and its performances having been decidedly superior, the credit of first *successfully* introducing anthracite coal for smelting purposes has been very justly claimed by the citizens of that place. For although the furnace at Mauch Chunk had overcome many difficulties, its abandonment so soon was by many regarded as *prima facia* evidence of failure—while the other has continued in operation, with short intervals, to a very recent period. It is still standing, and under a favorable aspect of the iron market, might probably be again worked with profit.

The Pottsville furnace was completed, and put in blast on the 26th of October, 1839, under the direction of the celebrated Mr. Perry. This gentleman, who had frequently visited Mr. Crane in Wales, and was familiar with the process adopted by him, declared that the performances of this furnace more than equalled those obtained by that gentleman. They were, therefore, esteemed as in the highest degree successful, and an intelligent iron master, (Hon. Dr. Eckert,) who had observed its workings, declared that it had triumphed over difficulties and accidents, during the first fortnight of its existence, which would have chilled up any charcoal works over and over again! The hearth was tapped night and morning, and the yield at each time varied from sixty to sixty-three pigs, equal to about three tons of metal. It is an all-important fact, that in charging the stack, nothing but pure anthracite coal and iron ore was used. Not a scrap of old metal, wood or charcoal was used, except for the mere purpose of first ignition.

The erection of this furnace was mainly accomplished through the efforts of Burd Patterson, Esq., who, from the earliest history of this region, has been identified with every measure of its onward progress. He is still a resident of Pottsville, and, as heretofore, stands connected with all new and praiseworthy enterprises.

In January, 1840, the furnace having now performed successfully for three months, a deputation, consisting of the late Nicholas Biddle, Thomas Biddle, Isaac Lea, Jesse Richards, J. M. Sanderson, and Dr. B. Kugler, visited Pottsville to inspect the iron-works, and to award a prize of five thousand dollars, subscribed by certain influential citizens of Pennsylvania, to be presented to the individual who would, within a specified time, succeed in smelting a certain amount of iron ore, with anthracite coal, &c. This prize was accordingly awarded to the proprietor of the Pottsville furnace, and therefore settles the question as to the person and place entitled to the credit of having first succeeded in this important enterprise.

The celebration of this event was a happy and brilliant affair, and it was not long ere the Union was filled with the importance of the achievement thus commemorated. The committee were invited to a dinner at the Mount Carbon House, and a toast, complimental to the distinguished gentlemen composing it, having been offered, Mr. N. Biddle responded to it in behalf of his colleagues, in a speech of great practical learning and profound eloquence, at the conclusion of which he offered the following toast:

" *Old Pennsylvania*—Her sons, like her soil, a rough outside, but solid stuff within; plenty of coal to warm her friends, plenty of iron to cool her enemies."

The Pottsville Furnace was soon followed by another in the vicinity, called the Valley Furnace. This was put in blast, September 17, 1841, and "succeeded admirably from the first moment of its action." It used the ore found upon the ground.

At the latter end of 1842, after the passage of the tariff act of that year, anthracite furnaces began very rapidly to multiply. In the following year they were found in full blast, and others going into operation, in almost every county in the State, where coal and iron ore were at all accessible. The number continued annually to increase, at an astonishing rate, until the passage of the present tariff law, which has thus far had a very disastrous effect upon this branch of American industry. It will not be long, however, before we recover all the strength that has been lost or impaired during the last few years, for such is the enterprise of our citizens that they *will* produce, notwithstanding the competition of their British rivals.

Until the year 1740, iron was made in England almost exclusively with charcoal, and prior to that period none of the iron stones of the coal region were used; but as soon as the iron manufacturers found it necessary to locate themselves in the coal region for the purpose of being convenient to the new kind of fuel they were about to adopt, they found the necessity of searching for ore nearer their works than the magnetic ores that they had been in the habit of using were; the result was, that an abundance of excellent ore was discovered in the coal regions in the immediate vicinity of their works, and although it did not yield so high a per centage of iron as the magnetic ores, they found it more profitable than transporting richer ores from a distance.

With regard to this region, a like result has been experienced; for it was not until after the erection of the furnace at this place, that any investigations had been instituted as to whether iron ore was to be obtained or not. But no sooner had explorations commenced than new and large deposits of iron ore were found daily, and the ore pronounced to be of an excellent quality. Mr. Benj. Perry, the intelligent anthracite founder, has visited several of these mines, and gave it as his opinion that any number of furnaces could be supplied with ore for an indefinite time.

In comparing the ores of this country with those of England and Wales, we find the average richness to be nearly the same; but we have a decided and important advantage in the thickness of the veins, many of which being upwards of three feet thick, and from that down to six inches. The average richness of the ores taken from the coal regions of England and Wales, is about 33 per cent. The average richness of eight specimens of ore, taken from the Pottsville coal region, was 33·18 of metallic ore. These specimens were analysed under the direction of Prof. Rogers, late State Geologist—some of them yielding 39, 38, and 37 per cent., and all taken from different veins. Prof. R. in his fourth annual report to the Legislature of this State, speaking of these ores, says: "Especial care has been taken to submit to chemical examination, such specimens *only* as represent the *average* character of their respective beds—choosing those freshly opened in the mines, or in some deep excavation, and *rejecting*, as far as possible, samples gathered from the outcrop, or found loose on the surface; as they invariably contain too high a per centage to prove a fair criterion," &c.

The presence of inexhaustible supplies of coal and iron ore, suggests an important advantage in the comparatively limited capital necessary to carry on iron works. For while iron masters in other sections of country are compelled, at all seasons, to keep on hand a large supply of coal and ore, no such necessity would exist here. Supplies could be procured in small quantities, as desired, for immediate use, and the necessity of buying large quantities at *high prices*, is thus entirely overcome. The same argument holds good, as regards means of transportation, and speedy and cheap access to market. While iron works at many places have no avenue to market during the winter, and are consequently compelled to retain a large stock of their manufactured product on hand—the manufacturer here could send it to market in such quantities, and at such times as the demand might justify.

We may next consider the *cheapness* of the fuel, as well as of the ores used. For the former, the fine refuse coal that has been crowding our mines and landings for years past, is now brought into use for generating steam and heating the blast, and besides answering admirably the purpose, it is afforded free of charge, and delivered to the furnace by the coal operators, so anxious are they to get rid of the large quantities annually accumulated about their premises. This, it will be granted, is an important consideration.

There is another consideration, with regard to those locations where the advantages, in some instances, consist merely or principally in being in the immediate vicinity of the ore. After the smelting of the ore into pig metal is accomplished, no more ore is required; but in the process of making bar-iron, about *four tons of coal* are necessary to manufacture one ton of the latter, so that, independent of the saving in the cost of making pig metal in the coal region, the saving in converting it into bar-iron, at a large rolling mill, would be immense.

The middle anthracite region, as we are assured by our friend, William F. Roberts, holds out inducements of the most favorable character for the investment of capital, in all the branches of iron making and iron manufactures. The coal is of superior quality, and may be mined at very low rates. Its iron ore is rich and in abundance, while it has other important facilities for iron-making establishments to operate with the greatest economy and profit.

The lands of the Dauphin Coal Company, we may add, are also admirably calculated to sustain extensive iron establishments. Taking in view the admirable outlets to market, and the peculiar character of the coal, and richness of the iron ore, they may be said to enjoy unequalled advantages for this branch of manufactures.

## PROCESSES OF MINING COAL.

We shall now resume the subject of mining, and briefly allude to some of the principal improvements lately introduced.

After the introduction of railways, there seems to have been little done in the way of improvements, to facilitate the operation of mining. But without tracing, in regular order, the introduction of each new feature, as the present is contradistinguished from the past, we shall at once proceed to explain the *modus operandi* of mining, as observed in the present day.

In the first place, it may be necessary to premise that the *range* of all the coal veins in the Schuylkill basin is east and west, converging to the eastward, and diverging westward, with such slight variation from the general rule, as not to be worthy of notice. The *dip* of the veins is to the south: and their angle of inclination from the horizon varies from 30° to 40° parallel, in all cases, with the surrounding strata. From 1833 the number of operations below water-level has annually increased, in a regular per centage with the increase of the trade. As they are the most extensive, and would, perhaps, prove most interesting to the stranger, we shall now describe the minutiæ of which they are comprised.

FIG. 52.—VIEW OF A COAL SLOPE.

When a vein of coal is being worked below water-level, a steam-engine and pumps are necessary to raise up the accumulated water in the mine; for *below* water-level means, simply, that the coal is being mined at some point *below* the bed of the adjacent river, creek, or rivulet. The first step to be taken at the commencement of an operation of this kind, is to ascertain where the vein *crops out* to the surface, or so near to the surface as to be easily found, from a previous knowledge of the range of the vein. A favorable location must then be selected, twenty or thirty feet to the northward of the crop of the vein, for the location of a stationary steam-engine. This must be where a sufficiency of water can be had for the supply of the steam-boilers; and if not near to a main railroad, prudence will dictate that it must be so situated that a branch or lateral road can be laid down near the place where the engine is to be erected. The descent into the mine is called a *Slope*, and thus those mines below water-level, called *Slopes*, are contradistinguished from those above water-level, called *Drifts*. Engines erected for the purpose of hoisting the coal up the Slope, and pumping the water out of the mine, are usually of the capacity of from forty, fifty, and sixty horse-power, nearly all horizontal high-pressure, and working with a slide-valve. They are generally built in a very neat, simple, as well as a strong and efficient manner, and invariably by the mechanics of the coal region.

The location of the engine being determined upon, a slope, or inclined plane, must be driven down in the vein, and consequently at the same angle of inclination. The thickness of the vein is usually excavated, and the slope must be sufficiently wide to admit of two railway tracks, from thirty-six to forty inches wide each, to be laid down; with room, also, for the pumps on one side, (and sometimes both sides) and travelling road on the other side (or sometimes *between*

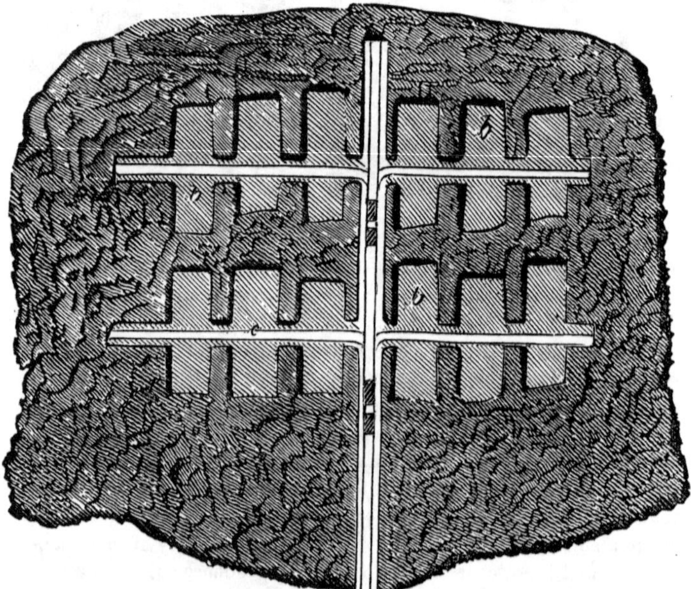


FIG. 53.—GROUND-PLAN OF A COAL MINE.

the two railway tracks) for the miners and laborers—the whole width of the slope being usually from eighteen to twenty-two feet. The slope is driven down about two hundred feet for the *first level*, at the bottom of which the *gangways* are commenced, running at right-angles from the slope, east and west in the vein, and are continued at distances discretionary with the operator, or to the extremity of his mining limits. The slope and gangways form a capital T. The gangways are frequently driven one, two, or three miles, with turnouts at intervals for trains to pass each other. They are made about seven feet high, and sufficiently wide to admit a railroad track to be laid down, on which a well loaded car, having from one to two tons of coal, may pass freely. (Fig. 53. The gangways are indicated by the letter *c*, on the left.) The cars are hauled to and fro by horses or mules; the latter being preferred, as well because of their diminutive size as for their stamina. The gangways being driven in a sufficient distance from the bottom of the slope, the next thing is to commence digging out or *mining the coal*. The coal in the vein is left undisturbed on each side of the slope to a distance of thirty or forty feet east and west, and extending all the way up the surface; the coal thus left, in mining phraseology, is called *pillars*, and is suffered to remain for the purpose of strengthening or supporting the slope; as in an extensive mine and in a good vein its use may be required for a great number of years. A *pillar* of coal of some twenty feet in width is also left all along the upper side of the gangway; and above this pillar, and up to the surface, all the coal is worked out. The plan of working adopted by miners is this: two miners and a laborer generally work a-breast, (the excavated squares, indicated by the letter *b*, are breasts where the coal is being worked out,) like the swarth of a *cradler* in the harvest field, usually from thirty to forty feet in width from the pillar above the gangway up to the surface. They make an opening from the gangway through the pillar above, about where the centre of the breast will be, of four or five feet wide, for a *shute;* after which the full extent of the breast is opened out, and the shute continued up the centre, down which the coal slides into a car in the gangway. When the coal is dug out, the roof is supported by *props of timber*, placed at a distance from each other, varying from six to eight or ten feet, as the roof may be found to be substantial or indifferent. The seams of coal vary from two to twenty-five feet in thickness, (rarely exceeding the latter figure). Those of from six to ten feet are considered best, as they can be worked with greater facility and profit. They can be so propped and roofed as to enable the miner to take out every particle of coal, without the slightest danger of accident; while those of greater thickness must be worked in *chambers*, and large *pillars of coal* left standing to support the roof; and even then the miner is exposed to danger from the detached pieces falling down continually.

From ten to fifty of these breasts are worked simultaneously up to the surface; after which, if the gangway is far enough extended, new ones are commenced, and the same operation repeated, until all the coal on that *level* is worked out. When this is done, the slope must again be driven down some two or three hundred feet; gangways again opened, railways laid down, and the same process of mining the coal continued. And thus the miner gradually gets deeper and deeper into the bowels of the earth, and to reward his industry and perseverance, nature has provided the purest and best coal low down, so that the farther down he ventures, the better and richer becomes his reward!

The deeper the mine, however, the more difficulty is experienced in keeping the works properly supplied with fresh and wholesome air; and nothing but long practical experience can furnish a thorough knowledge of this very important branch of the mining business. We shall reserve some remarks which we intend to offer on this subject, for the conclusion of the present article.

Going now to the shutes in the gangway, we find cars loaded with coal. A mule, which is in most cases used, will draw three or four of these loaded cars to the foot of the slope, where they are left, and empty cars hauled back to be loaded. One of the loaded cars is then pushed upon a turning platform, by a person stationed there for that purpose; he places the car fairly for the railroad track in the slope, attaches the chain to it, draws the pull of the bell as a signal to inform those above that "all is ready," and it is hoisted up the slope by the engine, while an empty car descends, at the same time, on the other track. The car of coal being now brought to the top, it is unhitched, pushed aside, and an empty car pushed into its place, hooked to the chain, and, a loaded car being now attached on the *other* track, the bell is again rung, and the empty car descends and the loaded one ascends, as before. This hoisting and lowering of cars is always going on with despatch during the day-time, and sometimes during the whole night, there being often two sets of hands and miners, one for the day and the other for the night. The time usually occupied for bringing up a car is about one minute, which includes attaching to and detaching the car from the chain, &c. Where from one to two hundred tons of coal are prepared and shipped daily, (besides the refuse and accumulated rubbish of the mine, which must be brought up,) it will be seen that it forms one of the most important features in mining.

The next feature in mining is that of preparing the coal for market, that is, cleaning it from the slate and earthy matter that sometimes is mixed with it, and breaking it in suitable sizes for the various purposes required. The coal dirt, consisting of small particles of coal

FIG. 54.—COAL BREAKER.

and slate, besides various kinds of earthy matter excavated in the mine, is hauled out and deposited in heaps along the sides of the hills, where it sometimes forms large elevations. The loaded coal cars, which are here represented as coming from a drift, or a mine above water-level, are seen on the left, running, by a slight inclination, to the Coal Breaker, which is represented in fig. 54. To fully retain the *idea*, the loaded cars are seen coming out of the mine, and going directly to the Breaker works. The Breaker machinery is, of course, erected as near to the mouth of the mine as local circumstances will admit, and considerable elevation is necessary in order to break and prepare the coal at as little expense as possible. The side of a hill is therefore preferred, as a railroad to conduct the coal from the mouth of the mine to the Breaker can easily be constructed, and will thus avoid the expense of ropes for an inclined-plane, upon which to haul up the loaded cars.

The Breakers are all turned by steam, with but a few exceptions, where water is at hand. An engine of twelve or fifteen horse-power is requisite for driving the Breaker, and turning the circular screens, and they are built on the same plan as the larger engines used at Slopes for hoisting up the cars and pumping out the water. The Breaker-rollers are of cast-iron, placed in very strong, compact frame-work, and turned by means of a leather belt and gearing-wheels. The most approved rollers are those perforated between the teeth, being an improvement on the former solid periphery-rollers originally invented, inasmuch as there is thus less solid surface presented to the coal in breaking, and, consequently, less *crushing* and wastage of the coal. The loaded car being brought to the head of the Breaker, it is *dumped*, and the coal falls into a small shute, from which it flows into the Breaker. The coal thus passes between the revolving rollers, whose *projecting teeth* break it into pieces of all sizes. From the rollers, the broken coal falls into screens, which also revolve, and having four or five sections of net-work of different sizes, the different sizes of the coal are thus assorted, falling out of the interstices of the screen into shutes below, which are indicated in fig. 54, as hanging directly over the cars. The coal screen was one of the most important inventions of the day. Previously to the introduction of Breakers, the coal was screened by hand. The screen was from 5 to 8 feet long, and from $1\frac{3}{4}$ to $2\frac{1}{2}$ feet in diameter, and placed in a frame, slightly inclined. As the coal entered the more elevated end, the screen was turned round by hand, like a grindstone. When Breakers were introduced, the screens, as previously, were constructed of bar iron, riveted on frame work. But great trouble and expense were experienced, from their liability to break, and the difficulty of repairing them, the whole work being necessarily stopped until this was accomplished. Attention was soon attracted to the subject, and it was not long ere mechanical ingenuity suggested a remedy. A machine was invented by a citizen of Pottsville, by which the largest and thickest wire is wrought into shape suitable for weaving, which is done by very heavy and improved machinery. Wire as thick as an ordinary ram-rod is crimped by this process, which merely consists of a heavy hammer, suspended in frame-work, which is made to fall upon the wire, placed under it, upon a surface allowing it to receive the particular *bend* desired, after which it is woven into frames of about three feet square. These frames are then placed over a large wooden cylinder, and rounded, when two or more sections are pointed and riveted together, which completes their circular form. The screen, thus complete, is removed from the bench, and joined with another of the same dimensions, but of larger or smaller net-work. These screens are remarkably durable, and are not the least feature which has tended to bring coal Breakers into universal use.

This process for crimping thick iron wire has introduced several new and important objects of manufacture, such as iron wire portable bedsteads, fences and ornamental railings, chairs, sofas, &c.

FIG. 55.—IRON RAILING.

Walker & Son, at the corner of Sixth and Market streets, Philadelphia, have a very extensive establishment exclusively devoted to the production of these articles, which, at no distant day, will supersede most of the same articles now made of wood.

After the coal leaves the screen and falls into its appropriate shutes, railroad cars are hauled immediately along side the openings, which, being raised up like the wickets in a mill dam, the coal falls out into the car, and when a sufficient quantity is obtained, the shute is closed, and then the coal leaves forever the scenes of its past history, and is borne off to its future destiny.

The size of chain generally used for hoisting coal is three-fourths and seven-eighths of an inch; formerly smaller chains were used, and, in fact, smaller engines and lighter machinery; but long experience and heavy bills of repairs have taught the coal operators that engines, pumps, gearings, chains, &c., must be strong and substantial in order to withstand the incessant lifting and straining to which they are subjected.

We may now offer a few remarks in regard to drainage, and the plan of pumping the water out of mines. The capacity of the pump varies from ten or twelve to fourteen inches. The working barrel is placed a little above the turning platform at the bottom of the slope, from which pipes are connected up to the surface, or near enough to the surface to have the water carried off. Pump rods are attached to the bucket in the working-barrel, and extend, of course, to the top of the slope, and are connected, by means of a large pump-wheel, with the engine. Below the working-barrel of the pump, and below the turning platform at the foot of the slope, a *sump* is driven down, of the same dimensions as the slope, to the depth of thirty or more feet. This forms a basin into which the water of the mine collects from all the gangways and turnouts, and when the amount of water in the mine is not very great, it will be a considerable time in filling, during which there need, of course, be no pumping. In rainy seasons the water is sometimes rendered very troublesome in the mines, and it is therefore expedient to have the sump, and all connected with the pumping apparatus, in good order and constant readiness. Pipes are attached below the working-barrel and into the sump, and a connection being thus formed, the water is pumped out. The water is gen erally pumped out at times when the engine is not hoisting coal, though it is often necessary, however, to hoist and pump at the same time. At some of the collieries two engines are used, one for hoisting up the coal, and the other for pumping up the water. Several hogsheads of water are thrown up per minute, with great ease and regularity.

Drifts.—In working a coal mine above water-level, no engine or pumps are required. The drift is commenced on the surface, at the foot of a hill, where the vein crops out, and is driven through the vein in the same manner as described when below water-level. The mine being far enough in, gangways are extended to the right and left, and the coal worked out upon the same plan as in slopes, when it is hauled to the breaker by horses and mules. As the gangway is above water-level, with a slight inclination towards the drift, of course the water will run out, thereby rendering engines, pumps, and pumping apparatus, wholly unnecessary. The mines, both of drifts and slopes, are substantially propped up with timber, indicated in the annexed figure, 56, at *a*, *b*, and *f*, *f*, *f*, *f*, which are the *slabs* of boards; *d* indicating the groove or canal through which the water flows out.

*Blasting* is frequently resorted to in mining, especially when working large veins. For this purpose the *safety-fuse* is used almost invariably, the coal being generally so wet and damp that the ordinary processes of blasting would not answer, even if preferable in an

FIG. 56.—VIEW OF THE TIMBER OF A DRIFT OR COAL MINE.

economical view, which they are not. The safety-fuse, too, is perfectly safe, which gives it not the least important advantage. It is a species of fire-cracker or cartridge, the principal part of the composition being powder, which is surrounded by a hempen fabric, and then covered with another composition, to render it water-proof, of which the greater part is pitch. In blasting coal it is difficult to keep the water from filling up the drill-hole, but by inserting a piece of safety-fuse, and then fastening it tightly, no other preparations are necessary, The match is applied, and following the powder through the fuse, produces the desired result, affording ample time for the miners to withdraw, whenever desirable.

The anthracite coal fields are, throughout, more or less faulty; the southern region more especially. The seams of coal having been heaved up, and at other places sunk down, their local positions, if we may so say, are very much, and in various ways, disturbed and contorted. A vein of coal may be followed for half a mile, when, gradually or directly, it is found to run out, and a mass of solid rock occupying its place, and rising up immediately *through it*. To get on the vein again, this rock must be tunnelled at an expense varying from ten to thirty dollars per yard, and without knowing, positively, how far the tunnel must be extended ere the end can be accomplished. In cases like this fortunes have been and are annually spent. Persons who have engaged in the mining business, and invested large sums in the erection of the necessary buildings, machinery, railways, etc., after getting fairly into operation, and while their success seemed complete, have struck these faults, and in a short time have been thrown into utter bankruptcy. All their machinery is rendered comparatively idle, their regular business suddenly checked and deranged, and thousands of dollars going out of their pockets. Impressed with the belief, which seems to be invariable in such cases, that a few yards of tunnelling will again place them on the vein, they labor assiduously from day to day, and from week to week, entirely realizing, though not in the literal sense, the lines of Pope:

> Hope springs eternal in the human breast;
> Man never is, but always to be blest!

There are, as we have said, various kinds of faults; in fact, although they may all have been produced by the same general agency, they vary in their particular character according to the different positions previously occupied by the strata. At some places a stratum of clay, or a combination of earthy substances, is interposed; while at others no such obstacles appear, but the vein is broken off, and the dismembered portion *sunk down,* just as we can suppose a piece of glass, laying on several small rollers, and then suddenly broken into irregular fragments; some pieces would be comparatively large, some would no doubt nearly and quite join each other, while others would occupy various relative positions to the mass.

The reader will agree with us that coal mining, under such circumstances, cannot but be an extremely hazardous and uncertain business; and, indeed, the experience of some of our most enterprising and intelligent operators affords substantial proof of the fact. There is no such thing as overcoming or avoiding, entirely, even with the best practical experience, the difficulties and dangers with which it is fraught.

## VENTILLATION OF MINES AND GASEOUS EXPLOSIONS.

We now approach perhaps the most important, abstruse, and not the least interesting branch of the subject of coal mining. In England, thousands of dollars have been expended in experiments to arrive at a general system of ventillation; and volume upon volume of practical experience, and theoretical essay have been published. But a few years since the columns of the English Mining journals were filled with these lucubrations; and the astounding number of persons annually destroyed in the mining regions, at last excited the attention of the government, and a series of investigations were instituted under its authority. The means proposed were various, and the wealth of a Girard might be squandered ere any one theory could be rendered practical, *per se*, in all coal mining regions. The plan of ventillation must always, in a great measure, be governed by the nature of the coal itself; by the situation and local position of the strata; by the arrangement of the workings, and various other considerations, which make one mine different from another.

From various data before us, and from the practical knowledge imparted to us in repeated conversations with friends in the coal region, we shall endeavor to present to the reader an *abstract view* of the whole subject, leaving for him, if his taste so incline, to add to the general stock speculations and theories of his own.

The whole subject was recently taken up and reviewed by Prof. Ansted, of England, in the course of a series of lectures at King's College, on the practical application of geological science to mining. Though the published reports of these lectures are of great length, we shall endeavor to present a brief outline of his remarks upon the subject under consideration, which, though referring particularly to the coal mines of New Castle, will nevertheless illustrate those of our own country.

He stated that he should now confine his remarks exclusively to the phenomena connected with those accumulations of gas in coal mines which caused explosions; and he selected this opportunity, partly because the subject arose naturally out of that of ventillation in coal mines, being also itself a distinct subject of the greatest possible importance, and also because public attention had been called to it, by the frequent explosions and the great loss of life which often accompanied them. The applicability of means to prevent these direful accidents was a subject of the greatest importance, as regarded the internal economy of the mine; and he should, therefore, have to consider the circumstances under which accidents of this kind might be expected to take place; the danger of explosion in coal mines arose, no doubt, from issues of gas proceeding from the coal itself. He had already had occasion to mention more than once the fact that gas was constantly given out by coal; not only when exposed to heat, or unusual compression, but also under ordinary atmospheric conditions. When any one went into a coal mine for the first time, they would probably notice a peculiar singing noise, which, though it was not easy to understand, there was no doubt arose from the issue of gas from the coal. It was not known with certainty, whether this arose from the bursting of certain small vesicles in the coal; or whether, in consequence of the pressure of the gas on the successive films of which the coal was made up, the singing noise resulted from the gas coming out, as it were, in the form of a thin plate. Certain it was, that in coal mines there was this unusual noise, and that it was connected in some way or other with the presence of gas. The noise was heard unceasingly in mines of certain kinds of coal, and in every such case, therefore, measures must be taken with regard to its results. The quantity of gas thus produced varied, however, very much according to the nature of the coal and the amount of atmospheric pressure, which appeared to be the two principal causes which regulated the issue of the gas from the fractured surface of the coal. In order to give some idea of the quantity of gas sometimes given out, he might mention, that in the Bensham seam, which was known to be a particularly fiery one, gas was thrown out so rapidly, and in such purity, that by boring a little hole in the mineral, and applying a light, a jet flame would be produced. In this case, the gas would be in too pure a condition to be explosive, because, to make it so, a certain admixture of atmosphoric air was necessary. The quantity of gas thus given out naturally "by singing," from four acres of coal, was ascertained, some time ago, to amount to ten thousand hogsheads per minute. Seams of coal, however, varied very much in this respect, and some contained scarcely any gas at all. Besides this constant issue by singing, there was another way in which the gas was met with, namely, in what was called "blowers:" these were puffs of gas, sometimes taking place at long cracks, or at faults, and at others at mere holes; they were sometimes very common, and produced as much gas as served to light certain parts of the mine—as, for instance, the principal passages; and this, indeed, was the safest possible way of getting rid of it. In the Killingworth mine there was a "blower," which had been burning for some years. In other instances, a fault was touched which gave out gas, and when the same fault had again been pierced, it had produced no gas. Sometimes, as in the case of the Great Jarrow explosion, when the works approached near a fault, the pressure of the gas pent up in it had forced forward the coal, which divided it from the mine, and involved all the workmen in certain destruction. These were all very difficult circumstances to deal with, as it was impossible to anticipate them. Besides these blowers, which occurred in the natural state of the coal, there were constantly dangerous accumulations in portions of the mines which had been worked out, or partially worked, and in which the roof had partly fallen down. It was impossible to prevent these accumulations where the roof had fallen, and in old workings; and they were always more or less dangerous, because the gas was here inevitably mixed with atmospheric air, and generally in such proportions as to be highly explosive; they were also not uncommon in faulted districts—where the beds were in a broken state, the result of contraction, or pressure, at some remote period. In all these cases, accumulations of gas took place, and the gas was liable to burst forth on the smallest atmospheric change. If, for instance, a fall in the barometer indicated that the pressure of the atmosphere had become less, its existence would not be so great, and a quantity of compressed gas would be forced out by expansion. Gas being thus present in the mines, either in the body of the coal itself, or in accumulated quantities, the danger arose from the fact, that it was impossible to conduct mining operations under ground without lights.

The most convenient way of lighting, according to the miners, was by candles; and this, partly, because they were easy to carry about, and partly from long custom. There was a great prejudice in favor of candles, which they always would use wherever it was possible. Now, it must be remembered, that generally part of the mine only was liable to the ordinary issues of gas, while other parts were subject to what might be called extraordinary issues, such as arose from unexpected accumulations. In the former case the use of open lights would be perfectly safe, provided the general ventillation was tolerably good. Thus, it might be considered that, under ordinary circumstances, such parts of the mine would be safe, and the other parts dangerous; and it was generally found, that if the workmen were allowed to use candles to the safe parts, they did not object to use lamps in the dangerous portions. But the condition of any district was always liable to be disturbed by new blowers commencing, or by the influx of gas from the dangerous districts; and, therefore, a vigilant superintendence of the state of the mine was indispensable. It was also necessary, if anything was to be done in the dangerous parts of the mine, that light should be obtained by safer means than by candles, because the smallest contact of flame was sufficient to explode certain mixtures of carburetted hydrogen gas and the atmospheric air. The explosive admixture was a mechanical one, and it was necessary to understand precisely of what proportions it was formed. If the gas came out pure, and a candle was thrust into it, the flame would be extinguished, because there was nothing to support combustion; the gas itself would take fire. If there were three parts of atmospheric air and one of gas, it began to be faintly explosive; and, when once explosive, it would set fire to other gas, which was too pure to explode. When the gas was as one to six, it became very explosive, and continued so until the proportions were one to ten, when it became less explosive. The danger, however, was not destroyed until the mixture became weaker than one part of gas to fourteen of air; after that it merely enlarged the flame of any light which might be brought into it, and turned it blue. The miner was thus enabled, by the lengthening of the flame of his light, and its appearance, to tell whether he was in a dangerous part of the mine, and the amount of gas that might be present in the atmosphere. If the flame reached to a certain height, the practical miner could tell at once how nearly the atmosphere approached an explosive combination; and with some other similar points affecting his safety, and that of his fellow-workmen, long habit had made him familiar.

There were several considerations of importance connected with these explosive admixtures of gas and air; as, for instance, the effect produced by the breathing of a number of men, which very much lessened the danger, by altering the proportions of nitrogen and oxygen in the atmosphere; and thus one portion of the mine would be less explosive than another, though both might contain the same proportion of carburetted hydrogen.

Every part of the mine which was capable of being worked, should, in all cases, be visited with guarded lights, before other lights were allowed to be taken; and the state of the mine should always be well ascertained before open lights are used. There was also a certain amount of danger in going into some portions of a fiery mine; and, until within the last thirty years, the only means of obtaining any degree of light in such cases, was by the use of an apparatus called the *steel-mill.* This was a very ingenious contrivance, by which a train of sparks was produced by pressing a flint against a revolving plate of steel, and this afforded sufficient light to move about, but not enough to work by. It was, moreover, a dangerous contrivance, for, every now and then, there could be no doubt, a flame was produced sufficient to explode gas under certain circumstances, and particularly if olefiant gas should be present, which, however, did not often happen to be the case in English mines.

About thirty years ago, a great number of serious accidents occurred in the mines of the north of England, following each other in alarmingly rapid succession. Very many lives were lost, and the public attention was much directed to the question whether or not some improvement could be discovered in the way of lighting the mines. Sir Humphrey Davy, then in the zenith of his reputation, was applied to by Mr. Buddle, a well known colliery viewer of that day, and invited to turn his attention to the subject. Humboldt had before attempted to overcome this difficulty; but his contrivance was only partially useful for visiting dangerous mines, as it would not burn longer than half an hour, the flame being supported by a reservoir of atmospheric air within the lamp. Dr. Clanny improved this lamp, admitting the external air freely in cases when it was used in an explosive state; but this lamp was practically useless, as the explosions which took place inside it soon put out the light.

The learned lecturer then proceeded to explain Sir. H. Davy's invention—the safety-lamp, figure 57. Its principle was founded upon the discovery, that the explosion of the mixture in question did not pass through small tubes; and after numerous experiments, Sir. Humphrey Davy found that the length of the tubes was of no consequence, but that wire gauze, the apertures being of the proper dimensions, answered the same purpose. By this means all necessity for an exterior glass was got rid of, and the new lamp might be carried into the *most explosive* admixtures without danger. Such was the Davy lamp; and he believed that, as it was the first, it was also by far the best real safety lamp that had been invented. It was, perhaps, not perfect, judging only by experiment, but it was decidedly the best for all practical purposes, as it was more manageable than any other, and not so easily put out of order. The gauze usually employed was made of iron wire, and it had seven hundred and eighty-four holes to the square inch. Sir Davy, having perfected his lamp, went down to the New Castle coal field, and with Mr. Riddle, traversed with impunity some of the most dangerous parts of the Bensham seam, at that time the most fiery known. The Davy lamp had been used with great success ever since, and though some accidents had occurred under circumstances in which no lights but those of Davy lamps were present, it was as safe as any such instrument could be. The superiority of this lamp over more recent inventions consisted in its producing a greater quantity of light, and being more portable with at least as much safety. Mr. George Stephenson, the engineer, had also invented a lamp, which was called a "Geordie," after the name of the inventor. It was, however, merely a modification of the Davy lamp, by the addition of a glass tube, which answered the double purpose of increasing the light and keeping the flame steady, by shielding it from currents of air. This, perhaps, gave it additional safety while perfect, as it was possible to drive an explosion of common street gas through the gauze of a Davy lamp. The glass, however, was liable to be broken, and then the instrument became a large and somewhat dangerous "Davy." In some of the Belgian mines a lamp called "the Muesseler lamp," was in use, but it was a very complicated affair, and for that purpose was inferior to Sir H. Davy's invention. No doubt it was safer theoretically, as by it the flame was extinguished the moment the lamp was taken into a dangerous atmosphere; but this very much lessened its utility, the main point being that the lamp shall give light with safety under such circumstances. The Davy lamp, with care, might be taken anywhere. He (the lecturer) had himself been in every description of atmosphere, and he had often seen explosions take place inside the Davy lamp by which he was lighted. This, indeed, was a circumstance which happened constantly to every viewer, over-man, and Davy-man in the New Castle coal field. The true danger of the Davy lamp was one that would apply equally to any other, and it arose from the fact, that its constant use made the workmen careless, and the more it was used, therefore, the greater chance was there of accidents. The learned lecturer

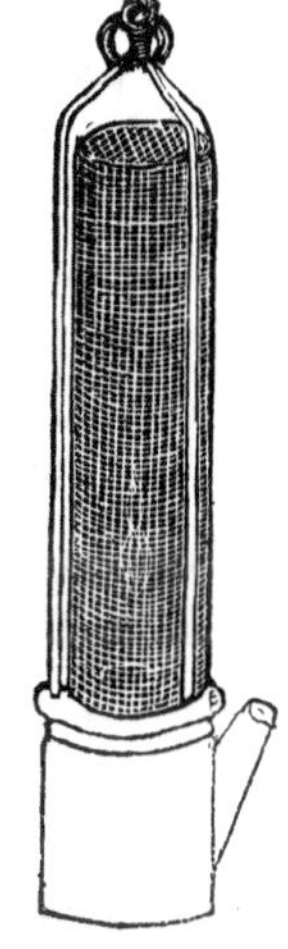

FIG. 57.

here exhibited a printed placard, setting forth the regulations which were adopted in several collieries in respect to Davy lamps, the most important being that which enacted that no man should be allowed to use his Davy lamp until it had been examined carefully by the Davy-man, and pronounced perfectly clean and in good order, nor until it was securely locked, so that the workman could not take it to pieces, and expose the flame in an unguarded manner.

With regard to actual accidents, the professor remarked that he should not say much beyond placing before them the circumstances of a few of the most disastrous, which would serve to explain the nature of the results of the explosions of which he had explained the cause. The most important, then, that he should mention, occurred in the Haswell colliery, on the 28th September, 1844, when ninety-five lives were lost; and the next was that at the Jarrow Pit, on the 3d of August, 1845, when forty-one persons were killed. In the same year thirty-six lives were lost at Killingworth; and in 1846, nineteen at Oldbury and thirty-six at Risca. By these accidents occurring within a period of little more than two years, upwards of two hundred persons lost their lives; and he had grouped these accidents together in order to give an idea of the extent to which they occurred. There were, then, two hundred lives lost in two years, in working coal mines, from accidents which, in the opinion of the juries who held inquests on the unfortunate sufferers, could not possibly have been prevented—the mines being in every case, except, perhaps, in that of the least considerable, in good condition at the time.

In the first of these accidents, that of the Haswell colliery, there were under ground at the time of the accident ninety-nine persons, of whom four only escaped. It was an important consideration (which also applied to other accidents), whether it was absolutely unavoidable, the catastrophe having occurred, that all these people should be killed. It might be that the whole matter was beyond human control, as in the case of a shipwreck. If we crossed the sea, the vessel, being made by human hands, was liable to accidents which might happen from stress of weather, which no one could help; or from carelessness, in respect of which some one would be culpable. If everything were done to render the vessel seaworthy, there would be no blame attaching in respect to its condition, though it should be wrecked in a storm; but if the vessel were sent to sea in an unsafe condition, there was blame. Just so was it with regard to mines; if the mine were in as good a condition as science, and the adoption of the best plans could make it, no blame might be attributed to any one for an unforeseen accident. But if it were possible so to manage the mine as to lessen the chance of accident, (he did not believe accidents could be entirely prevented); and if that were not done, then there would be blame. The sea must be crossed, despite of shipwrecks; and coal must be had, let it cost what it might; the point was to get it under the most favorable circumstances. In the case of each accident, therefore, it was important to consider whether everything was done by way of prevention and palliation which could be done. In the case before them, that of the Haswell colliery, the accident occurred in a part of the mine where the pillars had been removed, and the workmen were taking away the props. This was always a delicate operation, because the props being removed, the roof fell, and the accumulations of gas were disturbed, and often shifted. Before proceeding further with the particulars of this melancholy accident, it would be necessary to mention the effect of an explosion upon the atmosphere. The carburetted hydrogen, when exploded, became decomposed; the oxygen in the atmosphere mixed with the hydrogen and formed water, and nothing was left to breathe but pure carbonic acid gas. This was a most poisonous gas, and produced instant death by choking. At least, insensibility was instantaneous; and, although in such cases, when measures were taken immediately, recovery had taken place, death was generally inevitable. The result of an explosion, therefore, was to turn the atmosphere into pure carbonic acid gas; and every man in that district of the mine, in which the accident occurred, was doomed to certain death. The Haswell mine was divided into three great divisions, or three pannel workings; the explosion took place in the middle division, and thirty or forty persons, far removed from the scene of the accident, in another pannel, were poisoned by the after-damp. This could not have occurred had the ventillation of each pannel been as distinct as it might have been; and—he was, therefore, bound to say—as it ought to have been. The explosion at the Jarrow Pit, in August, 1845, by which forty-one persons were killed, was another instance where such a large destruction of human life was occasioned by the absence of two shafts; for the means of ventillation at the only one being destroyed by the accident, there was no means of restoring it previously to the mischief being done. Another interesting case occurred in the Killingworth colliery in 1845, which was the result of a fault. On one side of the district there was a long stone drift, at right angles to the main rolley-way of the pannel, in which the men were working; a fault was reached, and this gallery not prosecuted beyond it. The chief object was to drive through a mass of coal in order to get the ventillation completed, and for this purpose they were at work day and night. The fault was pricked in the course of the work in a succession of places, up to two or three days before the accident, which took place on a Thursday, without any unusual presence of gas being observed. On the Wednesday the fault was once more pricked, and no gas came out. On Thursday, one set of men had performed their allotted portion of work, and about two hours afterwards another gang of workmen descended; and it is singular that, though the gallery was considered so far dangerous that the men were working with Davy lamps, a boy was allowed to precede them with an open candle—a piece of carelessness most inexcusable, and for which they paid with their lives; before they reached the spot where the workings were, an explosion took place, and every person in the mine perished. Some idea of the force of these explosions might be deduced from the fact that on this occasion, a stopping consisting of thirty-six feet of rough materials, with an inch and a half brick wall on either side, was blown to pieces. In another instance, at the Jarrow mine, an accident happened, and on examining the mine afterwards, a cavity was found, in which there had been probably about two hundred cubic feet of gas in a very condensed state, and the side of the cavity being reduced by the working, it had given away, and let out the whole, which exploded with irresistible force. This was a kind of accident to which the miners were constantly liable, and which no vigilance could foresee or prevent. Accidents, and all their particulars, ought always to be recorded, and they could not be discussed too much. It was quite clear that, although it was impossible to avoid accidents, the risk was greatly diminished by good and effective systems of ventillation, and by unceasing care in the use of lights.

In continuation of this subject, we present the following, principally condensed from the work of the late Mr. Taylor, entitled "Statistics of Coal." The workmen of the Crouzot mine descended one morning, the one following the other, in rotation, into a shaft below, in which carbonic acid had accumulated during the night. Arrived at the level of the "bain," at a few yards from the bottom of the pit, the first fell, struck with asphyxia, without having time to utter a cry; the second followed immediately; the third saw his comrades prostrated on the ground, almost within reach of his arm; he stooped to seize them, and fell himself; another quickly shared the same fate, in his desire to save the others, and the catastrophe would not have been arrested had not the fifth been an experienced master miner, who obliged those who followed him to re-ascend.

The gases which result from the subterranean decomposition of the coal, have, besides carbonic acid, carbonic oxide, azote, sulphurous acid, and the carburets of hydrogen, which have a special odour. Before the coal takes fire, the interior air is already heavy and heated by the gaseous disengagements which are the precursors of ignition. As quickly as these symptoms are remarked, the coals already mined should be raised, and we should isolate from the surrounding air the region or the crevices which enclose the fire; employing at this work the laborers whose organization is known to be the best adapted to support the deleterious influence of these gases.

*Azote*, or nitrogen gas, is much less to be dreaded than the carbonic acid; because its action upon the animal economy is less energetic; besides, its production can only take place by the absorption of oxygen from the air, and it does not naturally exist in the fissures or cavities of the rocks. It has, then, no spontaneous disengagement; but if we penetrate into the works which have been a long time abandoned, and where there has been combustion, the azote will occupy, in consequence of its lightness, the higher parts of the excavations, while the carbonic acid will occupy the lower parts; the respirable air forming the intermediate zone. Azote is found isolated in certain mines, where there exists pyrites in a state of decomposition; the sulphurets changing into sulphates, absorb the oxygen and isolate the azote; the sulphuret of iron is, in this respect, the most active agent.

Azote manifests itself by the red color of the flames of the lamps, which ends by extinction; it renders respiration difficult, produces a heaviness of the head, and a hissing or singing in the ears, which seems to indicate a mode of action different from that of carbonic acid.

The ordinary lamp of the miner is extinguished when the air contains no more than fifteen per cent. of oxygen; [the atmospheric air is composed of twenty-one per cent. of oxygen and seventy-nine per cent. of azote,] it is also at this proportion of eighty-five per cent. of azote that asphyxia or suffocation is caused.

Proto-carbonated hydrogen, or inflammable air, designated by the French and Belgian miners under the name of *grisou*, is, of all the gases, the most dangerous; that which occasions the greatest number of accidents, not by asphyxia, which it can nevertheless produce when it is not mixed with at least twice its volume of air, but for its property of igniting when in contact with lighted flames, and of exploding when it is mixed, in certain proportions, with atmospheric air.

The *grisou* is more abundant in the fat and friable coals, than in the dry and meagre coals; it particularly disengages itself in the crushed places, *éboulements*, in the recent stalls whose surfaces are laid bare, and that so vigorously as often to decripitate small scales of coal and produce a slight rustling noise. The fissures or fractures of the coal, and even the clefts of the roof or the floor, give sometimes outlets to *soufflards*, or jets of gas. The action of this gas upon the flame of the lamps is the most certain guide in ascertaining its presence and proportion. The flame dilates, elongates, and takes a bluish tint, which can readily be distinguished by placing the hand between the eye and the flame, so that only the top of it can be seen. As soon as the proportion is equal to one-twelfth part of the ambient air, the mixture is explosive, and if a lamp be carried, it will produce a detonation proportionate to the volume of the mixture. When, therefore, a miner perceives at the top of the flame of his lamp the bluish nimbus which decides the presence of the fire-damp, he ought to retire, either holding his light very low or even to extinguish it.

The chemical effects of an explosion are the direct production of the vapors of water and carbonic acid, and the separation of azote. The physical effects are, a violent dilatation of gas and of the surrounding air, followed by a reaction through contraction. The workmen who are exposed to this explosive atmosphere are burned, and the fire is even capable of communicating to the wood work or to the coal; the wind produced by the expansion is so great that, even at considerable distance from the site of explosion, the laborers are thrown down, or projected against the sides of the excavations. The walls and timbering are shaken and broken, and crushing or falling down is produced. These destructive effects can be propagated even at the mouths of the pits, from which are projected fragments of wood and rocks, accompanied by a thick tempest of coal in the form of dust.

The evil rests not there; considerable quantities of carbonic acid and azote, produced by the combustion of the gas, become stationary in the works, and cause those who have escaped by the immediate action of the explosion to perish by *suffocation*. The ventillating currents suddenly arrested by this perturbation, are now much more difficult to re-establish, because the doors which served to regulate them are partly destroyed; the fires are extinguished, and often even the machines fixed at the mouths of the shafts to regulate the currents are damaged and displaced to such an extent, that it becomes impossible to convey any help to the bottom of the works.

## THE MINER'S DOOM.

'Twas evening, and a sweeter balm on earth was never shed,
The sun lay in his gorgeous pomp on ocean's heaving bed,
The sky was clad in bright array, too beautiful to last,
For night, like envy, scowling came, and all the scene o'ercast,
'Tis thus with hope—'tis thus with life, when sunny dreams appear,
The infant leaves the cradle-couch to slumber on a bier;
The rainbow of our cherished love, we see in beauty's eye,
That glows with all its mingled hues, alas! to fade and die!
'Tis dark, still night—the sultry air scarce moves a leaf or flower;
The aspen, trembling, fears to stir, in such a silent hour;
The footsteps of the timid hare, distinctly may be heard
Between the pauses of the song of night's portentous bird—

And in so drear a moment, plods the miner to his toil,
Compelled refreshing sleep to leave, for labor's hardest moil:
By fate's rude hand, the dream of peace is broken and destroyed—
The savage beast his rest can take, but man must be denied!
And why this sacrifice of rest?—did not the Maker plan
The darksome hours for gentle sleep, the day for work by man?
Yes—but the mighty gods of earth are wiser in their laws—
*They* hold themselves with pride to be their Creator's first great cause.
The miner hath his work begun, and busy strokes resound,
Warm drops of sweat are falling fast—the coal lies piled around.
And what a sight of slavery! in narrow seams compressed
Are seen the prostrate forms of men to hew on back and breast,
Fainting with heat, with dust begrimed, their meagre faces see,
By glimmering lamps that serve to show their looks of misery.
And oft the hard swollen hand is raised to wipe the forehead dews;
He breathes a sigh for labor's close, and then his toil renews.
And manly hearts are throbbing there—and visions in that mind
Float o'er the young and sanguine soul, like stars that rain and shine.
Amid the dreariness that dwells within the cavern's gloom,
Age looks to youth to solace him—waits for his fruits to bloom.
Behold! there is a careless face bent from yon cabined nook;
Hope you may read in his bright eye—there's future in his look;
Oh, blight not then the fairy flower, 'tis heartless to destroy
The only pleasure mortals know—anticipated joy!
Oh, God! what flickering flame is this?—see, see again its glare!
Dancing around the wiry lamp, like meteors of the air.
Away, away!—the shaft, the shaft!—the blazing fire flies!
Confusion!—speed!—the lava stream the lightning's wing defies!
The shaft!—the shaft!—down on the ground and let the demon ride
Like the Sirocco on the blast—volcanos in their pride!
The choke-damp angel slaughters all—he spares no living soul!
He smites them with sulphureous brand—he blackens them like coal!
The young—the hopeful, happy young—fall with the old and gray,
And oh, great God! a dreadful doom, thus buried to decay
Beneath the green and flowery sod whereon their friends remain—
Disfigured, and perchance alive—their cries unheard and vain!
Oh, desolation! thou art now a tyrant on thy throne,
Thou smilest with sardonic lip to hear the shriek and groan!
To see each mangled, writhing corpse to raining eyes displayed—
For hopeless widows now lament, and orphans wail dismayed.
Behold thy work! The maid is there, her lover to deplore;
The mother wails her only child, that she shall see no more;
An idiot sister laughs and sings—oh, melancholy joy!
While bending o'er her brother dead, she opes the sightless eye.

Apart, an aged man appears, like some sage David oak,
Shedding his tears, like leaves that fall beneath the woodman's stroke;
His poor old heart is rent in twain—he stands and weeps alone—
The sole supporter of his house, the last, the best, is gone!
This is thy work, fell tyrant!—this the miner's common lot!
In danger's darkling den he toils, and dies lamented not.
The army has its pensioners—the sons of ocean rest,
When battle's crimson flag is furled, on bounty's downy breast;
But who regards the mining slave, that for his country's wealth
Resigns his sleep, his pleasures, home, his freedom and his health!
From the glad skies and fragrant field he cheerfully descends,
And eats his bread in stenchy caves, where his existence ends.
Aye, this is he whom masters grind, and level with the dust—
The slave that barters life, to gain the pittance of a crust.
Go, read your pillar'd calendar, the record that will tell
How many victims of the mine in yonder churchyard dwell!
Hath honor's laurels ever wreathed the despot's haughty brow?
Hath pity's hallowed gems appeared when he in death lay low?
Unhonored in his memory, despised his worthless name—
Who wields in life the iron rod, in death no tear can claim!

A great number of accidents have taken place on *Monday mornings*, when the miners descend after having quitted the mine on Saturday. M. Bischof reports that, having visited a gallery which had been abandoned for several days, he found the gases liquated to such an extent that they were inflammable in every part of the area; detonating in the middle portion, while the almost pure atmospheric air filled the lower part.

It is very dangerous to allow these liquations to accumulate; it is necessary that the current of air be sufficiently active to produce immediately the diffusion of the gas in the air and its withdrawal out of the mine before the mixture has become explosive. But notwithstanding the precautions of ventillation—*aérage*—many mines would be completely unworkable if there had not been found the special means of guarding them from the fire-damp—*grisou*. The coal beds most dangerous—as has been previously stated—are those which are the most valuable for their good qualities; science and industry have therefore been called on to seek the means of combating the effects of the grisou, and we proceed to expose those which have been successively employed:

MEANS TO CHECK FIRE-DAMP.—The first idea which presented itself to the explorers was to disembarrass themselves of the gas by allowing the liquation to establish itself and by setting it on fire, so as to burn it, in the absence of the miners. For this purpose a workman, clothed in vestments of moistened leather, his visage protected by a mask with spectacles of glass, advanced, crawling on his belly, in the galleries where the fire-damp was known to exist, and holding forward a long pole, at the end of which was a lighted torch; he sounded thus the irregularities of the roof, the front of the excavations, and set fire to the grisous. This method, which has been employed, within twenty years, in the basin of the Loire, and even occasionally at the present day, in some of the English fiery collieries, has numerous inconveniencies. The workmen, whom they called *pénitents*, were exposed to dangers to such an extent, that a great number perished. When the gas, instead of being simply inflammable, was detonating, the solidity of the mine was constantly compromised by the explosions; the fire attacked the coal and the timbers; the gases, which resulted from the combustion, became stationary in the works, and menaced the workmen with asphyxia; at length it became necessary, in certain mines, to repeat, even three or four times a day, this perilous operation, and yet it in no respect obviated the rapid disengagements which caused these numerous accidents. This method was equally in use in the English collieries; only the penitent or fireman, instead of carrying the fire himself, caused it to be moved by means of a slider placed over a line of poles connected together, and directed by a system of pullies and cords. The danger was thus diminished for the fireman, who retired into a niche formed in a neighboring gallery; but in the meantime many were still overtaken, and besides, all the other inconveniencies remained.

The method called the *eternal lamps* was evidently better. It consisted in placing towards the top of the excavation, and in all the points where the fire-damp collected, lamps constantly lighted, which burned the grisou as fast as it was produced; the danger was diminished in a considerable degree, because there could not be formed such large accumulations of inflammable or detonating gas. This mode of proceeding was, however, renounced in a great number of mines, on account of the production of carbonic acid and of azote; a production the more sensitive, since, to facilitate the liquation of the gases, the air ought not to be very strongly agitated.

At length it was devised to profit by the property possessed by platina in sponge to facilitate the combustion of the hydrogen with which it was brought in contact, and pellets, composed of one part of platina and two parts of clay, were made, and placed near the points at which the grisou or fire-damp concentrated. But all these efforts, based upon the incited combustion of the inflammable gas, proved to be only dangerous and incomplete palliatives, which substituted for a great peril a series of other dangers, less imminent, doubtless, but equally distressing.

From that time all the well disposed continued to search for processes based upon another principle. Two only could conduct to a good result: 1. The withdrawal of the gases out of the mine; 2. A mode of lighting different from that which was in use, and which would suffice for the purposes of the miner without compromising his safety.

The principle of withdrawing—*entrainement*—of the gases by a rapid ventillation is, without contradiction, that which was the most natural to conceive; because it was already applied to all the other deleterious gases. Dr. Vèhrle proposed at first to effect the decanting of the gases by making the excavations (stalls?) communicate by ascending passages with a gallery embracing all the works, and uniting with an ascending shaft. But this project, otherwise impracticable, offered a remedy for only a part of these accidents; the execution alone of the necessary works could not have been made without the greatest danger, if these works had been undertaken in the coal; while, in the rocks of the roof, the expenses would have rendered them impracticable. But a good ventillation alone could not suffice to place the miners in security; it was an excellent auxiliary means, but it always left unsolved this important problem: *the prevention of the inflammation of the gases which disengage themselves from the surfaces of the stalls.*

The lighting alone could conduct to the solution of this problem, and numerous attempts had been made, under this head, when Davy discovered the safety-lamp. Before him, they had operated with a small number of lights, placed in the lowest positions, and at a distance from the stalls; the workmen kept these lamps in view, and when the blue nimbus, the indication of hydrogen, began to show itself, they extinguished them or withdrew, covering them with their hats. They made use of, also, in the most infected mines, various phosphorescent matters, and particularly a mixture of flour and lime formed from oyster shells, called Canton phosphorus, although the uncertain and ephemeral light which these materials produced, was but a very feeble resource. At length it was observed that the proto-carbonated hydrogen was somewhat difficult of ignition, and that the red heat was insufficient to accomplish it; thus it was practicable to carry a red coal, or a red hot iron into the fire damp without inflaming it, the white heat alone having the necessary temperature. They profited by this discovery by lighting the stalls by means of the steel-mill, previously described by Prof. Ansted, and such was the state of affairs when Sir H. Davy took up the subject, and commenced his brilliant experiments.

From the foregoing, the reader will have gleaned a true understanding of the nature of those accumulations in mines, which render ventillation a subject of such vast importance. The means adopted and suggested to prevent the explosion of these gases are innumerable; and notwithstanding that the discovery of the safety lamp, by which the presence of noxious gases can always be determined, has been a source of great moment and security, yet its use has not been effective in assisting to *expel* them, or of suggesting any certain method of ventillation, beyond the actual *discovery* of what *ought* to be expelled. It exhibits, infallibly, the actual presence of the gases; but it has not led to the means by which they can be dispersed and driven out; so that the very dangers which attended mining previous to its introduction, attend it now, and always will attend it until some means can be devised to drive them out of the mines as fast as they accumulate. A principle of ventillation, for example, that will answer the purpose during the summer season, would not answer at all (in many cases, at least in our region,) during the winter, notwithstanding the fact that a difference of but a few degrees occurs in the temperature of the mines, during the year. This, of course, arises from external causes, which must always govern the atmosphere of the mine. Thus, when the external atmosphere is varied, and suddenly changed by winds, lightning, frost, snow, and long continued rains, the gases in the mine will be found to accumulate, to disperse, or to be borne out of the mine, as the case may be. They are always regulated by the density of atmospheric air, and rendered explosive just in proportion as they unite with it; here the safety-lamp is useful, because, however explosive the combination may be, it will conduct the viewer safely through it; but, then, here is where *ventillation* fails, for the admixture cannot be disunited, or expelled, except in its own time and way.

The most general plan of ventillation adopted in this region is simply this: The atmospheric air is admitted at or near the mouth of the slope. After traversing the mine through every avenue, the current is drawn through an escape-hole, over which a furnace is erected, and a regular and intense heat kept up. The draft thus afforded (there being no other escape) is generally very strong, and as the current of air is borne along, it bears with it all the noxious gases in the mine. Whenever these gases accumulate in workings where the current of atmospheric air does not penetrate sufficiently, they are dispersed by the miners, by means of canvasses or banners; and when there is not sufficient air, boys are stationed with revolving fans, by which the air is kept comparatively clean. These, with like devices, varied as circumstances may suggest, are the means resorted to in the anthracite regions of this State. There is a viewer for each mine, who enters with a Davy-lamp, and always reports the actual condition of the mine before the miners go to work.

## MORAL CONDITION OF MINERS, ETC.

The mining population of our coal regions is almost exclusively composed of foreigners, principally from England and Wales, with a few Irish and Scotchmen. The former have a decided preference for working in small veins, and they can use the pick in the narrowest space, right and left, and in all positions. They cannot, of course, swing it over their shoulders, or give it the *force* which is deemed so necessary for effective work; but, holding it in front, and making short, quick strokes, the pick is still as effective in their hands, in a space of three or four feet, (or even less) as would be in less circumscribed limits.

This predilection of the English miners is principally attributable to the fact, that the coal veins of their mining regions are usually thin, and having entered them at a very early age, they have thus formed a preference for thin veins, and a prejudice against large ones, where it is necessary to blast, use ladders, &c. Every miner carries his lamp on his cap, to which it is *hooked*. (Fig. 58.) While pursuing their labors in the mines, they soon become thoroughly covered over with the black coal dust, and their clothes, which are of the coarsest fabrics, rudely patched together, are saturated with water. The mines are damp, and the floor usually full of coal-mud and water;

hence the miners and laborers wear heavy coarse shoes, with the soles covered with *tacks.* Although extremely healthy as a class, they are nevertheless, pale and somewhat delicate in the face, and their eyes may be said to be *prominent.* Their features are not regular, and they cannot justly lay claim to manly beauty. They know little but what pertains to their subterraneous employments; making that the subject of their discussions, their jests and their pastimes, they have little care for things concerning the *upper crust.* They are, to a certain degree, superstitious; even the most intelligent of them yield to it. For example, it is considered an evil omen when a stranger, in entering the mine, *begins to whistle.* It produces a certain effect among them, and destroys, in a measure, their good spirits. A miner *never whistles,* and when, occasionally, they hum a tune, it is more of a soft and plaintive character than the popular songs of the day. The employment seems well calculated to indulge thought—calm, complacent ideas. There is no wildness, no ambition; they seek only *contentment,* and are satisfied with their lot.

FIG. 58.—A COAL MINER.

Visitors to the mines are cordially received, and every attention is shown them by the workmen. As a stranger would derive little satisfaction from his visit, unless he placed himself in the care of some one thoroughly acquainted with the mine, the workmen observe the old established custom of requesting every visitor to *pay his footing,* that is, the present of twenty-five or fifty cents (or a dollar or five dollars, if you like) to the person or persons who "show him the elephant." This request is generally made when the parties are the greatest distance from the slope, and when the visitor would naturally apprehend some difficulty in finding his way out! But, inasmuch as the conductor is withdrawn from his labor, and the visitor, without him, would be liable to get in the way of the workmen, and perhaps meet with some accident—besides his disability to understand the operations—the payment of the "footing" should never be neglected nor begrudged. The ladies, of course, pass free—the only charge being a *smile* or so.

The moral condition of the mining population of the anthracite regions of this State, is vastly superior to that of the same class in any other country. They reside in rude cottages, it is true, and do not enjoy the same elegancies of life which are obtained in many other industrial pursuits; but they have abundance to eat, good clothes to wear, and *money in their pockets*. A more generous-hearted people, more devoted to their friends, and faithful in their domestic attachments, does not live. Upon their arrival here, where labor is usually plenty, the first fruits of their industry are carefully hoarded, and when a sufficient sum is gathered together, it is sent home to cheer and bless some kindly-remembered relative or friend. Thousands of dollars are thus annually sent off by the humble laborers of the coal regions; and the fact illustrates the golden trait of our nature, which must in all time to come hide a catalogue of sins.

Strange are the incidents which sometimes occur in virtue of this noble impulse. Not long ago, an honest and industrious miner, after several months unceasing toil, had laid by a sufficient sum to pay the passage of his wife and several children from England to this country, besides a handsome sum for necessary expenses. The passage had been secured, and the money forwarded. The wife and her children in due time arrived; but, alas! where was the kind husband and father? In his *grave*. A day or two previous to their arrival, he had fallen at his work in the mine—a victim to an explosion of *fire-damp*.

The career of the miner, repairing daily to his subterranean workshop is, indeed, full of unfortunate streaks; dismal as solitude; black as the earth he delves. The scarred chieftain knows less danger, and much more glory.

But although the lot of the miner is little to be envied, it must be said, to the eternal credit of our country and its institutions, that it is here stripped of the odious features which characterize it in other countries. To exhibit the contrast between the mining districts of England and our own country, we have prepared the following items from the report of the investigating commissioners, appointed by the British Parliament, a few years ago. The degrading practice of employing children and females in mines does not prevail here. Boys are employed to drive the horses, and to assort the coal as it descends into the shutes from the breaker; but these duties are light, and suited to their capacity. Females, old or young, have never been engaged in the mines of this country; (thank heaven, our countrymen appreciate them too highly not to offer them better *engagements*, and more pleasant and appropriate *employment!*)

In 1841 the commissioners proceeded to investigate the condition of the laborers, male and female, in the mines of Great Britain. Of the number of children employed in the iron, coal, tin, and lead mines, it is difficult to ascertain, or to form any nice estimate; but the number must be very large. In many pits they are set to work at a very early age, some at six years, and at all ages beyond that. According to the evidence of Dr. Mitchell, the proportion of men to boys in the iron-stone pits of Staffordshire is one hundred to seventy; in the coal pits it is one hundred to ninety. Many of these pits, especially the iron-stone, are low, and horses cannot be employed, which is the principal reason of there being so large a proportion of boys in comparison with the men, to push the skiffs or carriages to the foot of the shaft.

In some of the mines the improper and reprehensible practice of employing female children, to perform precisely the same kind of labor as that performed by the boys, prevails. The practice of employing females in coal pits, says one of the commissioners, is flagrantly disgraceful to a Christian, as well as a civilized country. * * * In the Flockton and Thornhill pits the system is even more indecent; for though the girls are clothed, at least three-fourths of the men for whom they hurry work, are in a state of almost complete nudity, and in this state they assist one another to fill the corves eighteen or twenty times a day. I have seen this done myself, not once or twice, but frequently. "Girls," continues the report, "from five to eighteen, perform all the work of boys. There is no distinction whatever in their coming up the shafts, or in going down—in the mode of hurrying or thrusting—in the weights or corves, or in the distances they are hurried—in wages or dues. They are to be found alike vulgar in manner and obscene in language; but who can feel surprised at their debased condition, when they are known to be constantly associated, and associated only with men and boys, living and laboring in a state of disgusting nakedness and brutality; while they have themselves no other garment than a ragged shirt, or in the absence of that, a pair of broken trowsers, to cover their persons!"

In the mining districts of Scotland, the employment of females in this description of labor, is generally considered to be so degrading, that "other classes of operatives refuse intermarriage with the daughters of colliers who have wrought in the pits."

The report of the collieries, &c., in the east of Scotland, by Mr. Franks, contains correct and authentic information as to the condition of the laborers employed in them. The descriptions are illustrated by drawings, exhibiting the operations and position of the children in the mines. The following extract from his report will enable our readers to form a conception of the places and kind of work devolving upon the children and young persons pursuing their several occupations:

"Many of the mines in the east of Scotland are conducted in the most primitive manner; the one horse gin to draw up the bucket, no separation in the shaft, the ventillation carried on in many places by means of old shafts left open, &c. The negligence of underground workings corresponds with the above, the roads being carelessly attended to, and the workings very irregularly carried on, so that the oppression of the labor is as much increased by the want of good superintendence as by the irregularity of the work-people themselves. The roads are, most commonly, wet, but in some places so much so as to come up to the ankle; and where the roofs are soft, the dripping and slushy state of the entire chamber is such that none can be said to work in it in a dry condition, and the coarse apparel the labor requires absorbs so much of the drainage of the water as to keep the workmen as thoroughly saturated as if they were working continually in water.

"The workings in the narrow seams are sometimes 100 to 200 yards from the main roads, so that the females have to crawl backwards and forwards with their small carts, in seems in many cases not exceeding 20 to 28 *inches in height.*"

In fact, says a very intelligent witness, (Mr. Wm. Hunter, the mining foreman of Ormiston colliery,) upon the occasion of being authorized to issue an order to exclude women and children from the colliery, "in fact, women always did the lifting, or heavy part of the work, and neither they nor the children were treated like human beings, nor are they where they are employed. Females submit to work in places where no man or even lad could be got to labor in; they work on bad roads, up to their knees in water, in a posture nearly double. They have swelled haunches and ankles, and are prematurely brought to the grave, or what is worse, a lingering existence." "In surveying the workings of an extensive colliery under ground," says Robert Bold, the eminent miner, "a married woman came forward, groaning under an excessive weight of coals, trembling in every nerve, and almost unable to keep her knees from sinking from under her. On coming up, she said in a plaintive and melancholy voice, 'Oh, sir, this is sore, sore, sore work! I wish to God that the first woman who tried to bear coals had broken her back, and never would have tried it again.'"

Now, when the nature of this horrible labor is taken into consideration, the extreme severity, its regular duration of from 12 to 14 hours daily, and sometimes much longer; the damp, heated, and deleterious atmosphere, in which the work is carried on; the tender age and sex of the workers; when it is considered that such labor is performed, not in isolated instances, selected to excite compassion, but that it may be regarded as the type of the every day existence of hundreds of our fellow creatures—a picture is presented of deadly physical oppression and systematic slavery, of which those unacquainted with such facts would not credit as existing in the British dominions.

We may add, as worthy of remark, that to this labor, which is at once so repulsive and severe, the girls are invariably sent at an earlier age than boys—from a notion very generally entertained amongst parents, that they are more acute and obedient.

Such is a sample of *British* slavery;—let us hereafter hear no more of the supposed horrors of negro slavery from that quarter! The stain, John Bull, is on your own hands!

## MISCELLANEOUS STATISTICS.

Coal, observes an anonymous writer, is evidently a result of the decomposition of the compound of bodies from which it is obtained. It consists of the greatest part of the earthy principle of these compound bodies, with which a part of the saline principle, and some of the phlogiston of the decomposed oil, are fixed and combined very intimately. Coal can never be formed but by the phlogiston of a body which has been in an oily state; hence it cannot be formed by sulphur, phosphorus, metals, nor by any other substance the phlogiston of which is not in an oily state. Every oily matter treated with fire in close vessels, furnishes true coal; so that whenever a charry residuum is left, we may be certain that the substance employed contains oil.

The inflammable principle of coal, although it proceeds from oil, certainly is not oil, but pure phlogiston, since coal added to sulphuric acid can form sulphur; to phosphoric acid, can form phosphorus, &c., and since oil can produce none of these effects till it has been decomposed and reduced to the state of *coal.* Besides, the phenomena accompanying the burning coal are different from those which happen when oily substances are burnt. The flame of charcoal is not so bright as that of oil, and produces no flame or soot.

All the phlogiston of coal is not burnt in the open air, particularly when the combustion is slow. One part of it exhales without decomposition, and forms a vapor, or an invisible and insensible *gas.* This vapor, (which is, or at least contains a great deal of fixed air,) is found to be very pernicious, and to affect the animal system in such a manner as to occasion death in a very short time. For this reason it is dangerous to remain in a close room or place where charcoal or any other sort of coal is burnt. Persons struck by this vapor are stunned, faint, suffer a violent headache, and fall down senseless and motionless. The best method of recovering them is, simply, timely exposure to the open air, and by making them swallow vinegar, and breathe its steam.

Among coals considerable difference is observable, which proceeds from difference in the bodies from which they are made; some coals, particularly, are more combustible than others. This combustibility seems to depend on the greater or less quantity of saline principle they contain; that is, the more of the saline principle it contains, the more easily it decomposes and burns. This difference in coal varies in about the same proportion that the difference in the properties of various kinds of wood varies when exposed to fire. The difference in coal, unlike that in wood, relates also to the localities where it is found; it is, therefore, rarely that the opposite extremes of its analytical properties are united in the same spot. We append an analysis of these two extremes—the first being that of the purest and best coal, and the latter the inferior and least valuable.

ANALYSIS OF ANTHRACITE.

| | | |
|---|---|---|
| 1 | Carbon, | 90 per cent. |
| | Volatile matter, | 6½ " |
| | Ashes, | 3½ " |
| | | 100 |
| 2 | Carbon, | 77 per cent. |
| | Volatile matter, | 11 " |
| | Ashes, | 12 " |
| | | 100 |

This difference in the quality of coal is again perceptible in reference to its *weight*. We append the following, which will exhibit the character of our anthracites according to the weight of each respectively, per cubic yard.

WEIGHT OF ANTHRACITE COAL.

FIRST, OR SCHUYLKILL REGION.

| *Localities proceeding from West to East.* | *Weight of a cubic yard in lbs.* |
|---|---|
| Lykens' Valley, | 2224 |
| Stony Creek, six mile openings, | 2244 |
| Big Flats, | about 2351 |
| Rausch Gap, | 2453 |
| Lorberry Creek, | 2484 |
| Pottsville, | mean 2504 |
| Tamaqua, Vein N. | 2700 |
| Lehigh, Mauch Chunk, | 2615 |
| " Nesquehoning, | 2646 |

SECOND, OR MIDDLE REGION.

| *Localities proceeding from West to East.* | *Weight of a cubic yard in lbs.* |
|---|---|
| West Mahanoy Coal, | 2313 |
| Hazleton, | 2615 |
| Girardville, | 2700 |
| Beaver Meadow, | 2700 |

The Pennsylvania anthracite appears to be altogether heavier than the European, as will appear from the following:

EUROPEAN.

| | |
|---|---|
| South Wales, (Swansea,) | 2131 |
| France, (Grenoble,) | 1809 |
| Black Spring Gap, | 2351 |

PENNSYLVANIAN.

| | |
|---|---|
| Wilkesbarre, (Baltimore co.) | 2484 |
| Pottsville, | 2649 |
| Tamaqua, heaviest, | 2808 |

The bituminous coals of the United States are also considerably heavier than those of Europe. With but one or two exceptions, all coals of the United States exceed one ton in weight to the cubic yard, while there is no instance of the bituminous coal of Europe reaching that weight.

Prof. Johnson, who conducted, by authority of Congress, a series of experiments on American coals, applicable to steam navigation, and to other purposes, in his report to that body, in 1844, says that the justly celebrated foreign bituminous coals of New Castle, Liverpool, Scotland, Pictou, and Sidney—coals which constitute the present reliance of the great lines of Atlantic steamers—are fully equalled, or rather surpassed in strength, by the analogous coals of Eastern Virginia; that they are decidedly surpassed by all the free-burning coals of Maryland and Pennsylvania, and that an equally decided advantage in steam-generating power is enjoyed by the anthracites over the foreign coals tried whether we consider them under equal weights or equal bulks.

Mr. Johnson remarks, that coal, "when sold by weight, and used on shore, the weight per cubic foot is a point of little moment. Space for storage is easily obtained. But, in steam navigation, bulk, as well as weight, demands attention; and a difference of twenty per cent., which experiment shows to exist between the highest and lowest average weight of a cubic foot of different coals, assumes a value of no little magnitude." This is obviously true, since, if other things be equal, the length of a voyage must depend on the amount of evaporative power afforded by the fuel which can be stowed in the bunkers of a steamer, always of limited capacity.

## GENERAL FEATURES OF PENNSYLVANIA ANTHRACITE.

The anthracites have specific gravities, varying from 1·39 to 1·61; retain their form when exposed to a heat of ignition, and undergo no proper intumescence while parting with the small portion of volatile matter which they contain; or, if changed at all, are only disintegrated into angular fragments. Their flame is generally short, of a blue color, and consequently of little illuminating power. They are ignited with difficulty, give an intense concentrated heat, but generally become extinct while yet a considerable quantity remains unburnt in the grate.

In experimenting with American coals, Prof. Johnson thus describes the differences between the red and white ash coals of our region:

FIRST, OF THE RED-ASH, PEACH MOUNTAIN.

The sample exhibited a deep jet-black color, an uneven splintery fracture; a lustre varying from dull to shining, according to the direction in which the fracture is made. Like all the other anthracites, it was wholly unaffected by atmospheric influences in a period of eighteen months, during which time they were in my charge.

This sample is more easily separated at the surfaces of deposition than most of the white-ash coals, but less so than that of Lykens' Valley. It has no exterior indication of impurity, such as discoloration from oxide of iron, or efflorescence of metallic salts. It has certain surfaces polished and minutely striated, appearing as if they had been subjected to friction under intense pressure. This feature is not, however, of so frequent occurrence in this, as in many other samples of anthracite.

Its specific gravity, determined by two specimens, was found to be 1·465 and 1·4632—the mean of which enables us to calculate the weight of a cubic foot of solid coal at 91·505 pounds. But the weight of 70 charges of two cubic feet each, in the state of lumps, gave 37·7939 pounds per cubic foot, proving that the actual weight in the market is but 0·5878 of the calculated weight in the mine. The same data prove that 42·64 cubic feet of space will be required for one gross ton.

| *Analysis* | | |
|---|---|---|
| | Carbon, | 86·087 |
| | Volatile matter, | 6·965 |
| | Earthy matter, | 6·948 |

SECOND, OF THE WHITE-ASH, SUMMIT HILL.

The aspect and character of this coal leaves no doubt that it will remain for any desired length of time, either under shelter or in the open air, without material change.

The coal was received generally in lumps, requiring to be reduced in order to be burned advantageously on the grate. Its aspect is that of most of the harder anthracites, possessing a deep black color, shining uneven and splintery fracture, with occasional exposure of conchoidal forms; a striated, rather grayish appearance, generally indicative of considerable portions of earthy impurity, marks certain surfaces. The seams of deposition are seldom followed by the fractures.

The specific gravity of two specimens was found to be 1·6126 and 1·5679, from which the calculated weight per cubic foot are 100·79 and 97·99 pounds respectively, or on an average, 99·39 pounds.

| *Analysis* | | |
|---|---|---|
| | Carbon, | 88·052 |
| | Volatile matter, | 5·235 |
| | Earthy matter, | 6·663 |

The foregoing tables will, doubtless, give a satisfactory exhibit as to the relative value of the different coals named for steam navigation, iron making, or for generating steam ordinarily. For stove use, the following experiment will determine the respective value of the white and red-ash varieties. Two rooms of nearly the same size, and having the same temperature, were selected to ascertain how many pounds of each kind would be required to heat them to a temperature of 65 degrees, during a period of fifteen hours, when the temperature out of doors, at 9 A. M., was at ten degrees below the freezing point. Two days were occupied in the trial, so that the red and white-ash coals might be used in *alternate* rooms. Fires were made at 9 A. M. and continued until 12 P. M. Two thermometers (one in each room,) were suspended at the greatest distance from the grates, and the temperature was carefully registered every hour. The result was as follows: thirty-one pounds, each day, of the Schuylkill *red-ash* coal, gave a mean temperature of 64 degrees; and thirty-seven pounds, each day, of the Lehigh *white-ash*, taken from a vein of high repute, gave a mean temperature of 63 degrees—thus making 2000 pounds of the red-ash to be equal to 2·387 pounds of the white-ash, or red-ash coal at $5.50 per ton, to be equal to white-ash ash at $4.61. This, says Mr. Taylor, settles the question between the two coals on the score of economy.

## COMPARATIVE VALUE OF ANTHRACITE AND BITUMINOUS COALS.

An impression has long prevailed that, for purposes of steam navigation, Cumberland coal was generally superior to anthracite, and it always enjoyed greater favor in the market. This impression must be destroyed—for a greater mistake never existed on the public mind. A series of experiments were recently instituted on several steamers of the United States Navy, under order of the government, the result of which is a triumphant vindication of anthracite coal, not only for steam vessels, but for generating steam under all circumstances—broadly establishing its vast superiority over bituminous and every other known coal. These experiments were brought about principally through the instrumentality of the Hon. James Cooper, our distinguished senator in Congress. The previous experiments of Prof. Johnson were unsatisfactory, because the tests were prosecuted on too small a scale to ascertain the real merits of anthracite; and in the manner of burning it, he committed a mistake which led him greatly to depreciate its value compared with Cumberland coal, in equal weight. These mistakes were apparent to Mr. Cooper, and he therefore took up the subject, and called the attention of Congress to it. A resolution offered by him was passed at once, requiring the Secretary of the Navy to institute experiments, the result of which we shall lay before the reader. Mr. Cooper resides at Pottsville, and his exertions in asserting, and thus establishing, with the broad seal of the United States government, the superiority of anthracite *over all other coals*, are deserving, and will receive, the lasting thanks of the thousands engaged in the trade of Pennsylvania.

*Experiments with Bituminous Coal, made with the boilers of the United States Steamer Fulton, at the New York Navy Yard, January,* 1852.

The temperature of the water in the boilers being at 38° F., and the temperature of the boiler room 18°, the fires were lighted at ten hours thirty minutes, A. M. At eleven hours forty minutes, A. M., the temperature of the water was 212° F., and steam began to be generated at the atmospheric pressure. Time raising steam, seventy minutes. The temperature of the boiler-room had now increased from 18° to 32° F. At eleven hours fifty-four minutes, A. M. the steam pressure in the boilers was thirty pounds per square inch above the atmosphere. Time of obtaining thirty pounds of steam, one hour twenty-four minutes from a temperature of 32° F. Up to this time there had been fed into the furnaces one thousand pounds of dry pine wood, equal to five hundred pounds of coal, and two thousand eight hundred and twenty-six pounds of (Cumberland) bituminous coal. Total, three thousand three hundred and twenty-six pounds.

The engine was now set in operation to work off all the steam, which the above amount of coal would generate, no more being fed to the furnaces. In fifty-three minutes the steam pressure was reduced from forty pounds to five pounds, and the number of double strokes of piston made from forty-one to seven, when the engine was stopped. During the time the engine was in operation, the steam was cut off at half-stroke.

The engine consisted of one cylinder, fifty inches in diameter, and ten feet four inches stroke. The space between the cut-off valve and piston, including clearance, to be filled with steam per stroke is 3·094 cubic feet. The calculation of the amount of water evaporated is made from the quantities of steam measured out by the cylinder, divided by the relative bulks of steam of the experimental pressures and the water from which it is generated.

The initial pressure of the steam in the cylinder is taken at one pound *less* than in the boilers. The space displacement of piston filled with steam, per stroke, is 70·448 cubic feet, to which must be added the above 3·094 cubic feet, making a total of 73·542 cubic feet.

| Time.—Minutes. | Pressure above atmosphere, per square inch, in pounds. | Number of double strokes of piston made. | Cubic feet of water evaporated. |
|---|---|---|---|
| 6 | 32½ | 41 | 10.326 |
| 5 | 25 | 34 | 7.311 |
| 5 | 22½ | 32 | 6.483 |
| 5 | 19¼ | 31½ | 5.879 |
| 5 | 16½ | 31 | 5.352 |
| 5 | 14 | 30 | 4.796 |
| 5 | 11¼ | 26 | 3.797 |
| 5 | 9 | 23 | 3.075 |
| 5 | 7¼ | 21 | 2.631 |
| 5 | 5½ | 12 | 1.393 |
| 2 | 4¼ | 7 | 0.868 |
| | | | 51.911 |

Taking the weight of a cubic foot of sea water at 64·3 pounds, the total weight evaporated is (51·911 × 64·3)=3337·877 pounds. The boilers of the Fulton contained 82·000 pounds of water at the initial temperature of 32° F., which was raised to 212° F., and 3337·877 pounds of it evaporated by three thousand three hundred and twenty-six pounds of coal.

Now it requires five times and a half as much caloric to evaporate a given bulk of water from a temperature 212° F., so as to raise it to that temperature from 32° F. The quantity of fuel, therefore, expended in raising the water from the initial temperature to that of 212° F., compared to that expended in evaporating the 3337·877 pounds from that temperature, will be as (82·000 × 180°)=14,760,-000 to (3337·877 × 990°)=3,304,498·23, or as 4·4666 to 1·000; consequently

$\frac{3326}{4.4666}$=744·6 pounds of coal were consumed in evaporating 3337·87 pounds of sea-water, or 4·483 pounds of water per pound of coal. It was intended to have made, on the following day, an experiment, under precisely the same circumstances as above, with anthracite; but it was found impossible from the presence of *ice* to work the engine, the experiment was therefore only made so far as regards the time of getting up steam, with the following results, viz:

The fires were lighted with the same *quantity* and *kind* of wood, and the same *quantity* of coal that had been used the day previous. At seven hours and twenty minutes, A. M. the temperature of the water in the boiler being 38° F., and that of the boiler room 32° F., with the natural draft, the temperature of the water at eight hours and five minutes was 212° (steam,) and the boiler room 43, F. Time to generate steam, forty-five minutes. At eight hours and twenty minutes the steam pressure in the boiler was thirty pounds per square inch. Time of obtaining thirty pounds of steam from water at 38° F. was *one* hour.

With the bituminous coal it will be seen that it required seventy minutes to obtain steam from water at the temperature of 32° F., while it only required forty-five minutes with the anthracite; being a difference of time in this respect of about thirty-six per cent. of the bituminous time.

The data for a comparison of the evaporative values of the coals was obtained by another experiment as follows:

*Experiments with White Ash Anthracite made with the boilers of the United States Steamer Fulton, in New York Bay, January* 1, 1852.

This experiment was made with the steamer under way, while steaming with steady pressure of steam and revolutions of the wheel, as follows:

Steam pressure (*initial*) in cylinder per square inch above the atmosphere, twenty-five pounds; double strokes of piston per minute, twenty-one and one-third; cutting off at from commencement of stroke, three-eighths; consumption of coal per hour, eighteen hundred pounds.

From the above data, there was filled per stroke 52·837 cubic feet of the space displacement of the piston, to which add 3·096 cubic feet of space comprised between the cut-off valve and piston, making a total of 55·931 cubic feet of steam of twenty-five pounds pressure, which would be per minute 55·931 $\times$ 42⅔=2386·39 cubic feet, and per hour 143,183·40 cubic feet. Dividing this last number by the relative bulks of steam of the pressure generated, and the water from which it was generated, we obtain $\frac{143183.40}{684}$=209·332 cubic feet of sea-water, which at 64·3 pounds per cubic foot, amounts to 13,460·047 pounds, evaporated by eighteen hundred pounds of coal, or seven thousand four hundred and seventy-pounds of sea-water per pound of coal.

3. *Experiment with White Ash Anthracite Coal made with the boilers of the pumping-engine at the United States Dry Dock, New York Navy Yard.*

A comparative experiment was made with the boilers of the pumping-engine at the New York Navy Yard, in October, 1851, on the comparative advantages of anthracite and bituminous coals; all the conditions were as nearly alike as practicable. With the anthracite coal a combustion of nine hundred and eighty pounds per hour, evaporated a sufficient quantity of water to supply the engine with steam of twelve pounds pressure above the atmosphere, per square inch, for four hundred and twenty-five double strokes of piston per hour, the steam pressures being alike in both cases; the economical values of the coals will be represented by the number of double strokes of piston made, divided by the quantity of fuel per given unit of time; or will be, anthracite $\frac{425}{980}=0{\cdot}4337$: bituminous $\frac{294}{1100}=0{\cdot}2673$, or the anthracite is better than the bituminous in the proportion of $\frac{0{\cdot}4337}{0{\cdot}2637}=1{\cdot}623$ to $1{\cdot}000$.

It is proper to remark that these boilers were expressly designed for burning *bituminous coal.*

## COMPARISON.

The coals used in these experiments were the kinds furnished by the agents of the government for the use of the United States Navy Yard and Steamers, and was taken indiscriminately from the piles in the yard without assorting.

The bituminous was from the "Cumberland" mines. The anthracite was the kind known as "White Ash Schuylkill."

From the preceding data, it appears that in regard to the rapidity of "getting up" steam, the anthracite exceeds the bituminous thirty-six per cent.

That in economical evaporation per unit of fuel, the anthracite exceeds the bituminous in the proportion of 7·478 to 4·483 or 66·8 per cent.

It will also be perceived, that the result of the third experiment on the boilers of the pumping-engine at the New York Dry Dock, which experiment was entirely differently made and calculated from the first and second experiments, gave an economical superiority to the anthracite over the bituminous of 62·[illegible] per cent.; a remarkably close approximation to the result obtained by the experiments on the "*Fulton's*" boilers, (66·8 per cent.,) particularly when it is stated that the boilers and grates of the pumping-engine were made with a view to burning bituminous coal, which has been used since their completion; while those of the "Fulton" were constructed for the use of anthracite. The general characters of the boilers were similar, both having return drop flues.

Thus it will be seen, from the experiments, that, without allowing for the difference of weight of coal that can be stowed in the same bulk, the engine using anthracite could steam about two-thirds longer than with bituminous.

*These are important considerations in favor of anthracite coal* for the uses of the Navy; without taking into account the additional amount of anthracite more than bituminous that can be placed on board a vessel in the same bunkers, or the advantages of being free from *smoke,* which in a *war*-steamer may at times be of the utmost importance in concealing the movements of the vessel, and also the almost, if not altogether, entire freedom from spontaneous combustion.

The results of the experiments made last spring on the United States steamer "Vixen" were so favorable, that I recommended to the Bureau of Construction, &c., the use of anthracite for all naval steamers at that time having, or to be thereafter fitted with *iron* boilers; particularly the steamers "Fulton," "Princeton," and "Alleghany," the boilers for all of which were designed with a special view to the use of *anthracite*, and with the approval of that bureau.

The "Fulton's" bunkers are now filled with anthracite, and the consumptions referred to in the engineer's report on that steamer show, during the short time she has been at sea, that the anticipated *economy* has been fully realized.

In view of the results contained in this report, I would respectfully recommend to the Bureau of Yards and Docks, the use of anthracite in the several Navy Yards, and especially for the engine of the Dry Dock at the New York Navy Yard.

In conclusion, I desire the approval of the Bureau to make such investigations as my duties will permit, with regard to the *experience* of the durability of *copper* boilers, when used with bituminous or anthracite coal; which can be done without any specific expenditure.

The inquiry may prove highly important to the Navy Department, as the use of anthracite under copper boilers has been heretofore generally considered as more injurious than bituminous coal, and is consequently not used by government in vessels having copper boilers.

Respectfully submitted, by your obedient servant,

CHARLES B. STUART,

*Engineer-in-Chief, U. S. Navy.*

Commodore JOSEPH SMITH,

*Chief of Bureau of Yards and Docks.*

Some further remarks of Mr. Stuart, in reference to this subject, may be found in his report to the Secretary of the Navy, dated May 2, 1852. This portion of the Report combats the experiments of Prof. Johnson, and exhibits the mistake he made, as a practical test, in declaring the comparative qualities of the two coals for steam-marine purposes—the experiments having been, upon the whole, too hasty, incomplete, and limited, to properly develope the real strength and value of the anthracite, while they were sufficiently large to exhibit the bituminous coal to the best advantage.

## ANTHRACITE COAL TRADE OF PENNSYLVANIA.

The following table exhibits the quantity of anthracite coal sent to market from the different regions in Pennsylvania, from the commencement of the trade in 1820 to 1851, inclusive, together with the annual increase:

| Years. | Schuylkill | Lehigh. | Lackawanna. | Other regions. | Aggregate | Annual increase. |
|---|---|---|---|---|---|---|
| 1820 | | 365 | | | 365 | |
| 1821 | | 1,073 | | | 1,073 | |
| 1822 | | 2,240 | | | 2,240 | |
| 1823 | | 5,823 | | | 5,823 | |
| 1824 | | 9,541 | | | 9,541 | |
| 1825 | 6,500 | 28,393 | | | 34,893 | 25,352 |
| 1826 | 16,767 | 31,280 | | | 48,047 | 13,154 |
| 1827 | 31,360 | 32,074 | | | 63,434 | 15,837 |
| 1828 | 47,284 | 30,232 | | | 77,516 | 13,082 |
| 1829 | 79,973 | 25,110 | 7,000 | | 112,083 | 34,567 |
| 1830 | 89,984 | 41,750 | 43,000 | | 174,734 | 62,551 |
| 1831 | 81,854 | 40,966 | 54,000 | | 176,820 | 2,086 |
| 1832 | 209,271 | 70,000 | 84,600 | | 363,871 | 187,051 |
| 1833 | 252,971 | 123,000 | 111,777 | | 487,748 | 123,877 |
| 1834 | 226,692 | 106,244 | 43,700 | | 376,636 | decrease. |
| 1835 | 339,508 | 131,250 | 90,000 | | 560,758 | 184,122 |
| 1836 | 432,045 | 148,211 | 103,861 | | 682,428 | 121,670 |
| 1837 | 523,152 | 223,902 | 115,387 | | 881,476 | 199,048 |
| 1838 | 433,875 | 213,615 | 78,207 | | 739,293 | decrease. |
| 1839 | 442,608 | 221,025 | 122,300 | 11,930 | 819,327 | 80,034 |
| 1840 | 452,291 | 225,318 | 148,470 | 15,505 | 865,414 | 46,087 |
| 1841 | 584,692 | 143,037 | 192,270 | 21,463 | 958,899 | 93,485 |
| 1842 | 540,892 | 272,546 | 205,253 | 57,346 | 1,108,001 | 149,102 |
| 1843 | 677,295 | 267,793 | 227,605 | 86,000 | 1,263,539 | 155,538 |
| 1844 | 839,934 | 377,002 | 251,005 | 127,993 | 1,631,669 | 368,130 |
| 1845 | 1,083,796 | 429,453 | 273,435 | 188,401 | 2,023,052 | 391,783 |
| 1846 | 1,237,002 | 523,002 | 320,000 | 205,075 | 2,343,992 | 320,940 |
| 1847 | 1,583,374 | 643,973 | 388,203 | 299,302 | 2,982,309 | 638,317 |
| 1848 | 1,652,835 | 680,746 | 437,500 | 256,627 | 3,089,238 | 106,929 |
| 1849 | 1,605,126 | 801,246 | 454,240 | 303,730 | 3,242,541 | 153,403 |
| 1850 | 1,712,007 | 722,622 | 543,353 | 276,339 | 3,254,321 | 11,780 |
| 1851 | 2,184,240 | 989,296 | 788,495 | 415,099 | 4,377,130 | 1,122,809 |
| | 17,367,628 | 7,562,128 | 5,080,721 | 2,264,792 | 32,755,232 | |

The next table gives the actual cost of the works mentioned in it, but does not include the running-stock or fixtures, which would probably swell the amount at least twenty to twenty-five millions additional. Thus, the three principal improvements, including equipments, have a capital investment of nearly $35,000,000; viz.: the Reading Railroad, $17,000,000—the Schuylkill Navigation, $10,200,000, and the Lehigh Navigation, $7,500,000.

## CANAL AND RAILROAD SYSTEM CONNECTED WITH THE COAL TRADE.

| | Miles. | Cost. |
|---|---|---|
| Lehigh Slackwater Navigation, | 37½ | $4,455,000 |
| Lehigh and Susquehanna Railroad, | 20 | 1,354,000 |
| Mauch Chunk and Summit Railroads, | 40 | 400,000 |
| Delaware Division of Pennsylvania Canal, | 43 | 1,735,958 |
| Beaver Meadow Railroad, | 26 | 360,000 |
| Hazleton Railroad, | 10 | 120,000 |
| Buck Mountain Railroad, | 4 | 40,000 |
| Summit Railroad, | 2 | 20,000 |
| Delaware and Hudson Canal, | 108 | 3,250,000 |
| Morris Canal, | 102 | 4,000,000 |
| Schuylkill Canal Navigation, | 108 | 5,785,000 |
| Philadelphia, Reading and Pottsville Railroad, | 98 | 11,500,000 |
| Little Schuylkill Railroad, | 20 | 500,000 |
| Mine Hill Railroad, with Laterals, | 55 | 550,000 |
| Danville and Pottsville Railroad, (44m. unfinished) | 30 | 680,000 |
| Mount Carbon Railroad, | 7 | 155,000 |
| Mount Carbon and Port Carbon Railroad | 2½ | 120,000 |
| Schuylkill Valley Railroad, | 14 | 300,000 |
| Mill Creek Railroad, | 6 | 120,000 |
| Lykens' Valley Railroad, | 16 | 200,000 |
| Wisconisco Canal, | 12 | 370,000 |
| Swatara Railroad, | 4 | 20,000 |
| North Branch Canal and Extension, (unfinished) | 163 | 2,790,310 |
| Dauphin Company's Railroads, (unfinished) | 52 | 1,500,000 |
| Railroads by Individuals, (estimated) | 120 | 250,000 |
| Railroads under ground, (estimated) | 200 | 250,000 |
| Railroads proposed or under way, in the coal regions, | 100 | 300,000 |
| Union Canal—(estimated for coal trade), | — | 1,000.000 |
| Total length and cost, | 1400 | $42,034,268 |

The Philadelphia *North American*, a few months since, published a series of articles on the *future* of the coal trade and the canal and railway system connected with it, which have some permanent points of interest, and we shall here introduce some extracts:

The Reading Railroad Company alone has invested in its works a sum about equal to one-half the capital of the late bank of the United States, with annual earnings to the amount of $2,314,340, a net annual profit of $1,125,393, and an annual tonnage of coal shipped from its port at Richmond *greater than the whole foreign tonnage of the city of New York*. Surely such an establishment deserves to be better known than it is, not only to capitalists, but to citizens generally.

The history of the Reading Railroad, in its main features, is like that of almost all our public works, except, perhaps, the Pennsylvania Railroad, now in progress. It was commenced on a capital not sufficient to complete it. The property acquired by the original stockholders has, therefore, been pledged successively to bondholders and holders of preferred stock, to obtain the necessary means of carrying out the original design. These several investments may be summed up briefly as follows:—

| | |
|---|---|
| Common stock, - - - - - - | $4,159,832 |
| Preferred stock, - - - - - | 1,551,800 |
| Sinking Fund stock, - - - - - - | 393,700 |
| Bonds - - - - - - - | 10,794,183 |
| Total, - - - - - - - - | $16,899,515 |

In obtaining the large amount of money represented in these figures, during a period of sixteen years, from 1836 to 1852, no doubt many sacrifices have been made. Had the original proprietors entered upon the business with $17,000,000 cash in hand, it is but reasonable to suppose that all the property now owned by the company might have been acquired with a much less expenditure of money. Without professing to be entirely familiar with the history of its financial operations, this much we suppose may be taken for granted. The important practical question, however, still remains, namely, is the property, represented by this $16,899,515, capable of perpetuating itself, and at the same time of producing a fair annual dividend on the whole amount?

Before proceeding to answer this question, it is proper to remark that the work of the company is now fairly complete. The road itself is made, with a double track for the entire distance, from tide water at Richmond to the mines, and with the numerous sidings and turnouts necessary for such a vast business. The company have obtained, by purchase, the city terminus of the State Railroad, extending three and a half miles from Broad and Willow to the junction on the other side of the Schuylkill. They have secured a site for a depot of the largest dimensions at the junction of Broad street and Willow, besides constructing an office with ample accommodations, for the general business of the company, in Fourth street. They have, not only their immense wharves at Richmond, but sufficient additional wharf room secured in the same vicinity, to supply any additional increase to their business. They have their iron foundry, brass foundry, and steam tilt-hammer shop, for supplying and repairing the necessary machinery at the shortest notice and the cheapest rates.

The only increase to their permanent investment, therefore, that seems at all possible, is in the running power. The company have at present invested in this item, that is, in engines and cars, $2,276,576. With this they can perform the general business of the road, and bring to market 2,000,000 tons of coal annually. The utmost capacity of the road, suppose it to be covered with a continuous line of cars, running night and day, is estimated at 4,000,000 tons. Should the business of the road ever reach its utmost capacity, as it is not at all impossible it may, an additional investment in locomotives and cars would be necessary. But such an increase of business, all other expenses remaining comparatively stationary, would be accompanied with such an enormous increase of profits as to make the procuring of the necessary capital the easiest thing imaginable. Indeed, it is not improbable, in such an emergency, that the additional machinery would be obtained out of the excess of profits, and without increasing the capital investment.

Be that as it may, one thing is perfectly obvious, namely, that the work of the Reading Railroad Company is now substantially completed. Their last large investment was made in 1850 and 1851, when they bought the city terminus of the State Railroad, the additional wharf room in Richmond, and the real estate in Fourth street, and at the corner of Broad and Willow. They now have access to a port of their own (Richmond) for all coal intended for the foreign or coastwise market, and direct access, on a road of their own, to the very heart of the city, for all coal intended for home consumption. They have the necessary offices, depots, wharves, workshops, and running power, all in a state of complete efficiency. More than even this, they have adopted a settled policy of laying aside from their annual earnings a sum sufficient to renew the wear and tear of the road and machinery, and to perpetuate the property in its present condition. This sum, (3 cents per 100 tons a mile,) known as the renewal fund, has been fixed upon after an experience of ten years, and a comparison with the experience of all other roads in the world engaged in the same kind of business, and has been found amply sufficient.

We repeat, therefore, the only question to the capitalist is, whether this vast machinery, a railroad of a hundred miles, with its multiplied appurtenances, is capable of earning for its owners a steady and sufficient income on the capital invested in it.

The last year (1851) was a trial one. The Company was put to a more earnest trial of strength, probably, than it will ever be again. There was at the beginning of the year a general misunderstanding among the parties interested in the different coal regions. A severe and determined competition was the result. The Lehigh Company and the various New York companies engaged in the Lackawanna region, particularly the latter, put their coal at greatly reduced prices. To prevent the withdrawal of the trade from the Schuylkill region, the Reading Railroad and the Schuylkill Navigation Companies both reduced materially their rates for transportation. The railroad reduced its freight on coal from $1,53½ to $1,22½, or 31 cents per ton. This had the desired effect. The road not only maintained its accustomed amount of business, but largely increased it. What is more, even at these reduced rates, while the stock of its main rival, the Delaware and Hudson Canal Company, fell during the competition $50 per share, the Reading Railroad earned more than six per cent. on its whole investment, after paying all expenses, and after laying aside the annual sum provided to renew the works and keep them in repair. The gross receipts of the road were $2,314,330. The total expenses $1,188,936. Add $109,847, (the sum set apart for the renewal fund) and we have still a net balance of $1,015,547, or a small fraction over six per cent. on the entire investment of $16,899,515.

The effect, therefore, of the keen contest of 1851, has been to show that the Reading Railroad can afford to transport coal at rates that would be ruinous to all the other companies. The other companies, in view of this clearly established result, have now changed their policy, leaving the railroad free to raise its freight without loss of business. The small increase of ten cents per ton, which is proposed, and which is now entirely practicable, would, without any increase of tonnage, make an addition of $165,000 to the receipts of the company; or, in other words, an increase of four per cent. in the profit of the common stock.

But there is a prospect, perhaps we might say a certainty, of a largely increased business. The coal transported on the road in 1851, was one million six hundred and fifty thousand two hundred and seventy tons, being an increase of two hundred and ninety-eight thousand seven hundred and sixty-eight tons over the year previous. Nor was this at the expense of the rival companies. All the companies had an increased business. The whole amount of coal brought to market from the Schuylkill, Lehigh, and Lackawanna regions, during 1851, was one million one hundred and one thousand and fifty-one tons more than in any previous year. Yet, with all this increase, the market has not been glutted, the price has at no time been such as not to yield profitable results to the producer, and the principal markets are at this moment bare. There is not coal enough now in the city of New York to last the winter out.

Taking the increase of last year (two hundred and ninety-eight thousand seven hundred and sixty-eight tons) as the basis for a calculation of the business of the Reading Railroad for the year 1852, and supposing no special causes to exist ensuring even a greater increase, the amount of coal which it will have to transport the present season will be one million nine hundred and forty-nine thousand and thirty-eight tons. But there are reasons why even a greater increase may be expected. In the first place, every season thus far has opened some new method for using coal profitably in mechanical and commercial business. In the second place, the extraordinary severity of the present winter has nearly doubled the amount consumed for domestic purposes. The increased consumption from this source alone, according to the estimate of those familiar with the subject, is not less than three hundred thousand tons. This will make the business unusually brisk during the whole of the next season. Under all the circumstances, it can hardly be doubted that the railroad during 1852 will be freighted to the full capacity of its present engines and cars, say two million tons.

We have said nothing here of the increase in other items of business. The merchandise and passenger business has thus far increased *pari passu* with that of coal. As it constitutes, however, only about one seventh of the whole business of the company, it is not material to the present argument. So far as these items are to be accounted at all, they only go to swell the general force of our conclusions.

In estimating the bearing of these conclusions upon the value of the company's stock, it should be remembered that this large increase of profits is not to be divided equally among the whole sixteen or seventeen millions invested.

Nearly eleven millions of this investment consist of bonds bearing a definite interest of five and six per cent. All the net savings of the company, after paying the five and six per cent. interest on, say, eleven millions of bonds, are to be distributed among $6,105,332 of stock.

Now suppose an increase of twenty-five per cent. in the business of the company, which was the rate of increase last year. The gross receipts last year were $2,313,330. Add twenty-five per cent. to this, and we have, as a probable conjecture, $2,891,662 as the gross receipts for 1852. The net profits for 1851, (after deducting the sum laid aside as a renewal fund,) were 43 8-10 per cent. of the gross receipts. The recent successful experiments in the use of anthracite coal for locomotives will no doubt lessen materially the working expenses, and consequently make the net profit bear a still larger ratio to the gross income. But, supposing the ratio to be the same, 43 8-10 per cent. on $2,891,662 gives us $1,266,548 as the net profits of 1852. This is on the supposition that the business will continue to be conducted on the same terms as in 1851. But it will be recollected, ten cents per ton additional freight is to be paid on coal, which on two million tons would make a net increase to the income of $200,000 additional, or $1,466,548 in all. The interest on the bonded debt is about $647,650, reckoning it all at six per cent. There would be left $818,898 to be divided among $6,105,332 of stock, or a clear dividend of more than thirteen per cent. This is a hypothetical statement, exhibiting what may be regarded as a picture of probabilities. At least it is so viewed by the friends of the road.

There is still another feature of the company which tends to impart stability to its operations, and which, at this particular time, is supposed to give additional value to the stock. They commenced, three years ago, laying aside $100,000 annually from their earnings as a sinking fund. This sum is expended in buying up and cancelling their own bonds, thus gradually and surely reducing their indebtedness. At the same time, to give the stockholders the full benefit of the earnings of their property, this $100,000 is distributed among them in the form of stock. Thus, should the net earnings of the road for any one year be 10 per cent., the shareholders would receive about 7 per cent. in cash and 3 per cent. in stock. The stock thus created we have called, for convenience, the sinking fund stock. The plan has been in operation for the last three years, during which time a stock dividend of 9½ per cent. has been declared and set aside, but not actually distributed. This accumulation of 9½ per cent. sinking fund stock, together with whatever of the same kind shall be created the present year, (making, say, 12 per cent. altogether,) is to be distributed to the stockholders at the close of 1852, besides a cash dividend such as the earnings during the year may warrant.

If to all these considerations we add the almost indefinite increase of business before the company in coming years, it seems difficult to divine a reason why its stock should not ultimately be among the choicest investments in the market. If it has not, like the Camden and Amboy Railroad Company, a legal monopoly, it has, what is far better than any monopoly of man's making, physical advantages that enable it effectually to distance all competition. These advantages arise from the nature of its business and the character of the region in which it is conducted. It is of the very nature of a heavy coal business, that the cars or boats carrying it to market have to return empty. Hence a most serious waste in the use of that which is of necessity the main item of expense—the motive power. The hauling is all one way. Now in this respect the Reading Railroad has a peculiarity which distinguishes it from all other railroads in the world. Such is the gradual slope of the Schuylkill region, through which the railroad runs, that the track has on an average a descending grade of about six feet to the mile, the whole distance from the mines to tide water. In other words, a great part of the actual transportation is effected by the mere force of gravity. The engine which brings a loaded train down is only required to be powerful enough to take the empty cars back. In this respect it has an advantage over every other railroad in existence for this specific business—an advantage which depends, not upon any act of legislation, but upon the unrepealable laws of nature, as its main business does upon one of the most imperative of the wants of man.

The anthracite coal trade has thus far doubled itself every five or six years. In reference to the future, the smallest rate of increase that we have ever heard named, as among the probabilities of the case, is that of 20 per cent. per annum. Taking this ratio as the basis of conjecture, we have the following exhibit of the probable consumption of coal for the next three years :

| | | Tons. |
|---|---|---|
| 1851 | (actual) | 4,383,899 |
| 1852 | (conjectural) | 5,269,678 |
| 1853 | do | 6,312,813 |
| 1854 | do | 7,577,375 |

What are the existing means for getting this large amount of coal to market? We say "existing," because, with the heavy amount of capital already invested in the business, and centered in the hands of four leading companies, it is not probable that any new means will be very soon called into existence. Four lines already established and in working order, and controlling a capital of not less than $45,000,000, render any new enterprise, of the nature of a competition, a very formidable undertaking. Indeed, we might almost dismiss as idle, all conjectures as to the establishment of new coal lines until the present lines are worked to the extent of their capacity.

The question then recurs, what is the capacity of our several lines for the transportation of coal.

We begin with the Delaware and Hudson Canal. The principal part of the coal transported by this company is that mined by themselves at or near Carbondale. This coal is first transported by railroad sixteen miles to Honesdale to what is called a "gravity" road. That is, the coal is raised a certain number of feet up an inclined plane, by means of a stationary engine. From the point thus gained the road is constructed for some distance with a descending grade sufficient to move the cars along by the force of gravity. Then a new inclined plane; and so on to the terminus on the canal. The Honesdale road is constructed throughout with a double track, one for the loaded cars, and one for the empty cars. The ascent of the inclined planes is made in some cases by means of water-power. Four cars of 3½ tons each are drawn up the inclined planes at a time. The average number of trips per day is about one hundred and twenty. The utmost capacity of the Honesdale road for transporting coal is given at five hundred and fifty thousand tons per annum. The Railroad of the Pennsylvania Company, from the neighborhood of Pittston to Hawley, (a port on the Delaware and Hudson Canal, eleven miles beyond Honesdale,) is also a gravity road, forty-four miles in length, with ten inclined planes on the empty car track, and twelve on the loaded track, each plane from seventeen hundred to two thousand feet in length. Three of these planes are worked by water-power, the others by steam. Most of these planes are supplied with three stationary steam-engines of thirty horse power each. There are, in all, upon this road, forty-eight steam-engines, of thirty horse power each. The road in some places is constructed on high tressel work. In one place it is from twenty-eight to thirty-five feet high for more than a mile. This tressel work was made in a hurry, is easily put out of order, and requires constant attention to keep it wedged in its place. The gravity part of the road is badly constructed, the grade in some places is too steep, and the wheels of the cars have to be spragged well with care. There is in every part much wear and tear. The maximum capacity of this road is four hundred and fifty thousand tons per annum.

The utmost capacity, therefore, of the two lines which supply the Delaware and Hudson Canal is 1,000,000 tons, which is just about the capacity of the canal itself. The canal has been already, and very recently enlarged, and, without still more enlargement, cannot carry more than 1,000,000 tons. This latter amount, therefore, must be taken as the measure of its present capacity, beyond which it cannot go materially without an enlargement of the canal, and without the construction of additional roads, or of additional tracks and motive power on the existing roads. Such an enlargement would not only require large additional capital to a concern that has already invested (with its auxiliaries) about $11,000,000, but a temporary and very embarrassing suspension of business. Such an enlargement may possibly take place at some future time, but certainly not till both itself and all the other lines are worked to their full capacity, and at remunerating prices.

The Lehigh Canal has some exit for its coal by the Morris Canal across New Jersey. But the difficulties of transfer from one canal to the other at Easton, as well as of the defective lockage along the line of the canal itself, have heretofore rendered the Morris Canal almost a nullity, so far as the coal trade is concerned. These difficulties have been to some extent removed, and the Morris Canal, it is hoped, will hereafter be a sharer in this important business. There are, however, no data on which any accurate estimate of its business can be made. Its friends suppose that it may be able to carry as much as 400,000 tons. It cannot, in any event, be sufficient to alter any of the conclusions to be derived from the general scope of our argument.

The principal, and at present the only measure of the capacity of the Lehigh, is the State Canal, 60 miles in length, from Easton to Bristol. The capacity of this canal is 900,000 tons. The Lehigh Company, therefore, by all its means, cannot possibly do a business exceeding 1,300,000 tons, without an enlargement of its own canal, as well as of the State Canal, throughout its entire length, which neither the Company nor the State is in a condition at present to undertake, without some imperative necessity.

The present capacity of the Schuylkill Navigation Company is 800,000 tons. This may be increased to 1,000,000 tons by putting on additional boats. Any increase beyond this requires more water, which can only be obtained by the construction of large reservoirs among the mountains, where water may be stored up in the wet season. The business of this Company, by means of these additional investments, may be increased, possibly, to 1,500,000 tons. Its condition, however, seems to preclude any prospect of such an increase, at least until all existing means are put to their fullest development.

The Reading Railroad with its existing cars, engines and wharves, can transport 2,000,000 tons, and by adding cars, engines, and wharves, as the demand may increase, without interrupting its business, or increasing its permanent investment, may increase its coal tonnage to 4,000,000. Beyond this it may go almost indefinitely by means of parallel tracks, every additional track giving an increase of between two and three millions. Such an additional track, however, would require an addition to the permanent investment, though it would make no interruption to the business of the company. We will therefore consider 4,000,000 as their ultimatum under existing circumstances.

Now, if we put these several facts together, we will find that most of the companies are already worked nearly to their full capacity. The business of the Lehigh Company was 988,296 tons in 1851, leaving a margin for increase of only 310,704 tons. The business of the Delaware and Hudson Company was 795,095, leaving a margin for increase of 204,905. The Schuylkill did in 1851 a business of 579,156 tons, leaving a clear margin of 220,844 tons without additional boats, and of 520,844 tons with such addition. The business of the Reading road was 1,605,084 tons, leaving room for an increase of 394,916 tons with their existing running power, and 2,394,916 tons with the addition of running power as it is needed.

This gives us the means of forming some legitimate conclusions as to the necessary distribution of the trade among the several companies during the next two or three years. If we suppose the Delaware, and Hudson, and the Lehigh Companies worked at once to their full capacities, which we see no good reason to doubt, and if we supposed that the Schuylkill Navigation Company should be able at the end of a year from this time, to command the additional capital necessary to put upon their canal the additional boats needed, so as after that time to be worked to its full capacity, the outstanding balance of the demand must of necessity fall to the Reading road. This distribution of the business may be exhibited to the eye as follows:

| | 1852. | 1853. | 1854. |
|---|---|---|---|
| Lehigh Company (full capacity.) | 1,300,000 | 1,300,000 | 1,300,000 |
| Delaware and Hud. (full capacity.) | 1,000,000 | 1,000,000 | 1,000,000 |
| Schuylkill Nav. (full capacity.) | 800,000 | 1,000,000 | 1,000,000 |
| R.R.R. (necessary to supply demand.) | 1,160,678 | 3,012,813 | 4,275,375 |
| Estimated demand. | 5,260,678 | 6,312,813 | 7,575,375 |

If, under these circumstances, all the companies do not make money, they have not the business sagacity for which they are generally credited. They are now all thoroughly organized and equipped for a successful prosecution of the business. The experience of the last year seems to have convinced them all that their true vocation is to carry coal at remunerating prices, instead of seeking to underwork each other. Indeed, with the prospect before them of all having a demand for as much as they can possibly do, harmony among them as to rates and prices, is as certain as any future event, dependent upon the laws of either mind or matter. The operations of the present year have, indeed, already commenced upon an amicable basis. Pursuing this line of policy, every addition to their business adds, in a rapidly increasing ratio, to their net returns. The main expense of the roads, canals, engines, boats, hands, and other permanent investments, having been already incurred, increase of business brings with it small comparative increase of cost. The addition to their income from increased tonnage is almost entirely profit. The ratio of net profits, to the gross income, is necessarily a constantly increasing one, until each line is worked to its full capacity.

Under these circumstances, we repeat, all the companies must necessarily do a profitable business. Even those lines already worked to their full capacity, will increase their profits, because it will be the interest of the other companies, equally with themselves, to raise their freights and prices above those of the last year. The Delaware and Hudson Company has already raised the price of coal 50 cents a ton, and the Reading Railroad has added 10 cents a ton to its charge for freight. The result to the consumer will be, without doubt, a slight increase in the price of the article, put to the great companies engaged in mining and transporting it; we see not how it can be otherwise than a full realization of those rich returns for which they have so long, and at some periods so disastrously labored.

In the above estimates, the Union Canal, extending from Pinegrove to Middletown, has been overlooked. The capacity of this improvement, since its enlargement is equal to some 800,000 tons; of which probably 200,000 will be passed over the line the present season. The Swatara coal district, it is well known, is one of the richest and best in Schuylkill county, and the amount of coal that will hereafter seek the line of the Union Canal must increase very rapidly hereafter.

# PART III.

## Wyoming.

Thou comest in beauty on my gaze at last,
" On Susquehanna's side, fair Wyoming,"—
Image of many a dream, in hours long past;
When life was in its bud, and blossoming,
And waters, gushing from the fountain spring
Of pure enthusiast thought, dimmed my young eyes;
And by the poet borne, on unseen wing,
I breathed, in fancy, 'neath thy cloudless skies,
The summer's air, and heard her echo'd harmonies.

Two courses, comprising the substantials of our little repast, have been set before the friendly guest; the third, a favorite poetic dish, with some "trimmings" of our own scattered around, shall now be spread out. Wyoming—sweet vale of Wyoming!—'tis of thee that we shall sing!

There are several stage-routes by which Wilkesbarre may be reached—the most eastern one of which commences at Easton, in Northampton, and passing through Monroe, enters Luzerne county at the Great Swamp, locally called the *shades of death.* This route is interesting for its scenery, and is generally selected by the New Yorkers in their travels to and from Wyoming. The route is sixty-four miles in length, and passes the villages of Effort, Nazareth, Jacobsburg, Wind Gap, (in the Kittatinny mountain), Taylorsburg, Shafer's, Mount Pocono, Soxville, Stoddartsville, (on the Lehigh river) Beaumont, and Bear Creek to Wilkesbarre. The traveller at Mauch Chunk can either fight his way up the Lehigh, by stage to Whitehaven, thence by railroad, or he may proceed direct by stage via Lausanne and Beaver Meadow, to Hazleton, at which place he will intersect the line at Tamaqua. For travellers from Philadelphia, Tamaqua is, as before stated, the most direct starting-point, while the line of stages is superior to the others, or used to be—each coach employing four horses. The distance from this place to Wilkesbarre is about forty-five miles, the road crossing the mountains of Locust and Mahanoy, in Schuylkill, and those of Nescopec and Wyoming in Luzerne, with their interesting valleys and coal districts. At the foot of this latter mountain, (which, further north, is called the Moosic mountain—what nonsense!) lies the borough of Wilkesbarre, and before it stretches out, in a north-east and south-west direction, the beautiful and far-famed valley of Wyoming. In the centre of the valley flows the Susquehanna, in one broad, clear, and tranquil sheet, and north of it rises the towering mountain range of the Nanticoke, the Shawnee, and the Lackawanna. The Susquehanna breaks through the mountain range near Pittston, where, entering the valley from the north, it receives the Lackawanna river, and strikes to the south-west, traversing the valley in a nearly straight line, until within a few miles of Columbia county, where it makes a graceful bend around Wyoming Mountain; pitches due south to that of the Nescopeck, and then again stretches out in a south-west course, and enters that county.

The county of Luzerne is plentifully supplied with streams. Excepting one or two of nearly equal area further west, it probably con-

tains more small streams and little lakes than any other county in the State. The principal ones are the Susquehanna, the Lackawanna, the Lehigh, the Wapwallopen, Nescopeck, and the north fork of the Lehigh—all south-east of the Shawnee Mountain. West of this mountain range are numerous small rivulets, where "trout most do congregate," as Huntingdon, Shickshinny, Harvey's, Toby's, Bowman's and others. Some ten miles north-west of Wilkesbarre, is Harvey's Lake,

a beautiful transparent sheet of water, much resorted to by fishing parties. There are several other small lakes, near at hand, all of which abound in trout, and other game fish, and are surrounded with the most splendid scenery which the unerring hand of nature could produce. No where in the United States is the beautiful brook trout found in greater abundance,—no where is it found in better positions for the excitements of fishing,—no where are its freshwater retreats found amidst finer scenery, or streams of clearer and purer water. Hear what the poet has to say of this delightful sport:

> Now, when the first foul torrent of the brooks,
> Swell'd with the vernal rains, is ebb'd away;
> And, whitening, down their mossy-tinctur'd stream
> Descends the billowy foam: now is the time,

While yet the dark brown water aids the guile,
To tempt the trout. The well dissembled fly,
The rod fine-tapering with elastic spring,
Snatch'd from the hoary steed the floating line,
And all thy slender wat'ry stores prepare.
But let not on thy hook the tortur'd worm
Convulsive, twist in agonizing folds;
Which, by rapacious hunger swallow'd deep,
Gives, as you tear it from the bleeding breast
Of the weak, helpless, uncomplaining wretch,
Harsh pain and horror to the tender hand.
   When with his lively ray the potent sun
Has pierc'd the streams, and rous'd the finny race,
Then, issuing cheerful, to thy sport repair:
Chief should the western breezes curling play,
And light o'er either bear the shadowy clouds.
High to their fount, this day, amid their hills,
And woodlands warbling round, trace up the brooks;
The next, pursue their rocky-channel'd maze,
Down to the river, in whose ample wave
Their little Naiads love to sport at large.
   Just in the dubious point, where with the pool
Is mix'd the trembling stream, or where it boils
Around the stone, or from the hollow'd bank
Reverted plays in undulating flow,
There throw, nice-judging, the delusive fly;
And as you lead it round in artful curve,
With eye attentive mark the springing game.
Strait as above the surface of the flood
They wanton rise, or urg'd by hunger leap,
Then fix, with gentle twitch, the barbed hook:
Some lightly tossing to the grassy bank,
And to the shelving shore slow-dragging some,
With various hand proportion'd to their force.
If yet too young, and easily deceiv'd,
A worthless prey scarce bends your pliant rod;
Him, piteous of his youth and the short space
He has enjoy'd the vital light of heaven,
Soft disengage, and back into the stream
The speckled captive throw. But should you lure
From his dark haunt, beneath the tangled roots
Of pendant trees, the monarch of the brook,
Behoves you then to ply your finest art.
Long time he, following cautious, scans the fly;

And oft attempts to seize it, but as oft
The dimpled water speaks his jealous fear.
At last, while haply o'er the shaded sun
Passes a cloud, he desperate takes the death,
With sullen plunge. At once he darts along,
Deep struck, and runs out all the lengthen'd line;
Then seeks the farthest ooze, the sheltering weed,
The cavern'd bank, his old secure abode;
And flies aloft, and flounces round the pool,
Indignant of the guile. With yielding hand,
That feels him still, yet to his furious course
Gives way, you, now retiring, following now
Across the stream, exhaust his idle rage;
Till floating broad upon his breathless side
And to his fate abandon'd, to the shore
You gaily drag your unresisting prize.

Besides the trout, there are numerous other fish, peculiar to the fresh water varieties, to be found in these mountain streams, as the rock fish, sun fish, eel, perch, salmon, pike, and shad. This latter, which is the largest fish frequenting the Susquehanna, is now very rarely caught so high up, being "headed off" by the dams erected in the Susquehanna, at various points below. It may still be found, however, though in nothing like the abundance with which it was caught some years ago. Mr. Miner, in his work on Wyoming, speaks of cases where many thousand were captured, near Wilkesbarre, in a single haul of the seine. The flesh of the shad is probably more delicate than any existing fish of our streams; and though it lacks the lusciousness and the glutinous fin of the turbot, it is preferred to it by many judicious epicures, notwithstanding the drawback occasioned by its innumerable and sharply-pointed bones. Shad fishing has heretofore been confined exclusively, on this river, to the seine and dip-net, but Mr. Herbert, in his recent work on fishing, says that much sport may be found in fishing for it, during its upward run in the spring, with a powerful trout-line, and a rich gaudy fly. The higher up the river they ascend the more readily they will take the fly, while, at the same time, the flavor of their flesh is greatly increased.

The mountainous character of this country, and the frequent disarrangements of the stratification, have occasioned numerous romantic water-falls, one of the most interesting of which is that of Falling Spring, at the head of the valley, near Pittston.

The first settlers of this county were originally from Connecticut, with a few Germans and Scotch-Irish from Pennsylvania. The Germans from the lower counties and from Europe have more recently filled up the southern part of the county, and a great number of Irish and Welsh miners are settled around the principal coal mines. The people of the Wyoming valley—says Mr. Day—and along the Susquehanna above, still retain the manners, the steady habits, the enterprise and intelligence, and even the pronunciation of their New England fathers; and the external aspect of things—the villages with tall spires and shaded streets; the neat white houses with green blinds, and broad front-yards fragrant with flowers and shrubbery; and in the country, the ancient red-painted or wood-colored frame-houses—all mark the origin and peculiar character of the people. Prof. Silliman, on his visiting the valley some years ago, made the following remarks:

"The severe and long-continued struggle for the possession of this country, which was sustained by the original Connecticut settlers from fifty to eighty years since, and the repeated attempts which were made to disposses them by arms, sufficiently evince the high estimation in which it was held by all the parties. The prize for which the settlers contended was worthy of all the heroism, fortitude, and long-suffering perseverance, which, during so many years, they displayed—an exhibition of moral courage rarely equalled and never surpassed. Believing themselves, both in a political and personal view, to be the rightful proprietors of the country, they defended it to the death; and no one who now surveys this charming valley can wonder that they would not quietly relinquish their claim.

"The first glance of a stranger entering at either end, or crossing the mountain ridges which divide it (like the happy valley of Abyssinia,) from the rest of the world, fills him with the peculiar pleasure produced by a fine landscape, combining richness, beauty, variety, and grandeur. From Prospect Hill, on the rocky summit of the eastern barrier, and from Ross' Hill, on the west, the valley of Wyoming is seen in one view, as a charming whole, and its lofty and well-defined boundaries exclude more distant objects from mingling in the prospect. Few landscapes that I have seen can vie with the valley of Wyoming. Excepting some rocky precipices and cliffs, the mountains are wooded from the summit to their base; natural sections furnish avenues for roads, and the rapid Susquehanna rolls its powerful current through a mountain gap, on the northwest, and immediately receives the Lackawanna, which flows down the narrower valley of the same name. A similar pass between the mountains, on the south, gives the Susquehanna an exit, and at both places a slight obliquity in the position of the observer presents to the eye a seeming lake in the windings of the river, and a barrier of mountains, apparently impassable. From the foot of the steep mountain ridges, particularly on the eastern side, the valley slopes away, with broad sweeping undulations in the surface, forming numerous swelling hills of arable and grazing land; and as we recede from the hills, the fine flats and meadows covered with the richest grass and wheat, complete the picture by features of the gentlest and most luxuriant beauty."

The valley itself,—says Hon. Charles Miner,—is diversified by hill and dale, upland and intervale. Its character of extreme richness is derived from the extensive flats, or river bottoms, which, in some places, extend from one to two miles back from the stream, unrivalled in expansive beauty; unsurpassed in luxuriant fertility. Though now generally cleared and cultivated, to protect the soil from floods a fringe of trees is left along each bank of the river; the sycamore, the elm, and more especially the black walnut,—while here and there scattered through the fields, a huge shell-bark yields its summer shade to the weary laborer, and its autumn fruit to the black and gray squirrel, or the rival plough-boy. Pure streams of water come leaping from the mountains, imparting health and pleasure in their course; all of them abounding with the delicious trout. Along those brooks, and in the swales, scattered through the uplands, grow the wild plum and the butter-nut, while, wherever the hand of the white man has spared it, the native grape may be gathered in unlimited profusion. I have seen the grape-vine bending beneath its purple clusters, one branch climbing a butter-nut, loaded with fruit; another branch resting on a wild plum, red with its delicious burden—the while growing in their shade, the hazel-nut was ripening its rounded kernel!

Such were the common scenes when the white man first came to Wyoming, which seems to have been formed a perfect Indian paradise. Game of every sort was abundant. The quail whistled in the meadow; the pheasant rustled in its leafy covert; the wild-duck

FALLING SPRING—WYOMING VALLEY.

reared her brood, and bent the reed in every inlet; the red deer fed upon the hills, while, in the deep forests, within a few hours walk, roamed the stately elk. The river yielded, at all seasons, a supply of fish—the yellow perch, the pike, the catfish, the bass, the roach, and, in the spring, myriads of shad.

The name of Wyoming was long supposed to mean, being interpreted, "*a field of Blood;*" but Mr. Heckewelder, perfectly versed in Indian language, says it is a corruption of *Maughwauwama,* by which it was designated by the Delaware Indians, being a compound of *Maughwau,* meaning large, and *wame,* signifying plains, so that it may be translated *the large plains.*"

The entire region of country surrounding Wyoming was long claimed by the Indians of the Six Nations, while the Shawnees, Delawares, Nanticokes, and other tribes, were the occcasional possessors of the valley, which seems, at all times, to have been a favorite abode with these children of the forest. The women cultivated corn upon the plains, the men travelled the surrounding mountains, covered with lofty forests, in pursuit of game, and the river supplied an abundance of fish. The Delawares were settled on the eastern side of the valley, nearly opposite to the village of the Shawnees, called Maugh-wau-wauma, or Wyoming. The Nanticokes occupied the lower

end of the valley, near the falls bearing their name. This is a rapid on the Susquehanna, almost precipitous at one place, where the river forces its passage through a narrow gorge of the mountains, and escapes through the beautiful valley where it had been lingering for upwards of twenty miles, into a region wild with rock and glen. The tribes thus situated, and each but tenants of the Six Nations, maintained a peaceable intercourse for a season, and enjoyed their wild scenes unmolested.

It was during this period that the soil of Wyoming was first trodden by the feet of a missionary of the Christian religion. The Moravians, or "United Brethren," had commenced their missions in the new world several years before—in Georgia as early as 1734. Their benevolent labors were extended to Pennsylvania and New York six years afterward. In 1742, says Col. Stone, their great founder and apostle, Count Zinzendorf, visited America, to look after their infant missions. He arrived at Bethlehem, near the forks of the Delaware, in the following year. Affecting representations of the deplorable moral condition of the Indians had reached the Count before he left Germany, and his attention was early directed to their situation, and their wants, while visiting the missionary stations along the Delaware. He made several journeys among the Indians deeper in the interior, and succeeded, not without difficulty, in establishing a friendly intercourse with various tribes. In one of these journeys he plunged through the wilderness into the valley of Wyoming, for the purpose of establishing a missionary post in the town of the Shawnees. It was here, during the autumn of that year, that one of those beautiful and touching incidents occurred which add a charm to the annals of the missionary enterprise. The Count had expected to be accompanied by an interpreter, celebrated in all the Indian negotiations for many years of that age, named Conrad Weiser, (of whom we have already spoken, in connection with Reading, Part I.) whose popularity was equally great among the Indians of all nations by whom he was known. But Weiser was unable to go. Inflexible in his purpose, however, the Count determined to encounter the hazards of the journey, with no other companions than a missionary, named Mack, and his wife. On their arrival in the valley, they pitched their tents on the bank of the river, a short distance below the village of the Shawnees—at that period the most distrustful and savage of the Pennsylvania Indians. A council was called to hear their errand of mercy, but the Indians were not exactly satisfied as to the real object of such an unexpected visit. They knew the rapacity of the white people for their lands; and they thought it far more probable that the strangers were bent upon surveying the quality of these, than that they were encountering so many hardships and dangers, without fee or reward, merely for the future well-being of their souls. Brooding darkly upon the subject, their suspicions increased, until they resolved upon the assassination of the Count; for which purpose executioners were detailed, who were instructed to carry their purpose into effect with all possible secrecy, lest the transaction, coming to the ears of the English, should involve them in a yet graver difficulty.

The Count was alone in his tent, reclining upon a bundle of dry weeds, designed for his bed, and engaged in writing, or in devout meditation, when the assassins crept stealthily to the tent upon their murderous errand. A blanket curtain, suspended upon pins, formed the door of his tent, and by gently raising a corner of the curtain the Indians, undiscovered, had a full view of the venerable patriarch, unconscious of lurking danger, and with the calmness of a saint upon his benignant features. They were awe-stricken by his appearance. But this was not all. It was a cool night in September, and the Count had kindled a small fire for his comfort. Warmed by the flame, a large rattlesnake had crept from its covert, and approaching the fire for its greater enjoyment, glided harmlessly over one of the legs of the holy man, whose thoughts, at the moment, were not occupied upon the grovelling things of earth. He perceived not the serpent, but the Indians, with breathless astonishment, had observed the whole movement of the poisonous reptile; and as they gazed upon the aspect and attitude of the Count, and saw the serpent offering him no harm, they changed their minds as suddenly as the barbarians of Malta did theirs in regard to the shipwrecked prisoner who shook the viper from his hand without feeling even a smart from its venomous fang. Their enmity was immediately changed into reverence; and in the belief that their intended victim enjoyed the special protection of the Great Spirit, they desisted from their bloody purpose and retired.* Thenceforward the Count was regarded by the Indians with the most profound veneration. The arrival of Conrad Weiser, soon afterward,

* This interesting incident was not published in the Count's memoirs, lest, as he states, the world should think that the conversions that followed among the Indians were attributable to their superstitions. Mr. Chapman, in his history of Wyoming, has preserved the story, having, as he says, received it from one who was a companion of the Count, and who accompanied him (the author) to Wyoming.

afforded every facility for free communication with the sons of the forest, and the Count remained among them a considerable time. Some time afterwards several of the Moravian brethren visited the valley, and formed an agreeable acquaintance with the Indians, especially with the Nanticoke tribe, one of whom, eighty-seven years old, was a remarkably intelligent man. The Missionaries frequently preached to them through their interpreter, and the result was the

MISSIONARIES PREACHING TO THE INDIANS.

establishment of a regular mission post there, which was successfully maintained for several years, and until broken up by troubles as extraordinary in their origin as they were fatal to the Indians involved in them.

The contention which so long subsisted between the citizens of Connecticut and Pennsylvania, says Mr. Trego, in his Geography of Pennsylvania, and which caused so much blood to be spilled at Wyoming, originated in an interference of the territorial claims of the respective parties. Strange as it may appear at the present day, this region was claimed by Connecticut as being within the limits of its charter as granted by the English government, and in 1753 a company was formed in that colony for the purpose of making settlements at Wyoming. In 1762, about two hundred persons from Connecticut arrived, and established themselves on the east side of the river, about the mouth of Mill Creek, a little above the place where Wilkesbarre now stands. They lived in friendship with the Indians, and

ATTACKING THE WYOMING SETTLEMEMT.

soon extended their settlements to the west side of the river. This state of peace was, however, of short duration, for the settlement was suddenly attacked by the savages; about twenty persons were killed, others wounded, and the rest fled to the mountains, making their way, almost destitute of provisions, through a wilderness of sixty miles, to the settlements at Easton, on the Delaware.

The proprietor of Pennsylvania, having purchased this territory from the Indians, granted the lands at Wyoming to certain persons, who in 1769 took possession of them, together with the improvements made by the Connecticut people, who had been driven away by the Indians. In the same year forty new emigrants from Connecticut arrived, who, after much contention and difficulty with the Pennsylvania settlers, were most of them arrested and taken to Easton, but were afterwards liberated. Soon after two hundred more came from Connecticut, who built a fort for their defence, and prepared to resist the authorities of Pennsylvania. A series of skirmishes, and at length of open warfare between the rival bodies of settlers succeeded; armed forces were sent by the proprietary government of Pennsylvania to dispossess the Connecticut people; a number were killed on both sides, and this civil contention lasted until the breaking out of the revolutionary war, which exposed both the belligerents to a common foe.

The valley of Wyoming was thus allowed a season of comparative repose. Both Connecticut and Pennsylvania had more important demands upon their attention. The census of the valley, at this time, is estimated by Mr. Miner, at about twenty-five hundred inhabitants. At the opening of the revolution, says he, the pulsations of patriotic hearts throbbed with unfaltering energy throughout Wyoming. The fires of liberty glowed with an ardor intense and fervent. The people erected, at their own expense, suitable defences against the encroachments of the enemy, and several military companies were organized, which joined the army of Washington, and participated in the bloody conflicts at Brandywine, Germantown, and other places. Wyoming was an exposed frontier bordering on the country of the Six Nations—a people numerous, fierce, and accustomed to war. From Tioga Point, says Mr. Day, where they would rendezvous, in twenty-four hours they could descend the Susquehanna in boats to Wyoming. Nearly all the *able-bodied* men of Wyoming, fit to wear arms, had been called away into the continental army.

It was to be expected, that the savages, and their British employers, should breathe vengeance against a settlement that had shown such spirit in the cause of liberty. They were also, beyond doubt, stimulated by the absconding tories, who were burning with a much stronger desire to avenge what they conceived to be their own wrongs, than with ardor to serve their king. The defenceless situation of the settlement could not be concealed from the enemy, and would naturally invite aggression, in the hope of weakening Washington's army by the diversion of the Wyoming troops for the defence of their own frontier. All these circumstances together marked Wyoming as a devoted victim.

The following sketch of the memorable battle of 1778 is condensed from the plea of the Wyoming delegation, drawn up by the Hon. Charles Miner, and intended to be delivered before the legislature of Connecticut—with some additional facts from "the Hazleton Travellers," and other sources.

Late in June, 1778, there descended the Susquehanna, Col. John Butler, with his own tory rangers, a detachment of Sir John Johnson's Royal Greens, and a large body of Indians, chiefly Senecas. The British and tories numbered about four hundred—the Indians about seven hundred. Jenkin's fort was at the head of the valley, just below the gorge. This fort capitulated on the 2d July to a detachment under Capt. Caldwell. Wintermoot's fort had been built near Jenkin's, by a Low Dutch family of that name, with a view, as afterwards appeared, to aid the incursions of the tories. As suspected, Wintermoot's fort at once threw open its gates to the enemy. Here the British and Indian force was assembled at dinner just before the battle. To defend the settlement against this force was a half-raised company of Capt. Deathic [Dœterick] Hewitt, consisting of forty or fifty men, and the militia, the remains merely, out of which the three companies above mentioned had been enlisted for the continental army. There were several forts at Wyoming—not regular fortifications, with walls, and embrazures, and great guns—but stockades, built by setting logs on end in ditches, close together, surrounding a space for the retreat of the women and children, with no other means of defence than the small-arms of the men, firing through loop-holes. In all Wyoming valley there was but one cannon, a four-pounder, without ball, kept at the Wilkesbarre fort as an alarm gun.

Against such a force as the enemy mustered, not one of these forts could have held out an hour, or kept the foe from reducing them to ashes. Some of the aged men out of the train-bands formed themselves into companies to garrison the forts, and yield to the helpless such protection as they could. Except at Pittston—which, from its position, was imminently exposed—no company of the Wyoming regiment was retained for partial defence. All the rest assembled at Forty Fort, on the Kingston side, prepared in the best manner they could to meet the enemy. They numbered about four hundred men and boys, including many not in the train-band. Old, gray-headed men, and grandfathers, turned out to the muster.

Col. Zebulon Butler happened to be at Wyoming at the time, and though he had no proper command, by invitation of the people he placed himself at their head, and led them to battle. There never was more courage displayed in the various scenes of war. History does not portray an instance of more gallant devotion. There was no other alternative but to fight and conquer, or die; for retreat with their families was impossible. Like brave men, they took counsel of their courage. On the 3d of July they marched out to meet the enemy. Col. Zebulon Butler commanded the right wing, aided by Maj. Garret. Col. Dennison commanded the left, assisted by Lieut. Col. George Dorrance. The field of fight was a plain, partly cleared and partly covered with scrub-oak and yellow-pine. The right of the Wyoming men rested on a steep bank, which descends to the low river-flats; the left extended to a marsh, thickly covered with timber and brush. Opposed to Col. Zebulon Butler, of Wyoming, was Col. John Butler, with his tory rangers, in their green uniform. The enemy's right wing, opposed to Col. Dennison, was chiefly composed of Indians [led on, says Col. Stone, by a celebrated Seneca chief, named *Gi-en-gwah-toh;* or *He-who-goes-in-the-smoke*].* It was between four

* Until the publication of the Life of Brant, [by W. L. Stone,] it had been asserted in all history that that celebrated Mohawk chieftain was the Indian leader at Wyoming. He himself always denied any participation in this bloody expedition, and his assertions were corroborated by the British officers, when questioned upon the subject. But these denials, not appearing in history, relieved him not from the odium; and the "monster Brant" has been denounced, the world over, as the author of the massacre. In the work referred to above, the author took upon himself the vindication of the savage warrior from the accusation, and as he thought, at the time, with success. A reviewer of that work, how-

and five o'clock in the afternoon when the engagement began, and for some time it was kept up with great spirit. On the right, in open field, our men fired and advanced a step, and the enemy was driven back. But their numbers, nearly three to one, enabled them to outflank our men, especially on the left, where the ground, a swamp, was exactly fitted for savage warfare. Our men fell rapidly before the Indian rifles; the rear as well as the flank was gained, and it became impossible to maintain the position. An order to fall back, given by Col. Dennison, so as to present a better front to the enemy, could not be executed without confusion, and some misunderstood it as a signal for retreat. The practiced enemy, not more brave, but, besides being more numerous, familiarized to war in fifty battles, sprang forward, raised their horrid yell from one end of the line to the other, rushed in with the tomahawk and spear, and our people were defeated. They deserved a better fate. One of the men yielding a little ground, Col. Dorrance, a few minutes before he fell, with the utmost coolness, said, "Stand up to your work, sir." After the enemy was in the rear, "See!" said an officer to Capt. Hewett, "the enemy is in force behind us; shall we retreat?" "I'll be d——d if I do!" was his reply; and he fell at the head of his men. "We are nearly alone," said Westbrook; "shall we go?" "I'll have one more shot first," replied Cooper. That instant a savage sprang towards him with his spear. Cooper stretched him on the earth, and reloaded before he left the ground. When the left was thrown into confusion, our Col. Butler threw himself in front, and rode between the two lines, exposed to the double fire. "Don't leave me, my children," said he; "the victory will be ours." But what could four hundred undisci-

ever, in the Democratic Magazine, who is understood to be the Hon. Caleb Cushing of Massachusetts, disputed the point, maintaining that the vindication was not satisfactory. The author thereupon made a journey into the Seneca country, and pushed the investigation among the surviving chiefs and warriors of the Senecas engaged in that campaign. The result was a triumphant acquittal of Brant from all participation therein. The celebrated chief, Captain Pollard, whose Indian name is *Kaoundoowand,* a fine old warrior, was a young chief in that battle. He gave a full account of it, and was clear and positive in his declarations that Brant and the Mohawks were not engaged in that campaign at all. Their leader, he said, was *Gi-en-gwah-toh,* as already mentioned, who lived many years afterward, and was succeeded in his chieftaincy by the late *Young King.* That point of history, therefore, may be considered as conclusively settled.

[*Col. Stone's History of Wyoming.*

plined militia effect against eleven hundred veteran troops? The battle was lost! Then followed the most dreadful massacre—the most heart-rending tortures. The brave but overpowered soldiers of Wyoming were slaughtered without mercy, principally in the flight, and after surrendering themselves prisoners of war. The plain, the river, and the island of Monockonock were the principal scenes of this horrible massacre. Sixteen men, placed in a ring around a rock, (which is still shown, behind the house of Mr. Gay, near the river,) were held by stout Indians, while they were, one by one, slaughtered by the knife or tomahawk of a squaw. One individual, a strong man, by the name of Hammond, escaped by a desperate effort. In another similar ring, nine persons were murdered in the same way. Many were shot in the river, and hunted out and slain in their hiding-places, (in one instance, by a near, but adverse relative,)* on the now beautiful island of Monockonock. But sixty of the men who went into the battle, survived; and the forts were filled with widows and orphans, (it is said the war made one hundred and fifty widows and six hundred orphans in the valley,) whose tears and cries were suppressed after the surrender, for fear of provoking the Indians to kill them; for it was an Indian's pastime to brandish the tomahawk over their heads.

A few instances will show how universal was the turn-out, and how general was the slaughter. Of the Gore family, one was away with the army, five brothers and two brothers-in-law went into the battle. At evening five lay dead on the field, one returned with his arm broken by a rifle-ball; the other, and only one, unhurt. From the farm of Mr. Weeks, seven went out to battle; five sons and sons-in-law, and two inmates. Not one escaped—the whole seven perished. Ander-

* During the bloody fight of the 3d, some of the fugitives plunged into the river and escaped to the opposite shore. A few landed upon Monockonock island, having lost their arms in the flight, and were pursued thither. One of them was discovered by his own brother, who had espoused the side of the crown. The unarmed whig fell upon his knees before his brother, and offered to serve him as a slave forever, if he would but spare his life. But the fiend in human form was inexorable; he muttered, *"You are a d——d rebel!"* and shot him dead. This tale is too horrible for belief; but a survivor of the battle, a Mr. Baldwin, confirmed its truth to the writer with his own lips. He knew the brothers well, and in August, 1839, declared the fact to be so. The name of the brothers was Pensil.—*Stone's History of Wyoming.*

son Dana went into battle with Stephen Whiting, his son-in-law, a few months before married to his daughter. The dreadful necessity of the hour allowed no exemption like that of the Jewish law, by which the young bridegroom might remain at home for one year, to *cheer up his bride.* The field of death was the resting-place of both. Anderson Dana, Jr., still living—then a boy of nine or ten years old—was left the only protector of the family. They fled, and begged their way to Connecticut.

Of the Inman family, there were five present in the battle. Two fell in the battle, another died of the fatigues and exposure of the day; another was killed the same year by Indians.

About two-thirds of those who went out fell. Naked, panting, and bloody, a few who had escaped came rushing into Wilkesbarre fort, where, trembling with anxiety, the women and children were gathered, waiting the dread issue. Mr. Hollenback, who had swum the river, amid the balls of the enemy, was the first to bring them the appalling news—"*All is lost—lost!*" They fled to the mountains, and down the river. Their sufferings were extreme. Many widows and orphans begged their bread, on their weary way home to their friends in Connecticut. In one party of near a hundred, there was but a single *man.* As it was understood that no quarters would be given to the soldiers of the line, Col. Z. Butler, with the few other soldiers who had escaped, retired the same evening, with the families, from Wilkesbarre fort.

But those left at Forty Fort? During the battle, (said the venerable Mrs. Myers, who, then a child, was there,) they could step on the river bank, and hear the firing distinctly. For a while it was kept up with spirit, and hope prevailed; but by and by it became broken and irregular, approaching nearer and nearer. "Our people are defeated —they are retreating!" It was a dreadful moment! just at evening a few of the fugitives rushed in, and fell down exhausted—some wounded and bloody. Through the night, every hour one or more came into the fort. Col. Dennison also came in, and rallying enough of the wreck of the little Spartan band to make a mere show of defending the fort, he succeeded the next day in entering into a capitulation for the settlement, with Col. John Butler, of the British forces, fair and honorable for the circumstances; by which doubtless many lives were saved. The capitulation, drawn up in the handwriting of Rev. Jacob Johnson, the first clergyman of the settlement, stipulated, "that the settlement lay down their arms, and their garrison be demolished. That the inhabitants occupy their farms peaceably, and the lives of the inhabitants be preserved entire and unhurt. That the continental stores are to be given up. That Col. Butler will use his utmost influence that the private property of the inhabitants shall be preserved entire to them. That the prisoners in Forty Fort be delivered up. That the property taken from the people called Tories be made good; and that they remain in peaceable possession of their farms, and unmolested in a free trade through the settlement. That the inhabitants which Col. Dennison capitulates for, together with himself, do not take up arms during the contest."

The enemy marched in six abreast; the British and tories at the northern gate, the Indians at the southern; their banners flying and music playing. Col. Dorrance, then a lad in the fort, remembered the look and conduct of the Indian leader—all eye—glancing quickly to the right, then to the left, with all an Indian's jealousy and caution, lest some treachery or ambush should lurk in the fort. Alas! the brave and powerful had fallen! No strength remained to resist, no power to defend!

The terms of the capitulation, however, were disregarded by the Indians, and it was soon felt that they afforded no security whatever The remaining widows and orphans, therefore, a desolate band, with scarcely provisions for a day, took up their sad pilgrimage over the dreary wilderness of the Pokono mountains, and the dismal "Shades of Death"—(a term bestowed on the solitary wilderness bordering the valley, which is still applied.) Most of the fugitives made their way to Stroudsburg, in Monroe county, where there was a small garrison. For two or three days, while pursuing their melancholy journey, they lived upon whortleberries, which a kind Providence seems to have furnished along the way in extraordinary abundance—the manna of that solitary wilderness. Mr. Miner in one of his papers, entitled the Hazleton Travellers, (originally published in the newspapers of Wilkesbarre,) says:

What a picture for the pencil! Every pathway through the wilderness thronged with women and children, old men and boys. The able men of middle life and activity were either away in the general service, or had fallen. There were few who were not in the engagement; so that in one drove of fugitives, consisting of one hundred persons, there was only one *man* with them Let the painter stand on some eminence commanding a view at once of the valley and the mountain. Let him paint the throng climbing the heights, hurrying on, filled with terror, despair, and sorrow. Take a single group: the affrighted mother, whose husband has fallen; an infant on her bosom; a child by the hand; an aged parent, slowly climbing the rugged way, behind her; hunger presses them sorely; in the rustling of every leaf they hear the approaching savage; the "Shades of Death" before them; the valley, all in flames, behind them; their cottages, their barns, their harvests, all swept in this flood of ruin; their star of hope quenched in this blood-shower of savage vengeance!

The Weekses, who fell in the battle, have been mentioned. Not one escaped; the whole seven fell, and the old man was left like the oak struck with lightning—withered, bare, blasted—all its boughs torn away.

> "Man cannot tell
> With what an agony of tenderness
> He turned him to the battle-field, where lay
> His hopes—his children—fondly, dearly loved."

The engagement was on Friday. On Sunday morning twenty Indians came to his house and ordered breakfast. They told Mr. Weeks he must go—he could not stay—he must clear out. "All my sons have fallen," said the old man, "and here I am, left with fourteen grandchildren, all young and helpless." After breakfast one of the Indian leaders stepped up to Mr. Weeks, took the hat from his head, and put it on; he then wheeled into the middle of the street a large rocking-chair with a cushion in it, sat himself down, and rocked himself. The tigers, gorged with food, blood, and plunder, for the moment paused, and rocked themselves into something like good nature. In sending the family into exile, they allowed them a pair of oxen and a wagon, to carry the children, a bed, and some food. They went up the Lackawanna to Orange county, New York.

Mrs. Jenkins, in her very interesting narrative, says that in those times of peril and suffering the women performed their part. While the men were out on duty, the women gathered, husked and garnered the corn. I speak now of other years, for little was saved in the melancholy and bloody '78. "We had not only to do this, but at times to make our own powder!" "Your own powder, Mrs. Jenkins!" I exclaimed. "Was it so? Had your people not only to find troops for the continental army, to build their own forts, to raise men for their own defence, to clothe them, to arm them, to feed them; but were they obliged to make their own powder? But how did you make it?" "O, we took up the floors, and dug out the earth—put it up and drained water through it, as we leech ashes—mixed weak ley, boiled them together, let the liquid stand, and saltpetre would rise in crystallizations on the top; then we mixed sulphur and charcoal. Mr. Hollenback went down the river and brought up a pounder."

When Forty Fort capitulated, (Mrs. Hewitt was there at the time,) Col. John Butler, as he entered the gate, saw Sergeant Boyd, a young man about twenty-five. He was an Englishman, had deserted from the enemy, was an excellent disciplinarian, and had been serviceable in training our men. "Boyd," said Butler, recognizing him, "go to that tree," pointing to a pine not far outside the fort. "I hope your honor will consider me a prisoner of war." "Go to that tree!" repeated Butler sternly. Boyd went, and was shot down.

In March, 1779, the spring after the battle, a large body of Indians came down on the Wyoming settlements. The people were few, weak, and ill prepared for defence, although a body of troops was stationed in the valley for that purpose. The savages were estimated at about four hundred men. They scattered themselves abroad over the settlement, murdering, burning, taking prisoners, robbing houses, and driving away cattle. After doing much injury, they concentrated their forces, and made an attack on the fort in Wilkesbarre; but the discharge of a field-piece deterred them, and they raised the siege. The house of Thaddeus Williams was also attacked by a party. The old man was sick in bed, and Sergeant Williams, his son, with a boy of thirteen, withstood the siege, killed a part of the assailants, and entirely drove off the others.—[*Hazleton Travellers.*

Soon after the battle, says Mr. Day, Capt. Spalding, with a company from Stroudsburg, took possession of the desolate valley, and rebuilt the fort at Wilkesbarre. Col. Hartley, from Muncy fort, on the West Branch, also went up the North Branch with a party, burned the enemy's villages, at Wyalusing, Sheskequin, and Tioga, and cut off a party of the enemy who were taking a boat-load of plunder from Wyoming. Most of the settlers had fled after the battle and massacre, but here and there a family had remained, or had returned soon after the flight. Skulking parties of Indians continued to prowl about the valley, killing, plundering, and scalping, as opportunity offered. It was at this time that Frances Slocum was captured. The story of her life fully illustrates the remark, that "truth is strange—stranger than fiction." The following, originally published in the North American, of Philadelphia, narrates the circumstances of this singular affair:

At a little distance from the present court-house at Wilkesbarre, lived the family of Mr. Jonathan Slocum. The men were one day away in the fields, and in an instant the house was surrounded by Indians. There were in it a mother, a daughter about nine years of age, a son aged thirteen, another daughter aged five, and a little boy aged two and a half. A young man, and a boy by the name of Kingsley, were present grinding a knife. The first

INDIANS CAPTURING THE CHILD OF MRS. SLOCUM.

thing the Indians did was to shoot down the young man and scalp him with the knife which he had in his hand. The nine year old sister took the little boy two years and a half old, and ran out of the back door to get to the fort. The Indians chased her just enough to see her fright, and to have a hearty laugh, as she ran and clung to and lifted her chubby little brother. They then took the Kingsley boy and young Slocum, aged thirteen, and little Frances, aged five, and prepared to depart. But finding young Slocum lame, at the earnest entreaties of the mother, they set him down and left him. Their captives were then young Kingsley and the little girl. The mother's heart swelled unutterably, and for years she could not describe the scene without tears. She saw an Indian throw her child over his shoulder, and as her hair fell over her face, with one hand she brushed it aside, while the tears fell from her distended eyes, and stretching out her other hand towards her mother, she called for *her* aid. The Indian turned into the bushes, and this was the last seen of little Frances. This image, probably, was carried by the mother to her grave. About a month after this they came again, and with the most awful cruelties murdered the aged grandfather, and shot a ball in the leg of the lame boy. This he carried with him in his leg, nearly six years, to the grave. The last child was born a few months after these tragedies! What were the conversations, the conjectures, the hopes and the fears respecting the fate of little Frances, I will not attempt to describe.

As the boys grew up and became men, they were very anxious to know the fate of their little fair-haired sister. They wrote letters, they sent inquiries, they made journeys through all the West and into the Canadas. Four of these journeys were made in vain. A silence, deep as that of the forest through which they wandered, hung over her fate during sixty years.

My reader will now pass over fifty-eight years, and suppose himself far in the wilderness of Indiana, on the bank of the Mississinewa, about fifty miles southwest of Fort Wayne. A very respectable agent of the United States, Hon. George W. Ewing, of Peru, Ia., is travelling there, and weary and belated, with a tired horse, he stops in an Indian wigwam for the night. He can speak the Indian language. The family are rich for Indians, and have horses and skins in abundance. In the course of the evening he notices that the hair of the woman is light, and her skin under her dress is also white. This led to a conversation. She told him she was a white child, but had been carried away when a very small girl. She could only remember that her name was Slocum, that she lived in a little house on the banks of the Susquehanna, and how many there were in her father's family, and the order of their ages! But the name of the town she could not remember. On reaching his home, the agent mentioned this story to his mother. She urged and pressed him to write and print the account. Accordingly he wrote it, and sent it to Lancaster in this State, requesting that it might be published. By some, to me, unaccountable blunder, it lay in the office *two years* before it was published. In a few days it fell into the hands of Mr. Slocum, of Wilkesbarre, who was the little two year and a half old boy when Frances was taken. In a few days he was off to seek his sister, taking with him his oldest sister, (the one who aided him to escape,) and writing to a brother who now lives in Ohio, and who I believe was born after the captivity, to meet him and go with him.

The two brothers and sister are now (1838) on their way to seek little Frances, just *sixty years* after her captivity. They reach the Indian country, the home of the Miami Indians. Nine miles from the nearest white they find the little wigwam. "I shall know my sister," said the civilized sister, "because she lost the nail of her first finger. You, brother, hammered it off in the blacksmith-shop, when she was four years old." They go into the cabin, and find an Indian woman having the appearance of seventy-five. She is painted and jewelled off, and dressed like the Indians in all respects. Nothing but her hair and covered skin would indicate her origin. They get an interpreter, and begin to converse. She tells them where she was born, her name, &c., with the order of her father's family. "How came your nail gone?" said the oldest sister. "My older brother pounded it off when I was a little child in the shop!" In a word, they were satisfied that this was Frances, their long-lost sister! They asked her what her Christian name was. She could not remember. Was it *Frances?* She smiled, and said "*yes.*" It was the first time she had heard it pronounced for sixty years! Here, then, they were met—two brothers and two sisters! They were all satisfied they were brothers and sisters. But what a contrast! The brothers were walking the cabin, unable to speak; the oldest sister was weeping, but the poor Indian sister sat motionless and passionless, as indifferent as a spectator. There was no throbbing, no fine chords in her bosom to be touched.

When Mr. Slocum was giving me this history, I said to him, "But could she not speak English?" "Not a word!" "Did she know her age?" "No—had no idea of it!" "But was she entirely ignorant?" "Sir, she did'nt know *when Sunday comes!*" This was, indeed, the consummation of ignorance in a descendant of the Puritans!

But what a picture for a painter would the inside of that cabin have afforded! Here were the children of civilization—respectable, temperate, intelligent, and wealthy, able to overcome mountains to recover their sister. There was the child of the forest, not able to tell the day of the week, whose views and feelings were all confined to that cabin! Her whole history might be told in a word. She lived with the Delawares, who carried her off till grown up, and then married one of their number. He either died or ran away, and then she married a Miami Indian, a chief, as I believe. She has two daughters, both of whom are married, and who live in all the glory of an Indian cabin, deer skin clothes, and cow skin head-dresses! No one of the family can speak a word of English. They have horses in abundance, and when the Indian sister wanted to accompany her new relatives, she whipped out, bridled her horse, and then, *a la Turk*, mounted astride, and was off. At night she could throw a blanket around her, down upon the floor, and at once be asleep.

The brothers and sister tried to persuade their lost sister to return with them, and, if she desired it, bring her children. They would transplant her again to the banks of the Susquehanna, and of their wealth make her home happy. But no. She had always lived with the Indians; they had always been kind to her; and she had promised her late husband, on his death-bed, that she would never leave the Indians. And there they left her and hers, wild and darkened heathens, though sprung from a pious race. You can hardly imagine how much this brother is interested for her. His heart yearns with an indescribable tenderness for the poor helpless one, who, sixty-one years ago, was torn from the arms of her mother. Mysterious Providence! How wonderful the tie which can thus bind a family together!

Frances' second husband was known among the tribe as the *Deaf Man*, and the village where she lived was called Deaf-man's village. The United States by treaty has granted her a rich reserve of land. Her son-in-law, Capt. Brouillette, is a half-breed, of French extraction, and one of the noblest looking men of his tribe. The whole family are highly respectable among their nation, and live well, having a great abundance of the comforts of Indian life.

In the summer of 1779, Gen. Sullivan passed through Wyoming, with his army from Easton, on his memorable expedition against the country of the Six Nations. As they passed the fort, amid the firing of salutes, with their arms gleaming in the sun, and their hundred and twenty boats arranged in regular order on the river, and their two thousand pack-horses in single file, they formed a military display surpassing any yet seen on the Susquehanna, and well calculated to make a deep impression on the minds of the savages. Having ravaged the country on the Genesee, and laid waste the Indian towns, General Sullivan returned to Wyoming in October, and thence to Easton. But the expedition had neither intimidated the savages nor prevented their incursions. During the remainder of the year they seemed to make it their special delight to scourge the valley—they stole into it in small parties, blood and devastation marking their track.

Early in the spring of 1784, says Mr. Day, the settlers of Wyoming were compelled again to witness the desolation of their homes by a new cause. The winter had been unusually severe, and on the breaking up of the ice in the spring, (see engraving, page 265) the Susquehanna rose with great rapidity—the immense masses of loose ice from above continued to lodge on that which was still firm at the lower end of the valley; a *gorge* was formed, and one general inundation overspread the plains of Wyoming. The inhabitants took refuge on the surrounding heights, many being rescued from the roofs of their floating houses. At length a gorge at the upper end of the valley gave way, and huge masses of ice were scattered in every direction, which remained a great portion of the ensuing summer. The deluge broke the gorge below with a noise like that of contending thunderstorms, and houses, barns, stacks of hay and grain, cattle, sheep, and swine, were swept off in the rushing torrent. A great scarcity of provisions followed the flood, and the sufferings of the inhabitants were aggravated by the plunder and persecution of the *Pennamite* soldiers, quartered among them. Gov. Dickinson represented their sufferings to the Legislature, with a recommendation for relief, but in vain. This was known as the *ice flood;* another, less disastrous, which occurred in 1787, was called the *pumpkin flood*, from the fact that it strewed the lower valley of the Susquehanna with the pumpkins of the unfortunate Yankees.

After the peace with Great Britain, and the danger from the Indians having been, in a great measure, removed, the surviving inhabitants returned to their possessions at Wyoming, and being joined by

ICE-FLOOD ON THE SUSQUEHANNA.

many others, their settlement again flourished, and the village was rebuilt. They still refused, says Mr. Trego, to acknowledge the authority of Pennsylvania, or to be governed by her laws; and on the application of that State to Congress, a board of commissioners was appointed to determine the dispute between Pennsylvania and Connecticut concerning the jurisdiction of the territory in question, who, after a deliberation of five weeks, unanimously decided that the state of Connecticut had no right to the land in controversy. But though

WYOMING MONUMENT.

the Connecticut settlers were now willing to acknowledge the jurisdiction of Pennsylvania, they refused to yield up their farms and improvements to the Pennsylvania claimants, and a scene of trouble and contention between the different parties ensued, in which resort was again had to arms, and a number of persons were killed. The civil authorities of Pennsylvania were resisted, and the armed companies sent to sustain them were met and repelled by armed bodies of the Connecticut settlers. At length, after a long and harrassing contention, a compromise was effected; seventeen townships being granted to the Connecticut people, on condition of their relinquishing all claims to any other lands within the purchase of the original Connecticut Susquehanna Company, and compensation being made to the Pennsylvania claimants. Thus, at last, ended the Wyoming controversy; the New England settlers and their descendants became industrious and valuable citizens of their adopted state, and having now become blended with the general family of the commonwealth, they enjoy, in their blooming, beautiful and busy valley, the blessings of peace, plenty and prosperity. They are not, however, forgetful of the perils and sufferings by which their fathers established themselves in that favored spot, and have erected a monument on the battle ground of the "Massacre of Wyoming," over the bones of the unfortunate sufferers in that melancholy tragedy, to commemorate the deeds of that eventful day, and to show to future generations the spot where their forefathers fought, bled, and died in defence of their families and homes.

# LOCOMOTIVE SKETCHES,

WITH

## PEN AND PENCIL.

FROM

## PHILADELPHIA TO PITTSBURG.

JUNIATA SCENERY.

# LOCOMOTIVE SKETCHES

WITH

# PEN AND PENCIL.

## Philadelphia to Pittsburg.

No more we sing, as they sang of old,
 To the tones of the lute and lyre,
For lo! we live in an Iron Age—
 In the age of Steam and Fire!
The world is too busy for dreaming,
 And hath grown too wise for War:
So, to-day, for the glory of Science,
 Let us sing of the *Railway Car!*
The golden Chariots of ancient Kings
 Would dazzle the wondering eye,
And the heads of a million slaves might bow
 As the glittering toy rolled by;
But this is the *Car of the People,*
 And before it shall bow all Kings:—
Be they warned when they hear the shrieking
 Of the Dragon with Iron Wings!

AND I have a long journey before us—three hundred and sixty-three miles! But we have an iron horse, and his fiery breath never fails. The first mile of our journey is confined to the widest street in Philadelphia, and to a railroad constructed at its expense. Messrs. Bingham & Dock, forwarding merchants, having leased the railroad belonging to the State, have erected a handsome depot at the corner of Schuylkill 5th and Market sts. Passengers for Columbia, &c., take the cars here. We next pass the City Gas Works, situated on the Schuylkill. They

PHILADELPHIA AND COLUMBIA RAILROAD DEPOT.

are the most complete in their structure and arrangement of any similar establishment in the country. They were commenced in 1835, with some doubt as to the practicability of the undertaking, in an economical view. The entire area now occupied by the works is nearly eight acres, with a front on the Schuylkill of 800 feet. A high standard of illuminating quality of gas has been adopted, by the use of the proper varieties of coal, and the admixture of resin when they fall below the prescribed standard. The fat bituminous coals of our own State are principally used, with some samples of pure Cannel coal, some of it imported and some obtained from the extensive beds of Virginia, as well as those lately discovered in our own State, which, it seems, has almost every variety of coal known anywhere on the globe. The standard quality of the gas is that of twenty candle light —in other words, the light of an argand burner, consuming four feet of gas per hour, is equal to that of twenty sperm candles, moulded six to the pound. This high quality is attained by the use of our own coals alone, without the addition of resin or any other bituminous matter. After the bitumen is extracted from the coal, the coke is sold. The quantity of gas consumed by the city proper (exclusive of the adjoining districts, two of which have extensive gas works of their own,) for the year 1851, was over one hundred and eighty-two millions of cubic feet. The maximum production of the works is about 887,000 cubic feet every twenty-four hours—or, converted into gallons, something over seven and a half millions, which is about the same amount of water daily consumed during the summer season.

The extent of main-pipes laid down is equal to ninety-five miles. The number of meters is nearly ten thousand, and the whole number of customers about the same—employing a total number of lights of over one hundred and fifteen thousand, exclusive of about sixteen hundred in the streets, squares, and market-houses of the city. The total length of the pipes is about one hundred and twenty-six miles. A larger gas establishment is now about being erected a short distance from the present works, to which these will be transferred, as their productive capacity is too small to supply the increasing demand of the city, which exclaims with Gœthe, "Light—more light!" The new works will be erected at Point Breeze, on the Schuylkill, and will embrace an area of some seventy acres.

MARKET STREET BRIDGE.

A few yards from the gas works we pass over the Market Street Bridge, one of the finest and most substantial structures in the United States. One side of it is appropriated to the exclusive use of the Railroad, and has a double track laid down upon it. All the freight and passenger-cars passing over the railroad, are hauled to and from the city by horses, for which purpose a large number are employed. The other side of the bridge is used for the passage of horses and vehicles—while, for foot-passengers, there is an outside-walk, from which a fine view of the Schuylkill is obtained. A short distance below we can see the Blockley Almshouse, a large and elegant establishment, whose external appearance indicates anything else than the abode of poverty and misfortune. But distance, the cheat, "gives enchantment to the view," which an inspection of its interior features would soon dispel. The building consists of two thousand feet in front, three stories high, ornamented in the centre with a stately Tuscan portico, supported by six massive columns. It has accommodations for a vast number of inmates—there being now within its walls not less than two thousand, supported at an average weekly expense of 104 cents each. The house of employment connected with it yields annually about $22,000 worth of manufactures, and the farm nearly $15,000 worth of produce. The children's department of this institution embraces, at present, three hundred and sixty-five boys and over fifteen hundred girls—supported at an annual cost of near $70,000. The entire annual expense of the whole establishment may be stated at $165,000—including the insane department, in which there are some three hundred and fifty patients. The whole cost of supporting the poor of the city and county of Philadelphia, for the year 1851, is stated at $228,977.

> Were half the power that fills the world with terror—
> Were half the wealth bestowed on *rum* and courts—
> Given to redeem the human mind from error,
> There were no need of *poor-houses* and *forts!*

Upon emerging from the bridge, we enter the borough of West Philadelphia, with its mud and dust, and jim-crack cottages. It has a large and rapidly increasing population, which is principally composed of those who conduct business in the city, but do their sleeping out here—hence the dull, drowsy appearance of the place.

Hitched to our "loco-snorter," we wind along, for a short distance, the western bank of the Schuylkill, catching, here and there, a glance at some attractive object in the distance, as the Girard College, Fairmount Water-works, &c. A deep cut finally obstructs the view, and the railroad soon after plunges into the midst of the narrow patches, the sweet little gardens—(blooming with roses and cabbages)—and cozy cottages, so characteristic of the suburban districts of populous cities; and then pursues its way among the broad rich acres of the farmer. Rich!—no. Not rich, nor yet sterile;—but rather like a dismissed Irish servant, without a "cha-ractur." The soil is too clayey to be fertile, and the swelling fields and grazing cattle deny its barrenness. That it is not well drained is certain—that such soils greatly need it, we will attach our hand and seal. The truth is, there is a superabundance of manure expelled from the city, and it naturally finds its way here, where it is needed; and thus, for ten or more miles, we traverse a region of agriculture not supported by the usual agricultural economy—but principally devoted to vegetables and fruits, for which it receives from the *city* the means to uphold its limited productive capacity.

Eleven miles from Philadelphia—(three hundred and fifty-two from Pittsburg,) is White Hall—(which, by the way, was brown,—but now, in fact, is neither white nor brown, for it burned down some years ago, and has not yet been rebuilt.) It was a fine hotel, much resorted to by Philadelphians—*why*, we cannot tell. Pure air! Fudge! Talk of pure air in a clay-flat like this—as well talk of raising potatoes in a snow-bank. Go to the mountains for pure air—go to Pottsville, six or eight hundred feet above the Delaware, or go with us to the cloud-capped summits of the Alleghany, and you shall taste pure air, fresh from heaven.

Oh! let us go and breathe our woe
  In Nature's kindly ear,
For her soft hand will ever deign
  To wipe the mourner's tear;
She mocks not, tho' we tell our grief
  With voice all sad and faint,
And seems the fondest while we pour
  Our weak and lonely plaint.
Oh! let us take our sorrows
  To the bosom of the hills,
And blend our pensive murmurs
  With the gurgle of the rills;
Oh! let us turn in weariness
  Towards the grassy way,
Where skylarks teach us how to praise,
  And ringdoves how to pray;

And there the melodies of Peace,
  That float around the sod,
Shall bring back hope and harmony
  With the sweet voice of God.

The Eagle Hotel, (seventeen miles from Philadelphia) is chiefly remarkable as being one of the oldest inns in the State. The turnpike from Philadelphia to Lancaster, sixty-two miles in length, was commenced in 1792, and finished two years after, at a cost of nearly $500,000. It is probably the first improvement of the kind ever commenced in the United States. The system of turnpikes rapidly extended a few years after the completion of this enterprise, which was subsequently extended to Pittsburg, as well as beyond the State line into Ohio, and also in the east into New Jersey—thus forming a continuous turnpike road of nearly four hundred miles. A large number of similar roads, radiating from the main thoroughfare, soon after followed, so that the State was placed in admirable condition for travelling by stage, as well as for transporting, in Conestoga teams, the merchandize

CONESTOGA TEAM AND STAGE-COACH.

destined for the interior. The extensive travel thus created and concentrated on this once splendid thoroughfare largely increased the number of the inns. Among these, several were particularly noted in their day for the extent of their business and the style with which they were conducted, among which were the Eagle, already mentioned, the Paoli, kept by the late Gen. Evans, and the Ship, by the late John Bowen. These hotels had, and still have, large and splendid farms con-

nected with them, and were conducted with great profit to the proprietors. There were a large number of other inns intended for the accommodation of the wagoners—most of which, after the commencement of the main line of railway and canal, were discontinned—the ponderous Conestoga team being entirely superseded by the canal boat and railway car. The turnpike, previous to the loss of its trade, presented a busy scene—an almost unbroken procession of these wagons, each of them drawn by six large strong horses, and many of the teams having a row of bells hanging over the collar of each horse. The wagoners got up a song upon the loss of their "occupation," a verse of which is all we can "recommember"—

Oh, its once I made money by driving my team,
But now all is hauled on the railroad by steam.
May the devil catch the man that invented the plan,
For its ruined us poor wagoners, and every other man.

The "every other man" referred to were probably the innkeepers aforesaid. But really the railway *did*, at the outset, entail a serious loss along the principal line of turnpikes, and even now it exhibits a wreck that is rather mournful to contemplate. Not only have the Conestoga teams disappeared, but the stage—alas! the stage-horn no longer is heard—the bounding wheels no longer rattle over the white compact road.

We hear no more the clanking hoof,
  And stage-coach rattling by;
For the steam-king rules the travelled world,
  And the pike is left to die.
The grass creeps over the flinty path,
  And the stealthy daisies steal
Where once the stage-horse, day by day,
  Lifted his iron heel.

No more the weary stager dreads
  The toil of the coming morn;
No more the bustling landlord runs
  At the sound of the echoing horn;
For the dust lies still upon the road,
  And the bright-eyed children play,
Where once the clattering hoof and wheel
  Rattled along the way.

No more do we hear the cracking whip,
  Or the strong wheels' rumbling sound;
And ah! the water drives us on,
  And an iron horse is found!
The coach stands rusting in the yard,
  And the horse has sought the plough;
We have spanned the world with an iron rail,
  And the steam-king rules us now!

The old turnpike is a pike no more—
  Wide open stands the gate;
We've made a road for our horse to stride,
  Which we ride at a flying rate;
We have filled the valleys and levelled the hills,
  And tunneled the mountain side;
And round the rough crag's dizzy verge,
  Fearlessly now we ride!

On—on—on—with a haughty front!
  A puff, a shriek, and a bound:
While the tardy echoes wake too late
  To babble back the sound:
And the old pike road is left alone,
  And the stagers seek the plough;
We have circled the earth with an iron rail,
  And the steam-king rules us now!

The PAOLI, twenty-one miles from Philadelphia, was kept for many years by the late Gen. Joshua Evans, who formerly represented the county of Chester in Congress. He was truly a fine "old gentleman of the olden school." During the revolutionary movements in this vicinity, the house was occupied by Gen. Washington as his head-quarters. About one mile and a half west is the house in which Gen. Anthony Wayne* was born. About the same distance, and nearly

GENERAL WAYNE'S RESIDENCE.

in the same direction, is the field of the memorable Paoli massacre. Wayne, himself, had charge of the American forces thus slaughtered

*Gen. Anthony Wayne was born in the township of Eastown, Chester Co., (about one and a quarter miles south of the Paoli tavern,) on the 1st Jan., 1745. He received a thorough education, and was particularly skilled in the mathematics. After leaving school he became a surveyor, and also paid some attention to astronomy and engineering, by which he attracted the attention of Dr. Franklin, who became his friend and patron. At the opening of the revolution he was a prominent member of the provincial Legislature. He entered the army in 1775 as colonel of a corps of volunteers, and was afterwards active on the northern frontier at Ticonderoga. Here he was made brigadier-general on the 21st Feb., 1777. In the battle of Brandywine he commanded the division of

with more than barbarous ferocity, which occurred on the night of the 20th September, 1777. Soon after the battle of Brandywine the two contending armies again met, on the 16th of September, near this place, and were about to engage in hostile proceedings, when a severe rain storm came on, materially injuring their powder, and otherwise rendering an attack from the Americans impracticable. Washington thereupon withdrew to the Schuylkill, some five miles northeast, and sent Gen. Wayne, with 1500 men, to join Gen. Smallwood, and annoy the rear of the enemy, who was posted near a Welsh Church, not far off, called Tredyffim. Wayne had encamped in a very retired position, near the present monument, and at some distance from the public roads. The British General, receiving information from traitors who knew every defile in the neighborhood, and every movement of the republican troops, detached Gen. Gray, a brave but desperate and cruel officer, to cut off Wayne's party. Stealing his way through the

Chadsford, resisting the passage of the column under Knyphausen with the utmost gallantry until near sunset, when, overpowered by superior numbers, he was compelled to retreat. At the battle of Germantown he evinced his wonted valor, leading his division into the thickest of the fight.

In all councils of war he was distinguished for supporting the most energetic measures. At the battle of Monmouth, he and Gen. Cadwallader are said to have been the only two general officers in favor of attacking the enemy. His conduct on that occasion elicited the special applause of Gen. Washington. His attack upon the fort at Stony Point, in July, 1779, an almost inaccessible height, defended by a garrison of six hundred men, and a strong battery of artillery, was the most brilliant exploit of the war. At midnight he led his troops with unloaded muskets, flints out, and fixed bayonets, and without firing a single gun, completely carried the fort, and took five hundred and forty-three prisoners. In the attack, he received a wound, from a musket ball, in the head, which, in the heat of the conflict, supposing to be mortal, he called to his aids to carry him forward, and let him die in the fort. In the campaign of 1781, when Cornwallis surrendered, he bore a conspicuous part; and he was afterwards actively engaged in Georgia. At the peace of 1783, he retired to private life. In 1789 he was a member of the Pennsylvania Convention, and strongly advocated the adoption of the Constitution of the United States. In 1792, after Harmar and St. Clair had been repeatedly unsuccessful, Wayne took the command on the northwestern frontier, and by his wise and prudent measures, his excellent discipline and bravery, he gained the decisive battle of the Maumee, and concluded the war by the treaty of Greenville, in 1795. A life of peril and glory was terminated in Dec., 1796, in a cabin at Presqu'isle, then in the wilderness, and his remains were deposited, at his own request, under the flag-staff of woods, and up the narrow defile below the Paoli, he drove in the American pickets, and rushed upon the camp. The assailants were received with several close and destructive fires, which must have done great execution, but the American troops were compelled by superior numbers to retreat. The number of Americans killed and wounded in this action, amounted to one hundred and fifty. Gen. Gray, it is said, had ordered his troops to give no quarter. Many victims were massacred with ruthless and savage barbarity, after resistance, on their part, had ceased. The cry for quarter was unheeded; the British bayonet did its work with unpitying ferocity. It is said by some that the enemy set fire to the straw in the camp, thus torturing many sick and wounded victims who were unable to escape the flames. The whole American corps must have been cut off, if Wayne had not preserved his coolness. He promptly rallied a few regiments, who withstood the shock of the enemy, and covered the retreat of the others. When this attack commenced, Gen. Smallwood was already within a mile of the field of battle; and had he commanded troops to be relied upon, might have given a very different turn to the night. But his raw militia, falling in with a party returning from the pursuit of Wayne, instantly fled in confusion. The neighboring farmers decently buried the dead, numbering fifty-three persons, in one common grave, at a spot immediately adjoining the scene of action. Some thirty-five years ago, a military company of Chester County, aided with the

the fort on the margin of Lake Erie. His remains were removed in 1809 by his son, Isaac Wayne, to Radnor churchyard, in Delaware county.

By direction of the Pennsylvania State Society of Cincinnati, an elegant monument was erected, of white marble, of the most correct symmetry and beauty.

*South Front.*—In honor of the distinguished military services of Major General ANTHONY WAYNE, and as an affectionate tribute of respect to his memory, this stone was erected, by his companions in arms, the Pennsylvania State Society of the Cincinnati, July 4th, 1809, thirty-fourth anniversary of the Independence of the United States of America; an event which constitutes the most appropriate eulogium of an American soldier and patriot.

*North Front.*—Major General ANTHONY WAYNE, was born near the Paoli, Chester county, State of Pennsylvania, A. D. 1745. After a life of honor and usefulness, he died in December, 1796, at a military post on the shore of Lake Erie, Commander-in-chief of the army of the United States. His military achievements are consecrated in the history of his country, and in the hearts of his countrymen. His remains are here interred.

PAOLI MONUMENT.

individual subscriptions of the citizens, erected a monument over the remains of the gallant men. It is composed of white marble, and is a pedestal surmounted by a pyramid. Upon the four sides of the body of the pedestal are appropriate inscriptions. The grounds enclosed, embrace about thirty-six acres, rising to a gentle elevation, and presenting the form of the letter L. The monument is situated in the angle of the plot, surrounded by a heavy stone-wall, and shaded with stately trees.

The neighborhood of Paoli is full of interesting incidents connected with the revolution. A few miles to the right is Valley Forge, where Washington and his army were encamped during the severe winter of 1777. Here, half-naked, hungry, and sick, a large number of the soldiers died. The general aspect of the revolution then seemed dark and gloomy, and scarcely a ray of hope for future success was left. The subsequent campaign, however, dispelled many of these dark clouds, and after the affair at Monmouth, new hopes were instilled into the hearts of the patriots.

The two stations between Paoli and Downingtown are without interest. DOWNINGTOWN, thirty-three miles from Philadelphia, though a small village, is one of the oldest settlements in the State. The place and the vicinity was originally settled by English emigrants from Birmingham. The present occupants live, for the most part, upon property that has been in their families for many generations. The brick house a few rods west of the railroad depot was erected in 1728. It was then the first house (excepting a few log cabins) of the pioneers. Downingtown derives its name from Thomas Downing—*not*, as might be inferred, from the celebrated Major Jack Downing, of Downingville. Thomas Downing bought the land from the earlier settlers, in 1730, and soon after built a mill and several other improvements upon it, in virtue of which it assumed his name. It is a village of neat cottages and green foliage, having a strictly rural aspect. Its length is over a mile, stretching along the turnpike, the houses standing a few yards back from the road-side, surrounded with neat gardens, trees and shrubbery. Even the stores, and other places of business, are partially hidden amidst the profusion of foliage. The people are plain and sober-minded, but though very intelligent, are by no means remarkable for enterprise. Nevertheless, there is a considerable number of mills and factories in the vicinity, deriving support entirely from the great agricultural resources surrounding them, for Downingtown, situated in what is called *the great valley*, is in the heart of the richest agricultural region in Pennsylvania. The great valley lies between two ranges of hills, running nearly parallel with each other, from the Schuylkill river to a point near the western boundary of the county. Extensive quarries of limestone are opened at points all along the valley, for the supply of lime to the adjacent country. In some parts of its range, this limestone is light-colored or white, semi-crystalline or granular, affording, where the layers are sufficiently thick and compact, a splendid marble for architectural and ornamental purposes. A large portion of the marble used in the construction of the Girard College buildings was obtained in this vicinity. There is a quarry a short distance from Downingtown, where excavations have been made beyond a hundred feet in depth. The east branch of the Brandywine creek, a beautiful stream meandering along its grassy banks in the valley, passes through Downingtown, and furnishes the driving power to a large flour mill and to iron works. The railroad crosses this stream by the bridge a short distance below the town.

Eleven miles above Downingtown, and forty-one from Philadelphia, is Coatesville. This place is, in many respects, similar to Downingtown, though its situation is rather more picturesque—having, on its northern side, the range of hills which border the great valley. The railroad here crosses the west branch of the Brandywine, over a bridge towering eighty feet in the air, and stretching across a chasm nearly nine hundred feet in width. This stream affords a fine water-power, which is extensively used for flour, paper, and other mills, as well as factories of various descriptions, in the vicinity of Coatesville. Coatesville was originally settled by a family bearing that comfortable name, and dates its origin as far back as 1725.

Four miles above is PARKESBURG, which derives all its importance from the machine shops erected here by the State, for the repair and manufacture of the running machinery of the railroad.* A large number of hands are employed in these shops, which imparts an active appearance to the place.

PENNINGTONVILLE, forty-eight miles from Philadelphia, is the last station in Chester County. The place is a growing one, surrounded with an industrious farming population.

The territory now included in Chester County, together with much land lying in other counties, was honorably purchased of the Indians by William Penn, and was conveyed in several distinct deeds. The first, bearing date June 25, 1683, and signed by an Indian called Wingebone, conveys to William Penn all his lands on the west side of the Schuylkill, beginning at the first falls, and extending along and back from that river, in the language of the instrument, "so far as my right goeth." By another deed of July 14th, 1683, two chiefs granted to the proprietary the land lying between the Chester and Schuylkill Rivers. From Kikitapan he purchased half the land between the Susquehanna and Delaware, in September, and from Malchalola, all lands from the Delaware to Chesapeake Bay, up to the Falls of the Susquehanna, in October; and by a deed of July 30th was conveyed the land between Chester and Pennypack Creeks. This last instrument is a quaint piece of conveyancing, and will show the value attached by the natives to their lands:

"This indenture witnesseth that we, Packenah, Jackham, Sikals, Portquesott, Jervis Essepenaick, Felktrug, Porvey, Indian kings, sachemakers, right owners of all lands from Quing Qingus, called Duck cr., unto Upland, called Chester cr.,

* This road commences at the Market Street Bridge, in Philadelphia, and pursues a western course, by Downingtown and Lancaster, to Columbia, on the Susquehanna River, where it connects with the Eastern Division of the Pennsylvania State Canal. Formerly it had two inclined planes, one at Philadelphia and the other at Columbia; but both have been avoided by laying down new tracks, with but slight gradients—the highest not exceeding thirty-five feet to the mile. The motive-power on this road, as also on the Alleghany Portage Road, is furnished by the State, for which a charge is made in addition to the road toll. This road was among the earliest completed in the country—having been opened in April, 1834. Cost, $3,983,302.

all along the west side of Delaware river, and so between the said creeks *backwards as far as a man can ride in two days with a horse,* for and in consideration of these following goods to us in hand paid, and secured to be paid by William Penn, proprietary of Pennsylvania and the territories thereof, viz: twenty guns, twenty fathoms match coat, twenty fathoms stroud water, twenty blankets, twenty kettles, twenty pounds of powder, one hundred bars of lead, forty tomahawks, one hundred knives, forty pair of stockings, one barrel of beer, twenty pounds of red lead, one hundred fathoms of wampum, thirty glass bottles, thirty pewter spoons, one hundred awl blades, three hundred tobacco pipes, one hundred hands tobacco, twenty tobacco tongs, twenty steels, three hundred flints, thirty pair of scissors, thirty combs, sixty looking-glasses, two hundred needles, one skipple of salt, thirty pounds of sugar, five gallons of molasses, twenty tobacco boxes, one hundred jewsharps, twenty hoes, thirty gimlets, thirty wooden screw boxes, one hundred and three strings of beads—do hereby acknowledge, &c., &c. Given under our hands and seals, at New Castle, 2d of the eighth month, 1685."

Chester County received its name in the following manner: When Wm. Penn first arrived at Upland, now old Chester, turning round to his friend Pearson, one of his own society, who had accompanied him in the ship Welcome, he said, "Providence has brought us here safely. Thou hast been the companion of my perils. What wilt thou that I should call this place?" Pearson replied, "*Chester*, in remembrance of the city from whence I came." Penn also promised that when he divided the territory into counties, he would call one of them by the same name. In the beginning of the year 1683, the governor and council established a seal for each of the counties, assigning to Chester the *plough*—the device still indicative of the thrifty agricultural character of the inhabitants.

Before the close of the year 1682, no less than twenty-three ships had arrived in Pennsylvania from Europe, conveying more than two thousand souls. They were principally Friends, who had purchased allotments, and came to occupy them. Many were of opulent families, whom no common consideration could have prevailed upon to leave their homes; and whom, perhaps, nothing but the goad of unceasing persecution could have driven entirely away. All were industrious, discreet, and prudent, and every way fitted to render a colony prosperous, flourishing, and happy. Not an inconsiderable number of these settled in Chester County. Some had taken the precaution to bring with them frames of houses and other conveniences; some, who arrived early, were enabled to erect temporary cabins of logs, and some were compelled to pass the winter in rude shanties, or caves dug in the side of a hill.

At the time the European emigrants first settled in the county, it was principally overshadowed by forests, with here and there a small patch cleared by the natives for the purpose of raising corn. Owing to the Indian practice of firing the woods once or twice in a year, the small bushes and timber were killed in their growth, and of course the forests were but thinly set. One of the first settlers said, that at the time of his first acquaintance with the country, he could have driven a horse and cart from one of its extremities to the other, in almost any direction without meeting with any material obstruction.

The early settlers of Chester County were from different parts of Europe, England, Wales, Ireland, Holland, and Germany. Of these, the English, as they arrived first, seated the southern parts adjoining the Delaware, and a few took up lands bordering upon the Maryland line. They were principally from Sussex, (the residence of Wm. Penn,) Cheshire, Derbyshire, Leicestershire, and Northamptonshire. The Welsh occupied the eastern parts, and settled in considerable numbers. The oppression which they suffered in their native country from the tyranny of the nobles first determined their emigration, and the happy consequences resulting to the first adventurers, from their change of situation, induced many to follow them. Soon after their arrival here they generally joined the society of Friends, and established meetings. Wm. Penn once paid them a visit, but as they neither understood his language, nor he theirs, they could only enjoy the satisfaction of seeing him. It is said, however, that they were highly gratified with this mark of his attention and good-will, and took even their little children with them to the meeting which he attended, that they also might have a sight of the great proprietor. Rowland Ellis was one of their most conspicuous characters.

The Irish emigrants located the north and western sections of the county. Those who first arrived were generally men of some standing and character, and were welcomed as an accession of virtue and intelligence to the little community. They were almost all Protestants, and many of them Friends. The Dutch and Germans, who are now the principal land-holders in many of the northern townships, are not the descendants of the original settlers of those parts. Within the memory of those now living, they formed the smallest portion of the population in those very districts where they are now the most numerous. Their untiring industry and stubborn perseverance seem to have peculiarly qualified them to become successful tillers of a soil such as obtains there—fertile, indeed, but hard of cultivation; and the posterity of the Irish, who are not so remarkable for the patient qualities of character, seem to have gradually relinquished to them the possession of the land.

For a number of years the improvements in those parts of the county seem to have been much in the rear of those in other parts of the county. The log cabins of the early pioneers were still prevalent as late as 1760. This was partly owing to the uncertain tenures by which the real estate was held.

*Soul-Drivers.*—This was a name given to a certain set of men who used to drive redemptioners through the country, and dispose of them to the farmers. They generally purchased them in lots consisting of fifty or more, of captains of ships, to whom the redemptioners were bound for three years' service, in payment for their passage. The trade was brisk for a while, but at last was broken up by the numbers that ran away from their drivers. The last of the ignominious set disappeared about the year 1785. A story is told of his having been tricked by one of his herd. The fellow, by a little management, contrived to be the last of the flock that remained unsold, and travelled about with his master. One night they lodged at a tavern, and in the morning the young fellow, who was an Irishman, rose early, sold his master to the landlord, pocketed the money, and marched off. Previously, however, to his going, he used the precaution to tell the purchaser, that though tolerably clever in other respects, he was rather saucy, and a little given to lying; that he had been presumptuous enough at times to endeavor to pass for master, and that he might possibly represent himself as such to him!

Chester County is remarkably rich in its mineral resources, and for *variety* and general usefulness for chemical purposes, it is probably not surpassed by any other region, of equal extent, in the United States. We have elsewhere observed that, at various points on the Schuylkill, above Valley Forge, there are extensive deposits of the ores of copper and lead. The formation containing them traverses nearly all the counties of Pennsylvania east of the South Mountain. Out-crops occur at various places between the Schuylkill and the Susquehanna, and mining operations are now being prosecuted with considerable spirit and vigor. Indeed, the copper had been worked in the northern slope of the Mine Ridge, in Lancaster County, for some time previous to the revolution, and the old shafts are now being cleared out with the view of again extending them. In Delaware County arrangements have also been made to mine copper, and that county may be said to be literally a copper region. The "barrens" of York county, bordering on the Susquehanna, contain no inconsiderable quantity of chrome, being a continuation, in detached basins, of the same formation so long and so successfully worked in Lancaster County, near the Maryland line. A portion of this extensive chrome region also extends into Chester County,—but its proximity to navigation in Lancaster has enabled the operators to drive a splendid business in raising it. The mineral, we may add, is sought after to obtain from it the chromic acid, for the preparation of the beautiful chrome-yellow used in painting and dyeing. Lancaster County is, we believe, the only spot in the United States where it is found to any extent, and large quantities of it are annually shipped to Europe. There is a variety of other valuable minerals found in parts of Chester, among which may be mentioned asbestus, magnesites, amethyst, jasper, garnet, schorl, chalcedony, agate, sapphire, beryl, etc. etc.

The early settlers of Chester, we have already mentioned, were Quakers, and the county is still under the influence of their principles and social habits. Their religious creed has a tremendous influence upon the social economy of the people—regulating not only their course of action, but also their mode of thinking. There is little actual difference between their religious creed and that of Protestants generally;—but standing upon the broad platform that the greatest enemy of true religion is *pride*, and that, if not stoutly combated, it usurps the moral nature and sentiments of man, they wage a ceaseless war against this evil, and fortify themselves in every shape and form from its insidious approaches.

> Pride still is aiming at the blest abodes,—
> Men would be angels, angels would be gods.

Wm. Penn, in his work entitled "No cross, no crown," lays down the principles which he professed, and combats, in good plain English, the follies and wickedness of the church, all having their origin in *pride*. Splendid church edifices, fashionable dress, and ostentatious show, high living and voluptuous ease, the swell of music and the excitements of worldly amusement—all are denounced as nourishing a family of evils which finally overwhelm the true fountain of revealed knowledge. He is right, no doubt of it. Pride—pride lies at the bottom of nearly all our social evils, and it is sheer folly to deny or attempt to palliate it. Pride governs the church—ergo, the church is corrupted by it.

The Quakers, therefore, opposing themselves to this monster, observe a simplicity in all their movements through life which requires the greatest self-denial. They are eternally at war with the flesh. Their houses, their churches—their dress, language, thought—all show the paramount object in view—simplicity. They do not care about churches—it matters not where they worship. They address themselves to the inward spirit which God gave them—it is that which moves them—the flesh—the body is the mere earthly tabernacle—the temporary dwelling-place of the immortal spirit. Thus quietly, and without the assistance of music or worldly machinery, they sit and worship; and no doubt it is the most rational and spiritual way.

The following engraving illustrates the simple but substantial character of their meeting-houses. It represents one of their oldest places of worship near the Brandywine. The interior consists of plain wooden benches, with high backs, cleanly scoured, and destitute of paint.

QUAKER MEETING-HOUSE.

The benches, with a large ten-plate stove, comprise the entire furniture of the house!

As we leave Chester County, and pass through the range of hills called the Mine Ridge, the great county of Lancaster, in all its glory, expands before the eye. An intelligent Englishman called this county the "garden of America," and a view of it from this position will fully justify the propriety of the compliment. It is, without doubt, the garden of this glorious Union, and there are few spots in this wide, wide world, which could present a nobler scene to the eye than is here afforded. The Mine Hill winds around the county from the Susquehanna in a north-east and south-west course, while the South Mountain, or Conewago hills, border it on the north—the district between presenting one broad basin of fertility, with numerous subordinate elevations, rolling out one after the other, with intervening valleys and streams. The broad fields, when laden with the ripening harvests, swell to and fro with the sweeping gales, like the dark-green waters of the ocean.

> Heavens! what a goodly prospect spreads around,
> Of hills, and dales, and woods, and lawns, and spires,
> And glittering towns, and gilded streams, till all
> The stretching landscape into smoke decays!

The whole broad scene gradually sinks into the dim, blue vapory outlines of the bold Kittatinny, which skirt alike the landscape and the overarching firmament. The entire region of country between the Mine Ridge and the Conewago Hills, and between those hills and the Kittatinny, (called the great Cumberland valley,) presents one

continuous and almost unvaried scene of agricultural prosperity. The soil is naturally rich, as, in addition to the debris deposited over the surface by the decomposing matter of the higher lands, it is traversed at various points by extensive beds of limestone. The whole country is, therefore, in the highest state of cultivation; and in the economy which characterizes the general agricultural system, there is probably not a more prolific region in the United States. The farms are generally small—averaging about eighty acres each—and by a judicious division of the land, and rotation of crops, their high state of fertility is constantly maintained. The farms in the interior of the counties of Lancaster, Berks, Lebanon, and those of the eastern portion of the State generally, are more extensive than those situated along the lines of improvement—because, denied the ready access to market which is afforded to the latter, they are compelled to feed cattle during the winter, and thus consume the grain which could otherwise not be disposed of so profitably and conveniently. The cattle, or the great bulk of them, are purchased from western drovers during the autumn months, and being fattened in the winter, are sent to market in the spring, before western fat cattle arrive, and thus fair prices are usually realized. By this means an abundance of manure is secured to the farm, and a fair price for the grain consumed is realized.

The farmers living near the city, and on the lines of railroad, turn their produce, for the most part, into the dairy, or dispose of their grain by the bushel. They also raise a larger proportionate amount of poultry, vegetables, and floral and horticultural products. These latter, however, do not receive the attention they deserve; and we think if some of our farms were exclusively devoted to the production and improvement of poultry, (a thing very much needed, and, at the same time, very easily effected,) in the same manner as the most of them now are devoted to the fattening of cattle, they would prove more profitable than under the usual monotonous routine of farming. The same may be said of vegetables, as well as of floral and horticultural plants generally. They are all too much neglected; and one reason is, probably, that the mass of the farmers are not competent to raise them in perfection, because they require more cultivated taste and scientific principles in their production than the ordinary grass and cereal crops.

Farming is, in fact, throughout Pennsylvania, little less than systematic labor—well organized, it is true; but still only a monotonous routine of physical toil, too seldom relieved by mental exercise or enjoyment. This is unfortunate. It is the result of old established prejudices, deeply-rooted in our German population, who, resisting every modern innovation, hold fast to the time-honored principles, precepts and examples of their forefathers, and regard it as a moral and social duty to "follow in their footsteps." They, therefore, plough, plant, and reap, pretty much in the old way, without deviating to the right or left, but by industry, frugality, and close attention to their affairs, generally gather a competency, which is finally distributed amongst their children, who in turn travel over the same beaten track of agricultural life.

The system of cropping varies in different districts; the following, which is given by Mr. Trego, is one of the most common in this section of the State: A field which has been in pasture is ploughed up for Indian corn late in the fall, or, more usually, early in the spring. The corn is planted in the beginning of May, and well dressed and tended through the early part of summer. The corn is planted in straight rows, about two feet or more apart, and is hoed or ploughed twice, to keep down the weeds. Sometimes pumpkins are sown with the corn. About the last of October the corn ripens and is gathered, yielding from thirty to fifty bushels per acre, and on rich soils frequently more. About the first of the following April the same field is again ploughed, and sown with oats, which is harvested towards the end of July, producing from twenty-five to fifty bushels to the acre. The oats' stubble is then ploughed in, and the field, being well manured, is sown with wheat in the latter part of September. Rye is frequently sown instead of wheat, where the soil is light and thin, or where it is not manured; and many farmers sow both wheat and rye. In February or March, clover or other grass seeds are sown on the wheat and rye, which grow among the grain until harvest. The wheat and rye are generally fit to cut early in July, and commonly yield from twenty to thirty-five bushels per acre. The field is by that time covered with young clover, which is left until the following summer, when it is cut for hay in June, and a second crop is used for pasture or gathered for seed in September. The field may be re-mown the following year, or pastured until it comes again in course for Indian corn. Some farmers prefer sowing their wheat on a field freshly broken up from the grass sod; some omit the crop of oats between the corn, and the wheat or rye; others take off the Indian corn early, and sow wheat or rye immediately after it. The mere order of succession in the different crops is not very important, provided that the farmer is careful not to exhaust his land by too frequent repetitions, or by neglecting to plough, manure and dress his fields in the best manner.

Besides the grains already mentioned, buckwheat, flax, barley, potatoes, turnips, beets, and many other articles are cultivated. Different modes of culture and tillage are practiced in different parts of the State, according as the variety of soil, climate, or situation renders it necessary or expedient. In the more elevated and colder districts, the cultivation of Indian corn is not very successful; but grass, oats, and potatoes thrive admirably. Scarcely a farm is to be found in any portion of the Atlantic Slope, without its apple orchard of choice and selected varieties. Pears, peaches, plums, cherries, and other fruits are abundant, and though many farmers are careful to obtain the finest kinds, yet there is scarcely sufficient attention bestowed in selecting and grafting fruit. The demand for good fruit, particularly apples, is annually increasing, and it will probably not be long before it is abundantly supplied—especially as every quarter of the State appears to be well calculated for their growth. Heretofore the great bulk of the apples raised has been converted into cider, and on nearly all large farms the cider-press will be noticed as among their most prominent features. In Pennsylvania, among the Germans, particularly, there is a description of sauce called *apple butter*, and it is principally in the manufacture of this article that the cider and apples are consumed. In the rural districts apple butter is extensively used by every family—in fact, throughout the State, except in a few localities, its use is universal, and may be said to rank as one of the necessaries of the table. The cider is boiled in large kettles, holding from thirty to forty gallons, into which apples, properly paired and quartered, are thrown—say two bushels of prepared apples to twenty-five gallons of cider. After six to eight hours boiling, during which the liquor is constantly stirred, it begins to thicken, and when reduced to a tolerable paste, it is taken from the fire, deposited in earthen jars, and after standing a few weeks, is of good flavour for use. Boiling apple butter, in the counties of Lebanon, Berks, Lehigh, portions of Lancaster, and other German counties, is made the occasion of social celebration and interchange of neighborly courtesies. The young men and women of the neighborhood are invited to spend the evening, and it is here that, for the lack of better opportunities, and without expensive dress or ostentatious show, the substantial graces of the sex are exhibited.

Agriculture constitutes, by far, the most important interest in Pennsylvania, notwithstanding her immense beds of coal and iron, and extensive manufactures. Every other interest, however important, is merely subordinate to this, and it is a source of congratulation that such is the fact, not only to this State, but to the entire Union. Removed from the excitements, turmoils, and selfish intrigues of the city, the farmers are, upon the whole, purer in sentiment, more patriotic in feeling, and more industrious, honest, and straightforward in their course through life, than any other class of people. It is not to be disguised that, in some of the higher and nicer points of education, they are often lamentably deficient; but when we come to weigh their substantial virtues with the vices that usually accompany superficial intelligence, especially in populous places, the vast superiority of their condition, as Christian and virtuous citizens, is strikingly exhibited. Their retired and comparatively isolated position in the country, enables them to smother the spirit of pride and ostentatious show, which so often usurps good morals and supplants the better judgment of the town's people; and being thus rendered more simple-hearted and sober-minded, they are morally better men, and politically better citizens than any other class of people. The integrity of our agricultural population is to the political what the Alleghany Mountains are to the physical aspect of our glorious country—the *back-bone* of its prosperity. For while the one drains the country of its impurities, and pours forth its waters for the internal affairs of trade—purifying the atmosphere, and yielding metals which "subject all nature to our use and pleasure," the other regulates the political atmosphere, and saves it from the extremes into which excited and densely populated regions would be sure to embroil it.

In ancient times, the sacred plough employed
The kings and awful fathers of mankind;
And some, with whom compared your insect tribes
Are but the beings of a summer's day,
Have held the scale of empire, ruled the storm
Of mighty war; then, with unwearied hand,
Disdaining little delicacies, seized
The plough, and greatly independent lived.

It was the misfortune of republican governments, in more ancient times, that they had no agricultural population to rely upon—or, rather, that they lacked the powerful levers which we are now using so successfully to carry out our representative system, viz., the press and the post-office. In ancient Greece, for instance, where existed a complete democracy, the agricultural population was entirely proscribed for the want of these tremendous civilizers. To exercise the elective privilege the voter had to repair to the capitol—such a thing as voting at home was never dreamed of, because there were no means to enable the citizen to give an expression of his principles or to inform him of the nature of political affairs. The popular strength was therefore concentrated in the capitol, instead of being distributed, as it is here, over "our boundless continent." With these powerful instruments, the farmer in Oregon may exercise his political prerogatives with as much judgment and patriotism, as if he lived within a few miles of the capitol. The press is thus a conservator of intelligence, while the post-office is the distributer, and the two enable us to carry out the most admirable representative system which the world has ever known; and so nicely do all its details harmonize with the local position, feelings and principles of our people, that probably no other form of government, no matter how liberal, would promote our happiness and prosperity, or give one-fourth the strength and national grandeur which now belong to the people of free America:

Land of the forest and the rock,
  Of dark blue lake and mighty river,
Of mountain, rear'd on high to mock
The storm's career and lightning shock—
  My own green land forever!

In passing through this splendid agricultural region, the stranger will particularly observe the neatness and order which characterize the general aspect of the scene of farming operations, the good fences, the substantial and comfortable buildings, and especially the imposing appearance of the barn. Nearly every large farm has a cluster

GENERAL APPEARANCE OF A LANCASTER COUNTY FARM.

of buildings, the most promient of which is the barn, situated next to the mansion-house, around which are scattered wagon and carriage-sheds, corn-cribs, spring-house, wash-house, summer dining-house, etc.

with adjacent tenant house. The pride of a Pennsylvania farmer, however, is in his barn, and large sums of money are frequently expended in its erection. The structure is usually placed along side of a small hill, so that a four horse team may be driven into the barn floor without overcoming too steep a grade from the road, an arrangement equally desirable for other considerations. Barns are usually over one hundred feet in length, by about forty to sixty feet in depth —the loft and threshing-floor overarching by six or eight feet, the stables below forming a good shelter. Surprise is often expressed by strangers at the contrast generally presented between the appearance of the barn and the dwelling house—the former being comparatively more imposing than the latter. It is true the contrast often augurs unfavorably for the taste and personal convenience of the farmer; but there are circumstances governing the premises of the case which a due regard to economy will not allow him to overlook. Feeding a large number of cattle during the winter, as nearly every one does, he must provide accommodations of a corresponding character, ample in dimensions, and combining with neatness and durability of construction, spacious granaries, threshing-floors, hay-lofts, rooms for tools and implements, etc., besides stalls for six to a dozen head of horses.

The stock of horses in the eastern portion of Pennsylvania, and more particularly in Lancaster County, is worthy of remark. They are enormously powerful animals, bred entirely with a view to draught, and perfectly unfit for the saddle or light harness. Some thirty years ago, when racing was fashionable, the stock of horses embraced some splendid specimens of "blooded animals;" but as this amusement finally ran into gross licentiousness, the race-course was abandoned by respectable persons, and the quality of the stock, as far as swiftness is concerned, immediately deteriorated. The race-course near Lancaster, on the left side of the railway, between that place and Dillerville, was once the scene of some of the finest triumphs of the horse ever witnessed in this country. It was the pride and delight of many gentlemen of fortune, in those days, to enter the "stakes." The spirit of rivalry was carried to great lengths—and the horses themselves seemed

"To share with their masters the pleasure and the pride."

Fox-chasing, too, in the days of our "gran'-dads," was a favorite amusement, and many of those who declined to participate in the excitements of the race-course, warmly entered into the chase, for which swift hounds, as well as horses, were requisite. Fox-chasing is truly a splendid exercise, because in addition to its adventures and "hair-breadth escapes," it is made the occasion of social reunions amongst neighbors. The description of Thomson is no less applicable to old England than to young America at the time of which we are speaking:

——Give, then, ye Britons,
Your sportive fury, pitiless, to pour
Loose on the nightly robber of the fold!
Him, from his craggy winding haunts unearth'd,
Let all the thunder of the chase pursue.
Throw the broad ditch behind you; o'er the hedge
High-bound, resistless; nor the deep morass
Refuse, but through the shaking wilderness
Pick your nice way; into the perilous flood
Bear fearless, of the raging instinct full;
And as you ride the torrent, to the banks
Your triumph sound sonorous, running round,
From rock to rock, in circling echoes tost;
Then scale the mountains to their woody tops;
Rush down the dangerous steep; and o'er the lawn,
In fancy swallowing up the space between,
Pour all your speed into the rapid game.
For happy he, who tops the wheeling chase;
Has every maze evolv'd, and every guile
Disclos'd; who knows the merits of the pack;
Who saw the villain seiz'd, and dying hard,
Without complaint, though by a hundred mouths
Relentless torn: O glorious he, beyond
His daring peers! when the retreating horn
Calls them to ghostly halls of gray renown,
With woodland honours graced; the fox's fur,
Depending decent from the roof; and spread
Round the drear walls, with antic figures fierce,
The stag's large front: he then is loudest heard,
When the night staggers with severer toils,
With feats Thessalian Centaurs never knew,
And their repeated wonders shake the dome.
But first the fuel'd chimney blazes wide;
The tankards foam; and the strong table groans
Beneath the smoking sirloin, stretch'd immense

From side to side; in which, with desperate knife
They deep incision make, and talk the while
Of England's glory, ne'er to be defaced,
While hence they borrow vigor: or amain
Into the pasty plunged, at intervals,
If stomach keen can intervals allow,
Relating all the glories of the chase.
Then sated Hunger bids his brother Thirst
Produce the mighty bowl; the mighty bowl,
Swell'd high with fiery juice, steams liberal round,
A potent gale, delicious, as the breath
Of Maia to the love-sick shepherdess,
On violets diffus'd, while soft she hears
Her panting shepherd stealing to her arms.

The days of fox-chasing, in this quarter, have long since passed, and there are few yet living who shared the wild sport. Poor Reynard has escaped to wilder retreats in the mountains, from whose woody solitudes he can venture with more safety upon the lazy flocks ruminating in the green valleys below. Horse-racing, if not entirely abandoned, has degenerated into mere "scrub contests," in which "fast quarter-nags" run at a rate somewhat less than a mile in forty minutes and fifty-five seconds! Still the villagers of our agricultural districts enter into these little excitements with great satisfaction, and betting sometimes runs into large odds—

I'll bet my money on the bob-tail nag—
 Who will bet on the gray!—
Two-to-one on the Camptown brag—
 I'll take you, sir, on the bay!

As soon as the harvest is gathered, there is a succession of celebrations, in which the whole rural population take part—young and old of both sexes repair to the village, and interchange friendly greetings, join in the dance, show each other "delicate attentions," make presents, promises, explanations, and so forth. Horse-racing but adds to the general interest of the day, and there is nothing in which the "gay gallants" exhibit more pride than in the spirit and equipage of their steeds. Each farmer's son, when he becomes "of age," has his horse, raised under his own care, on the farm. Young and spirited like the rider, the animal is superfluously rigged with a bright shining yellow saddle, two flaming red streaked girths, a broad cropper and martingale, two broad reins, with a bridle of corresponding gaudiness, while below, to give a still more imposing aspect to the paraphernalia, is a broad leather halter. The whole is tastefully surmounted with a red or yellow tassel. Thus equipped, the young horse feels and is taught to feel that he has a duty to perform. It is his business to convey the impression that he has stamina—he must bite the bit, jerk up his head betimes, and paw furiously with his feet, and evince, by every possible jesture, the untamed wildness of his nature. He must learn to recognize his master's pride—the dear object of his hopes, his love, and fears—failing in which, he must needs feel the spur's prickly bite, and dance and prance gaily over the road! It is thus that horsemanship often opens the door to a maid's affections—she sees, she admires—then, alas! she loves!

We have stated that the horses in the districts of agriculture are exceedingly large and awkward—fit only for the plough and team. It is only in these districts, however, that such horses are mainly used. In other sections of the State the stock is better blooded, in consequence of being more used under the saddle. Oxen are employed to some extent for labor, particularly in the newly settled and rougher parts of the country—while mules are generally preferred at iron and colliery works, being very hardy and long-lived, as well as capable of drawing heavy loads in the team.

The stock of milch cows has been greatly improved by the importation of many noble animals from England, and so apparent is the advantage which has accrued, that it will probably not be many years before the whole of the present inferior stock will have disappeared from every good farm in the State. The sheep have been also much improved by crossing with the Spanish and English varieties. Wool-growing, however, forms but a very small branch of the agriculture of the eastern counties of Pennsylvania. In the county of Washington, and throughout the northwest and mountain regions of the Alleghanies, it is probably the most important feature of husbandry, and it is there where our flocks have been greatly improved in the quantity and quality of the fleece. The following statistics from the census bureau, exhibit the extent of the productive resources of Pennsylvania, in the year 1850—a year which, by the way, was by no means remarkable for general prosperity:

| | |
|---|---|
| Number of acres of improved land, . . . . | 8,619,631 |
| Value of farming implements and machinery, . | $14,931,993 |
| Value of live stock, . . . . . . | 42,156,711 |
| Quantity of wheat grown, in bushels, . . | 15,482,191 |
| " Indian corn, " . . . | 19,707,702 |
| " Tobacco, in pounds, . . . | 857,619 |
| " Wool, . . . . . . | 4,784,367 |
| " Wine manufactured, in gallons, . | 23,839 |
| " Butter, in pounds, . . . . | 40,554,741 |
| " Cheese, . . . . . | 2,395,279 |
| " Hay, in tons, . . . . | 1,826,265 |
| " Hemp, dew rotted, in tons, . . | 173 |
| " " water rotted, " . . . . | 686 |
| " Flaxseed, in bushels, . . . | 43,627 |
| " Maple sugar, in pounds, . . . | 2,218,644 |

The first settlers of Lancaster County, as also those of Berks, Lebanon, and portions of others adjacent, were Germans, many of them belonging to peculiar religious denominations, as the Mennonists, Seventh-day Baptists, &c. A large portion of the present population of the county still adheres to the religious tenets of their forefathers; and whatever may be thought of them in other respects, their mode of life is simple and without reproach. Virtuous, honest, and industrious, they constitute decidedly the most substantial and respectable class of the people. Certain of them wear long beards, and coats without buttons—the fabric invariably of their own plain manufacture. Seeing these venerable patriarchs disposing of their little business affairs in a simple, straightforward, and unaffected manner, prepossesses one very much in their favor, and challenges the highest respect for their religious sentiments. They endeavor to follow, as nearly as possible, the way pointed out, in their own view, by the Saviour; and the more effectually to do this, shut themselves out from the world, as far as circumstances will allow, and exclude every feeling of personal pride or vanity; kind, charitable and hospitable they practise in their own life the closest self-denial. We know no class of men who come nearer, in their *works and actions*, to the standard of pure Christian principle; they are good to a fault, and virtuous in the broadest sense of the word. They have no salaried preachers, but every man does what he can to illustrate the purity of religion, and to scatter broadcast the blessings of its holy teachings. In their private life, the lines of young Pope are probably as applicable to them as any other people on the globe:

Happy the man whose wish and care
A few paternal acres bound,
Content to breathe his native air,
In his own ground.

Whose herds with milk, whose fields with bread,
Whose flocks supply him with attire,—
Whose trees in summer yield him shade,
In winter, fire.

Blest who can unconcern'dly find
Hours, days, and years glide soft away,
In health of body, peace of mind,
Quiet by day.

Sound sleep by night; study and ease
Together mixed; sweet recreation;
And innocence, which most doth please,
With meditation.

Thus let me live, unseen, unknown,
Thus unmolested let me die;
Steal from the world, and not a stone
Tell where I lie!

The Seventh-day Baptists, who are seceders from the *Dunkers*, established themselves at Ephrata, in 1730. This is a little village lying about seven miles north-east of Lancaster. Their life, at first solitary, was soon changed to a conventual one, and a monastic society was established in 1733. Monastic names and habits were assumed, and the cloister soon numbered upwards of one hundred persons, of both sexes. The number of outside members was about two hundred or more. The society was prosperous and increased, and it soon became necessary to erect larger buildings. These were erected sometime about 1740, and consisted of a sister's house, with a chapel attached, and a brother's house, containing a large meeting-room, with galleries, in which the whole society assembled for public worship. These buildings, which are still standing, and exhibited in the following engraving, were surrounded with numerous others, of small dimensions, and included a school-house, printing-office, paper-mill, &c. The buildings are singular, and of ancient style of architecture, all the outside walls being covered with shingles.

BROTHERS' AND SISTERS' HOUSES AT EPHRATA.

The two houses for the brethren and sisters are very large, being three and four stories high: each has a chapel for their night meetings, and the main buildings are divided into small apartments, (each containing between fifty and sixty,) so that six dormitories, which are barely large enough to contain a cot, (in early days a bench and billet of wood for the head,) a closet, and an hour-glass, surround a common room, in which each subdivision pursued their respective avocations. On entering these silent cells, and traversing the long narrow passages, visitors can scarcely divest themselves of the feeling of walking the tortuous windings of some old castle, and breathing in the hidden recesses of romance. The ceilings have an elevation of but seven feet; the passages leading to the cells, or "kammers," as they are styled, and through the different parts of both convents, are barely wide enough to admit one person, for when meeting a second, one has always to retreat;—the dens of the kammers are but five feet high, and twenty inches wide, and the window, for each has but one, is only eighteen by twenty-four inches; the largest windows, affording light to the meeting rooms, are but thirty-four inches. The walls of all the rooms, including the meeting-room, the chapels, the saals, and even the kammers, or dormitories, are hung and nearly covered with large sheets of elegant penmanship, or ink-paintings, many of which are texts from the Scriptures, done in a very handsome manner, in ornamented Gothic letters, called in the German, Fractur-schrifften. They are done on large sheets of paper, manufactured for the purpose at their own mill, some of which are put into frames, and which admonish the resident, as well as the casual visitor, which ever way they may turn the head. There are some very curious ones: two of which still remain in the chapel attached to Saron. One represents the narrow and crooked way, done on a sheet of about three feet square, which it would be difficult to describe—it is very curious and ingenious: the whole of the road is filled up with texts of Scripture, advertising the disciples of their duties, and the obligations their profession imposes upon them. Another represents the three heavens. In the first, Christ, the Shepherd, is represented gathering his flock together; in the second, which occupies one foot in height, and is three feet wide, three hundred figures, in the Capuchin dress, can be counted, with harps in their hands, and the heads of an innumerable host; and in the third is seen the throne, surrounded by two hundred archangels. Many of these Fractur-schrifftens express their own enthusiastic sentiments on the subject of celibacy, and the virtue of a recluse life, while others are devotional pieces.

The society owned a farm, a grist-mill, paper-mill, oil-mill, and fulling-mill. All the society's property was in common, and the labor of the members; but individual members were not compelled to relinquish private property which they might have held previous to joining the society.

They receive the Bible as the only rule of faith and code of laws for church government. No monastic vows were taken, nor had they any written covenant. They believe in the divinity of Christ and the Trinity of the Godhead, and rely on the merits and atonement of the Saviour, and that he died for all who call upon his name and offer fruits of repentance. They contend for the observance of the original Sabbath. They hold on to the apostolic baptism, and administer *trine immersion* with the laying on of hands and prayer, while the recipient is kneeling in the water. They celebrate the Lord's Supper at night, in imitation of our Saviour, washing at the same time each other's feet, agreeably to his command and example. Celibacy they consider a virtue, but never require it, nor do they take any vows in reference to it. They never prohibited lawful intercourse, but when two concluded to be joined in wedlock, they were aided by the society. Celibacy, however, was always urged as more conducive to a holy life. They do not approve of paying their ministers a salary, thinking the gospel was sent without money or price—but share their own supplies with them.

It is not one of their customs to wear long beards, as is frequently said of them: this is more the case with the Dunkers and Mennonists. They are often represented as living on vegetables,—the rules of the society forbidding meats, for the purpose of mortifying the natural appetite,—and also as lying on wooden benches, with billets of wood for pillows, as an act of penance. The true reason and explanation of this matter is, that both were done from considerations of economy. Their circumstances were very restricted, and their undertaking great. They studied the strictest simplicity and economy in all their arrangements: wooden flagons, wooden goblets, and turned wooden trays, were used in administering the communion; and the same goblets are still in use, though they have been presented with more costly ones. Even the plates off of which they ate were octangular pieces of thin poplar boards; their forks and candlesticks were of wood, and also every other article that could be made of that material was used by the whole community. After they were relieved from the pressure of their expensive enterprise in providing such extensive accommodations, they enjoyed the cot for repose, and many other of the good things of life; though temperance in eating and drinking was scrupulously regarded.

Although opposed to bearing arms, they opened their houses cheerfully to succor and comfort the distressed inhabitants of Paxton and Tulpehocken during the old French war—for which the government rendered them its acknowledgments, and Gov. Penn offered them a whole manor of land, but they would not receive it. During the revolution they were decided whigs. After the battle of Brandywine the whole establishment was open to receive the wounded Americans; their Sabbath-school house was converted into an hospital; great numbers of the sick were transported here in wagons; the camp fever broke out among them, and one hundred and fifty were buried on the top of Mount Zion. Here their remains reposed unnoticed—unhonored and unsung—until a few years ago, when a subscription was set on foot, through the instrumentality of Mr. Konigmacher, and a monument to their memory commenced on the 4th of July, 1843. The occasion was celebrated with much spirit. Thousands of people from all parts of the surrounding country were present, and participated in the ceremonies attending the laying of the cornerstone. Hon. Joseph R. Chandler, of Philadelphia, delivered an address,

MONUMENT AT EPHRATA.

detailing and commenting upon the historical incidents involved. The monument is still in an unfinished state, but hopes are entertained that sufficient funds will soon be accumulated to accomplish the object intended.

EPHRATA MOUNTAIN SPRINGS.

The Ephrata Mountain Springs, kept by Joseph Konigmacher, Esq., are delightfully situated on the Ephrata Ridge. Here the South Mountain is the dividing ridge, the waters on the south side run into the Chesapeake, and in the north to the Delaware. The water is pure, soft sandstone and slate; the temperature of the different springs is, some very cold, others more moderate, but for drinking or bathing, and its restorative qualities to health for diseased or frail constitutions, it is but seldom excelled. As a delightful summer resort it is extensively patronized; the superb variety of landscape scenery, and the lovely drives affording a pleasant succession of novel and beautiful views.

Passing two or three unimportant stations, (including that of KINZER'S, where there is a branch railway to STRASBURG, three miles distant) we reach the city of LANCASTER. This city has recently made much progress in improvement, and its population has been correspondingly increased. It is now nearly fourteen thousand, whereas, only a few years ago, it was but eight thousand. It is an old town, having been laid out in 1730 by Andrew Hamilton, proprietor of the land, and at that time one of the most influential men connected with the government. For the purpose of attracting population, the proprietor sold the lots at low and accommodating rates, subject to a small annual ground-rent. This had the effect of drawing together a large number of inhabitants, principally poor mechanics; and the town was composed almost entirely of small one-storied houses. The citizens, by their industry and frugality, gradually became the owners of the humble tenements they occupied, subject to the ground-rent mentioned. There were no large manufacturing establishments, but a large number of small ones, conducted solely on individual account. There were very few families of fortune, but these were unusually rich, and commanded an influence of corresponding extent. As one-storied houses always *paid better*, in view of the class of people to be accommodated, those who had money to invest in *improvements*, generally expended it in the erection of such buildings.

The original aspect of the town is still retained, to some extent; though there are now in full operation several of the largest cotton and iron establishments in the State. The erection of these magnificent industrial establishments has given an impulse and tone to the business of the place which it never knew before. Like many other county-seats, Lancaster long labored under the paralysing influences of a superfluous professional population—a population which, whatever its social merit, does nothing but *consume*, without contributing to the real productions or substantial wealth of a community. The place is still literally overrun with professional men—including a horde of smiling, friendly politicians, awaiting their "turn" for the suffrages of the "free and independent electors" of the "old guard." There is no county in the State—there is probably none in the Union—where more interest is manifested in political affairs; at the same time, it must be observed, there is none which has more offices to bestow! Lancaster has produced some of the most skilful practitioners in the

LANCASTER.

political arena. Indeed, any one who has graduated in its schools, may safely venture forth, relying on his "tactics." The learned professions, too, embrace some of the brightest ornaments in the country, while the population, as a whole, is intelligent and enterprising—though with the elements surrounding them, hardly as enterprising and public spirited as might reasonably be expected. Some of the citizens are very rich, and, as recent experiments have proved, could safely invest their capital in objects conceived in the spirit of taste and liberality. Lancaster, with half the talent and energy wasted in her political struggles, might readily become one of the principal workshops of Pennsylvania. Convenient to the anthracite coal beds, situated in a most magnificent agricultural region, with inexhaustible deposits of valuable minerals—as iron ore, copper, chrome, lead, limestone, &c.—these advantages added to her close proximity to the seaboard, and a superabundance of idle capital, where is the obstacle between her and future greatness and prosperity?

The Conestoga, a beautiful winding stream, passes along the south ern outskirts of the city, and empties into the Susquehanna at Safe Harbor, ten miles distant. The Conestoga is rendered navigable for boats of the largest class, by a series of dams and locks. The amount of business done is small; but it affords a splendid water-power for mills, which are plentifully distributed along its banks, as well as the streams emptying into it. Lancaster has a larger number of flouring mills than any other county of equal extent in the Union—the whole number exceeding three hundred, exclusive of clover, saw, and other mills. The Conestoga is connected at Safe Harbor with the Tide Water Canal, situate on the opposite side of the Susquehanna. The boats are towed across the river by a steamboat. It is at this place that the splendid rolling mills and furnaces of Messrs. Reeves, Abbott & Co., are located. These works were erected a few years ago at a probable cost of two hundred thousand dollars. Large beds of iron ore surround them, and the site is in all respects admirable for an establishment of this character. Most of the iron used on the Central Railroad was manufactured at this place, and we venture nothing in saying, that in all the essentials constituting a perfect rail, a more complete one never was laid down in the United States. Having passed over a large portion of the road on foot, we had an opportunity of observing its qualities. At some places, where landslides had occurred, the superstructure of the road was entirely removed out of its place, and the rails bent nearly double; but in no instance had a fracture occurred. Nor is the metal too soft or yielding; for notwithstanding the heavy pressure to which a portion of the road has been exposed, for several years past, we did not meet a single instance of splitting or splintery fracture, so common along the edges of rails on other roads. This is particularly the case on the New York and Erie Railroad, and is one reason why the travel over it is rendered so dangerous. At some places the metal is too brittle, and will not withstand the bouncing momentum occasioned by short curvatures. The rail, therefore, breaks in the most *dangerous places*, and this is an evil which can only be overcome by the substitution of a new and better quality of rail. The Erie Railroad Company, with the laudable desire of remedying this defect on other portions of their road, ordered rails of American manufacture—the other, of course, having been so-called cheap English iron. But in this case they were unfortunate in getting a rail *entirely too soft*, so that it is full of splintery fractures, the edges flattened down, and deep depressions at the place of jointure. This portion of the road, therefore, is just as dangerous as the other, and so palpable has this fact become, that a New York paper, some time ago, speaking of the case of a criminal sentenced to be executed, advised the Governor to commute his sentence to transportation over the Erie Railroad, so as to give the *poor fellow one chance in a hundred for his life!*

In the immediate vicinity of Safe Harbor is a place called Indiantown. It is the site of an old Indian village, formerly occupied by the *Conestagos*, and some of the most interesting conferences between them and the whites were held there. Many relics of the Indians have been picked up by the inhabitants, and several rocks, along the Susquehanna, bear their quaint markings. William Penn paid them a visit on one occasion, and was received with marked respect. As the whites gathered around them, the Indians gradually fell within the pale of civilization, and for many years those that remained pursued the peaceful ways of agriculture. The surrounding country is very rich, being an extensive limestone formation, which, towards the north, gradually sinks under the Turkey Hills. These hills, along the Susquehanna, rise up in immense cliffs, giving to the scenery an aspect of great wildness and sublimity.

Safe Harbor has always been a great fishing-place, but has been rendered especially so, of late years, owing to the construction of the

SAFE HARBOR IRON WORKS.

SAFE HARBOR IRON WORKS.

dam in the Susquehanna, to render it navigable for the steam-tow boats. The dam appears to arrest the progress of the fish, in their upward course, and the fisheries below are rendered correspondingly successful. Immense quantities of shad are caught with the seine, the process of which is indicated in the above engraving. The seine is generally one hundred yards in length, and from four to six feet in width, varying according to the depth of the water. One edge of the seine is heavily loaded with lead, while the other has numerous corks

SHAD FISHING IN THE SUSQUEHANNA.

or wooden buoys, so that it floats in the water in an upright position. The seine is carried two or three hundred yards above the fishing battery, when one end is taken in a boat, which is rowed out from the shore in a circular course, and hauled in at the battery, the other end in the meantime, arriving at the same place. The process of hauling in the seine is represented in the engraving. Shad generally run in schools; and in clear weather, with the aid of a spy glass, their approach may be descried from the battery. The hauls are, therefore, irregular; but when a school is effectually surrounded with the seine, an immense number is sure to be captured, as few are able to escape its delicate net-work. Shad fishing is somewhat laborious, and is followed entirely with a view to profit. It is, however, at times, very exciting, and the fishermen have a merry time of it. Before the erection of the dams in the Susquehanna, shad-fishing was a regular pursuit for many persons all along the Susquehanna, and some fisheries were among the most productive properties in the State, their annual profits yielding, in a course of years, large fortunes.

The Conestoga, while it is the only small stream in the State upon which steamboats run, is also the first upon which experiments for steam navigation were made. William Henry, of Lancaster, as early as 1760, instituted a series of experiments on this subject, and it is said partially succeeded, but in the midst of them his boat was accidentally sunk in the river, and he himself shortly after died. But while his efforts thus failed, it remained for another son of Lancaster to complete what had only been commenced. About this period Robert Fulton was born, in the township bearing his name. His parents shortly after removed to Lancaster, where Fulton received a good education. He subsequently went to London, and having early evinced a partiality for painting, placed himself under the charge of Sir Benjamin West, the celebrated American artist, then and for a long time afterwards, a resident of that city. He here became acquainted with the Duke of Bridgewater, and other distinguished scientific and practical men, and entered warmly into some of their projects, in reference to canals and internal navigations. He soon after obtained patents for an *inclined-plane* for transportation, and certain instruments for excavating canals. Removing afterwards to France, he made himself master of the French, Italian, and German languages, and formed a lasting friendship in Paris with the celebrated Joel Barlow, in whose family he resided. In the meantime he acquired a knowledge of the higher range of mathematics, chemistry, and physical philosophy, the result of which was several important improvements in the mechanic arts and submarine navigation, for two of which he obtained patents. He performed many experiments in the harbor of Brest, with his plunging-boats and torpedos, demonstrating the practicability of employing subaquatic explosion and navigation for the destruction of vessels. This invention, remarks Mr. Day, attracted the attention of the British government, and overtures were made to him by the ministry which induced him to go to London, with the hope that they would avail themselves of his machines; but a demonstration of their efficacy which he gave the ministry, by blowing up a vessel in their presence, led them to wish to suppress the invention rather than encourage it; and accordingly they declined patronizing him. During this period he also made many efforts to discover a method of successfully using the steam-engine for the propelling of boats, and as early as 1793 made such experiments as inspired him with great confidence in its practicability. Robert R. Livingston, Esq., chancellor of the state of New York, and minister of the United States to the French court, on his arrival in France, induced him to renew his attention to this subject, and embarked with him in making experiments for the purpose of satisfying themselves of the possibility of employing steam in navigation. Mr. Fulton engaged with intense interest in the trial, and, in 1803, constructed a boat on the river Seine, at their joint expense, by which he fully evinced the practicability of propelling boats by that agent. He immediately resolved to enrich his country with this invaluable discovery; and on returning to New York in 1806, commenced, in conjunction with Mr. Livingston, the construction of the first Fulton boat, which was launched in the spring of 1807, from the ship-yard of Charles Brown, New York, and completed in August. It was one hundred feet long, twelve feet wide, and seven feet deep. In 1808 it was lengthened to one hundred and fifty feet, and widened to eighteen feet. This boat, which was called the Clermont, (from the seat of the Livingston family,) demonstrated on the first experiment, to a host of at first incredulous but at length astonished spectators, the correctness of his expectations, and the value of his invention. Between this period and his death he superintended the erection of fourteen other steam-vessels, and made great improvements in their construction.

THE CLERMONT.

"As I had occasion to pass daily to and from the building-yard," said Fulton, "while my boat was in progress, I have often loitered unknown near the idle groups of strangers gathering in little circles, and heard various inquiries as to the object of this new vehicle. The language was uniformly that of scorn, sneer, or ridicule. The loud laugh rose at my expense, the dry jest, the wise calculation of losses and expenditures, the dull but endless repetition of the "Fulton folly." Never did a single encouraging remark, a bright hope, or a warm wish, cross my path. Silence itself was but politeness veiling its doubts or hiding its reproaches. At length the day arrived when the experiment was to go into operation. *To me it was a most trying and interesting occasion.* I invited many friends to go on board to witness the first successful trip. Many of them did me the favor to attend as a matter of personal respect; but it was manifest they did it with reluctance, fearing to be partners of my mortification and not of my triumph. I was well aware that in my case there were many reasons to doubt of my own success. The machinery, (like Fitch's before me,) was new and ill made; and many parts of it were constructed by mechanics unacquainted with such work, and unexpected difficulties might reasonably be presumed to present themselves from other causes. The moment arrived in which the word was to be given for the vessel to move. My friends were in groups on the deck. There was anxiety mixed with fear among them. They were silent, sad, and weary. I read in their looks nothing but disaster, and almost repented of my efforts. The signal was given, and the boat moved on a short distance and then stopped, and became immovable. To the silence of the preceding moment now succeeded murmurs of discontent, and agitations, and whispers, and shrugs. I could hear distinctly repeated, '*I told you it was so; it is a foolish scheme; I wish we were well out of it.*' I elevated myself upon a platform, and addressed the assembly. I stated that I knew not what was the matter; but if they would be quiet, and indulge me for half an hour, I would either go on or abandon the voyage for that time. This short respite was conceded without objection. I went below and examined the machinery, and discovered that the cause was a slight maladjustment of some of the work. In a short period it was obviated. The boat was again put in motion. She continued to move on. All were still incredulous. None seemed willing to trust the evidence of their own senses. We left the fair city of New York; we passed through the romantic and ever-varying scenery of the Highlands; we descried the clustering houses of Albany; we reached its shores; and then, even then, when all seemed achieved, I was the victim of disappointment. Imagination superseded the influence of fact. It was then doubted if it could be done again; or if done, it was doubted if it could be made of any great value."

Fulton obtained a patent for his inventions in navigation by steam in February, 1809, and another for some improvements, in 1811. In the latter year he was appointed, by the Legislature of New York, one of the commissioners to explore a route for a canal from the great lakes to the Hudson, and engaged with zeal in the promotion of that great work. On the commencement of hostilities between the United States and Great Britain, in 1812, he renewed his attention to submarine warfare, and contrived a method of discharging guns under water, for which he obtained a patent. In 1814 he contrived an armed steamship for the defence of the harbor of New York, and also a submarine vessel, or plunging boat, of such dimensions as to carry one hundred men, the plans of which being approved by government, he was authorized to construct them at the public expense. But before completing either of those works he died suddenly, February 24th, 1815. His person was tall, slender, and well formed; his manners graceful and dignified, and his disposition generous. His attainments and inventions bespeak the high superiority of his talents. He was an accomplished painter, was profoundly versed in mechanics, and possessed an inventive faculty of great fertility, which was always directed by an eminent share of good sense. His style as a writer was perspicuous and energetic. To him is to be ascribed the honor of inventing a method of successfully employing the steam-engine in navigation,—an invention justly considered one of the most important which has been made in modern times, and by which he rendered himself both a perpetual and one of the greatest benefactors of mankind. Some of Fulton's relatives still reside near the spot where he was born; which is further interesting from the fact that the parents of the late Mr. Calhoun, of South Carolina, for a long time lived there; some persons allege that it was the place of his birth.

Lancaster is known, also, as the residence of the Hon. James Buchanan, who removed here from Franklin County upwards of forty years ago. Mr. Buchanan lives in a plain, but substantial brick house about one mile from Lancaster. It was lately the residence of Hon. W. M. Meredith, Secretary of the Treasury during General Taylor's administration. The estate, which is a delightful one, is surrounded by fine shade trees. It is called *Wheatland*, from the splendid agricultural district adjacent. The Hon. Thaddeus Stevens, one of the most brilliant lawyers in Pennsylvania, and at present the representative of this county in Congress, has also resided in Lancaster for many years past. He is a native of Caledonia County. Vermont. Personally, Mr. Stevens is one of the best men living;—politically, he is known for his extreme measures, which, were they less sectional in character, would render his position in Congress much more popular and commanding than it is under present circumstances. It is his splendid personal and intellectual qualities which sustain him; in the absence of these, with his strong sectional views, he probably never would have been heard at all in the councils of the nation.

Thomas M'Elrath, Esq., of the firm of Greely & M'Elrath, proprietors of the New York *Tribune*, lives in a handsome retired mansion, one mile north of Lancaster. Hon. Ellis Lewis, the eminent jurist, and one of the Supreme Judges of Pennsylvania, also lives at this place. He is regarded by many as one of the most learned exponents of the law in this country. His occasional contributions to our periodical literature have aided this reputation; these, although thrown off in his idle moments, may be regarded as the *poetic cream* accumulating upon the surface of a fine intellect, and which the more substantial elements of the law refuse to skim off. Judge Lewis has a daughter—the wife of James H. Campbell, Esq., of Pottsville—who also enjoys a conspicuous position in the literary circles of our country. Gliding, as she does, over the clear, transparent water of poesy, Mrs. Campbell has a mind and heart to realize all its varied beauties, and sings them in the purest and sweetest strains.

One mile northwest of Lancaster, the State Railway intersects that of the Lancaster and Harrisburg line, thirty-six miles in length. The former extends to Columbia, ten miles distant. Columbia is one of the most active and flourishing places in the State, delightfully situated on the Susquehanna River. It is in the midst of public improvements, radiating in every direction. The Pennsylvania canal commences here, following the valley of the Susquehanna to Northumberland, where it branches into two divisions, following respectively the north and west branches of that river. Below Northumberland, at Duncan's Island, another division, and by far the most important one, crosses the Susquehanna, and follows the Juniata to Hollidaysburg, situated on the eastern slope of the Alleghany Mountains. The Tide Water Canal meets the Pennsylvania Canal at Columbia, and follows the course of the river to Havre-de-Grace, in Maryland, thirty-six miles. The Baltimore Railroad extends to

York, where it unites with the main road from Harrisburg to Baltimore.

The principal trade of Columbia is in the descending lumber of the Susquehanna, in which a large amount of capital is invested. Some business is done, too, in the coal trade; but it is comparatively unimportant, confined to the more bituminous qualities for domestic purposes. One of the richest deposits of iron ore in the United States is situated within a few miles of the place, which is also surrounded with numerous furnaces and machine-shops, flour and grist mills, &c. The longest and most substantial bridge in the State, and probably in the Union, stretches across the Susquehanna. Its length is over one

VIEW OF THE SUSQUEHANNA ABOVE COLUMBIA.

mile and a quarter, and is adapted both for railway cars and vehicles, as well as for towing boats across the river. A short distance above Columbia, a bold and extensive ridge of white sandstone emerges from the surrounding formation, which, at the Susquehanna, towers several hundred feet in rugged, perpendicular cliffs, entirely overlooking the banks of the river below. This description of scenery, however, is peculiar to the Susquehanna, and is even wilder some ten miles below. About two miles above Columbia is the residence of Prof. Haldeman,

6

one of the most eminent of American Mineralogists and philosophers His residence is eminently worthy a gentleman of fortune and cultivated taste—being, probably, the most stately edifice in this part of the country, while its situation is altogether unsurpassed for bold, romantic profile, and delightful prospect. The view on page 61 is afforded from his spring-house, a short distance from the dwelling. The village of MARIETTA, one mile distant, is situated on the banks of the river.

Eleven miles from Lancaster is the village of MOUNT JOY, pleasantly situated in the heart of a beautiful agricultural region. Cedar Hill Seminary, near this place, is a well-known school for young ladies. The place is otherwise without general interest. ELIZABETHTOWN, eight miles further, is a village of some six hundred population. It is a short distance from the railroad, on the turnpike between Philadelphia and Pittsburg. Nearly all the villages along this once crowded thoroughfare have lost their former interest and prosperity, since the diversion of its trade to the lines of railroad and canal. After leaving this place, we enter the range of Conewago hills, some six miles wide, one of which is tunnelled. A splendid bridge crosses the stream a short distance beyond the tunnel, which

RAILROAD BRIDGE OVER THE CONEWAGO.

is nearly one hundred feet in height. This structure has just been completed, and is one of the most substantial of the kind in the State. The entire railroad, indeed, has recently been very much improved, and re-laid with a strong rail. The country between Elizabethtown and Middletown, nine miles, is strewn with huge blocks of trap rock, which constitute the characteristic feature of the Conewago range. Some of these blocks indicate, in their rounded structure, a deposition by drift, though it is more probable that they have been detached from their beds by the slow but powerful erosive agency of floods and rains, which, carrying off the *smaller detritus* associated with them, has left them thus isolated and exposed. These rocks are amongst the hardest which the varied state of the earth afford, and it is both curious and interesting to see them thus scattered over the surface of a recent formation, when they themselves belong to the earliest epochs of the earth's eventful history. The Susquehanna, no doubt, long battled with this range of hills before it was able to secure its final passage. Even now its waters are tossed into tremendous bounding waves, as they roll over the descending steps of its rocky bottom, while the channel is surrounded with bold jutting rocks, which form a great impediment to its navigation. The descending rafts of the Susquehanna pass through these falls with the rapidity of a railway train. The fall of the river, within a distance of little more than a mile, is probably not less than seventy feet. As the raft descends, it plunges, creaks, and bends in every direction—the high waves rolling and splashing frightfully—rendering the adventure at once exciting, novel, and perilous. We made the descent, several years ago, and can never forget the peculiar agitation of our nervous system, as our "long, low" craft made the first plunge into the troubled waters! Gracefully sank down the front platform, and furiously swept the eager water over our thirsty boards! That plunge over, another and another followed in quick succession. Looking round, we were quite bewildered with the real wildness and magnitude of the scene—the white-capped waves sweeping by with tremendous force, and dashing their empty furies against the sturdy rocks, which the men at the oars, with the most *desperate efforts,* were trying to avoid. They succeeded, and glancing back, as we were rapidly emerging from the scene of danger, a thrill of inexpressible delight burst up from our half-smothered "inner man." We shall never forget our passage through the Conewago Falls; it was a *pleasant trip!*

MIDDLETOWN, situated on the old turnpike, equi-distant from Lancaster and Carlisle, originally derived its name from that circumstance. It was "laid out" in 1775, having previously been the site of an Indian village. The Swatara here empties into the Susquehanna, and the Union Canal, elsewhere alluded to as the first improvement of the kind ever projected in the United States, unites with that of the Pennsylvania Canal. The section between this place and Pine Grove, the place of shipment of the coal passing over the work, has lately been enlarged so as to carry tonnage equal to the boats of the Pennsylvania Canals. The other section, from that place to Reading, where the canal unites with the Schuylkill navigation, is capable of floating boats of some thirty tons only. This section will also, no doubt, be enlarged, so as to make the tonnage uniform throughout the line. Middletown, as may be inferred from the number of rafts and piles of lumber strewn along the mouth of the Swatara, has an extensive trade in this business, as well as in coal. The place is a large

VIEW OF MIDDLETOWN.

and flourishing one—the citizens generally intelligent and enterprising. It is the residence of Gen. Simon Cameron, a distinguished statesman and financier, who has always commanded a large amount of influence in the political movements of the day. Gen. C. has served his native State with zeal and ability in the Senate of the United States, and has always, indeed, evinced a patriotic devotion to its interests.

Middletown covers a large area, and that part of it which is properly so called, is exhibited in the engraving. The village at the mouth of the Swatara is locally called Portsmouth being a mere offshoot of the larger and older town.

The first view we obtain of the Susquehanna is near Middletown, where it flows smoothly and tranquilly along its pebbled banks. It is all along strewn with pleasant islands, most of them overgrown with trees and vines, while some of the larger ones are cultivated and inhabited by the farmer. Indeed, there are some five or six of the most productive farms in Pennsylvania situated in this river, two or three of which have upwards of two hundred acres each. Most of the smaller islands are well adapted to the culture of tobacco, which is raised to a large extent. Had the same amount of money appropriated to the construction of the canals traversing the valley of this river been directly applied to the improvement of its navigation, there can be little doubt but that its trade would have been greatly increased. The Susquehanna drains over thirteen millions of acres of land, much of which is coal and valuable mineral land of different varieties. When this immense region is fully occupied and worked, our present canal and railway system will prove entirely inadequate to carry off its products, and we should not be surprised if not many years hence its improvement for steamboat navigation was seriously undertaken. And why not? There is really no serious impediment—the fall, upon the whole, is not too great to be overcome without dams or locks. Three or four millions of dollars, judiciously expended, would render it navigable, beyond doubt, for steamboats of the largest class. The river transports an immense, an incalculable amount of debris, all of which might be arrested in its progress, and deposited in such a manner as to form long narrow islands, thereby preventing its waters from spreading over too wide a surface, and rendering the channel much deeper, which, at its lowest stages, is seldom less than five feet. The falls, of which there are but five or six difficult ones, might readily be overcome without dams or locks. Their irregular rocky steeps could be blasted, or the force of the descent impaired by carrying the channel in a round-about direction.

The Susquehanna, between the Conewago hills and the Kittatinny Mountain, is crossed by the celebrated Cumberland Valley, sometimes called the Kittatinny Valley. This broad and fertile valley enters the State in Northampton County, on the Delaware River, and pursuing a nearly western course, crosses the Susquehanna, where, diverging in a more southerly course, it afterwards enters the State of Maryland—its entire length, in this State, being about sixty miles. The counties of Northampton, Lehigh, Berks, Dauphin, Cumberland and Franklin, are embraced within its range. It has numerous extensive layers of limestone, which, of course, is the principal source of its great fertility, while the soil is still much improved by the descending vegetable matter of the surrounding mountains. The valley, throughout its entire length, is well watered, and inhabited by an industrious thriving population. This valley, along the Susquehanna, has at some points a broad alluvial margin, which, however, is no less prolific for agricultural purposes. This may be observed by the traveller as he passes along in the cars—the numerous banks of sand and pebbles, embracing every variety of river stone, indicating the former flow or projecting arms of the river over them.

HARRISBURG, the capital of the State, is one hundred and seven miles north-west of Philadelphia, and two hundred and fifty-six miles south of Pittsburg. It lies three hundred and ten feet above tide-water. This is one of the handsomest towns in the United States, and it is deficient in nothing calculated to render it so. It lies on a broad alluvial flat, some twenty feet above the flow of the river, which gracefully winds around its western shores. The river is here again split with several beautiful islands, overgrown with a profusion of stately trees and rich wild foliage, which are reflected upon its clear unruffled surface. Two splendid bridges hang over it—one to accommodate the passage of horses and vehicles, the other for the Cumberland Valley Railroad, running to Chambersburg, in Franklin County. Both these bridges are over a mile in length, and are unsurpassed for their strength and architectural proportions. Some eight miles north the bold outlines of the Kittatinny Mountain are seen, traversing this portion of the State in an east and west direction—leaving abrupt, wild, and lofty coves along the Susquehanna. Back of the Kittatinny, further north, a succession of other mountains occurs—being the projecting spurs of the great anthracite coal formation, which, on approaching the Susquehanna, divides into two branches, one called the Lykens Valley, the other the Dauphin coal districts. The scenery around this place has, therefore, all the softness of a splendid agricultural valley, teeming with spirited little villages and imposing farm-buildings, agreeably contrasted with the soft green aspect of bold and

HARRISBURG.

lofty mountain ranges, through which the river tamely and serenely winds its peaceful way. It was wisely selected for the seat of government of this great commonwealth. The borough contains many beautiful public and private buildings—the latter, especially, far superior to those of most other towns in this State. The population is now upwards of eight thousand, having materially increased during the last few years. It contains several large iron, cotton and other manufactories, which furnish employment for a large number of operatives, of both sexes. The people are very intelligent, and the tone of the society is probably unsurpassed. The Legislature, at its annual session, attracts many of the better class of our citizens to the place, whose sojourn during the social excitements of the winter doubtless contributes much to the polish of the people. The Legislature, itself, we regret to say, is often composed of indifferent material. In selecting representatives, the question with the people too often is, "*who will accept?*" not "*who shall we send?*" Practical and sensible people —men of education, talents, and experience, all avoid it. The compensation is less than they are accustomed to receive—and the honor is not sufficiently great to tempt them from their regular pursuits. It is thus a difficult matter to get really good men into our Legislative sessions. There is no inducement for it; while there is seldom any question or principle of sufficient magnitude to call them forth on grounds of personal patriotism. Some years ago, at the outset of our splendid internal improvement system, and while the banking and educational policy of the commonwealth was still unsettled and incomplete, it was far otherwise. At that period some of the brightest intellects of the State illumined its Legislative Councils; men eminent for their private worth, their public spirit, and their comprehensive talents, constituted the representatives of the people. It was then regarded as an honor to mingle in its deliberations, because they were characterized with dignity, masterly eloquence, and practical wisdom.

Soon after the policy of the State in reference to internal improvements, banks, education, &c., had been marked out, a striking change occurred. The extraordinary interest awakened in these subjects, engulphed the State into a sea of political agitation, which has never had a parallel in its history. The appropriation of millions upon millions of dollars, for the prosecution of the State works, called forth thousands of political adventurers, who, like a pack of ravenous wolves, pounced upon the tempting feast, their mercenary appetites leaving nothing but the skeleton of their hospitable victim. Like leeches, their blood-thirsty appetites became sharper with the increasing weakness of the prey; and they gnawed, like a poisonous mineral, at its interior vitality, until the fretful clouds of bankruptcy hung over the feeble old Commonwealth, and ghastly *Repudiation* was about to lay its black seal upon her fallen credit! This was a gloomy—gloomy time! Nor have we yet altogether recovered from its paralysing effects. The secret of this unfortunate state of affairs—or, rather the *reason*, for it never was a secret—was simply this: The original intention of the friends of an internal improvement system contemplated but one main thoroughfare throughout the State, with one or two radiating branches. Setting out vigorously with the prosecution of this laudable enterprise, the work was shortly overwhelmed with difficulties entirely foreign to its own merits. While yet unfinished, a large number of branches were proposed, to construct which millions of dollars were unblushingly asked for. The friends of these local branches took common ground—they would vote for no more appropriations for the *main line*, without their proposed branches were provided for. The main line, which was already recognized as a matter of downright necessity, and cheap at any cost, was thus saddled with a host of unproductive branches, the construction of which only impaired and complicated its own value to the State. The State, thus embarrassed, had to borrow an immense amount of money, while the objects upon which it was expended failed to yield revenue enough to pay even the annual interest. In the meantime, many portions of the work were incomplete, and in this way, year after year, millions were squandered without the return of a dollar in the shape of profit! The men employed on the improvements controlled the elections; few were elected to office except hungry political gamesters, and, of course, they had everything their own way. The people, attributing most of the evil to the mismanagement of the works, and the political influences operating upon the Executive, in the appointment of their managers and superintendents, stripped him of these functions, and vested their whole charge in a board of commissioners, elected by their suffrages. This, however, effected little good, for the Canal Board, even now, is probably as corrupt as the most voracious political gourmand could desire. Elected entirely upon political considerations, the commissioners act under its influences, and render the works subordinate to its purposes. Instead of our State works being, as they ought to be, a system of trade and commerce, regulated solely with that view, they are a vast system of political corruption, poisoning the atmosphere wherever they penetrate. The only effectual way for the people to rid themselves of the whole tribe of partizan speculators, is to sell the works to the highest bidder, or bestow them on whatever parties will accept.

The whole length of completed canal, owned by the State, is about seven hundred and fifty miles, and of railroad one hundred and twenty miles. The following gives the route and cost of the several divisions: 1. The Delaware Division extends from Bristol to Easton, at the mouth of the Lehigh, connecting with the canal of that company—ascent, one hundred and sixty-four feet; length, sixty miles; cost, $1,374,744. 2. The Eastern Division commences at Columbia, (the termination of the State Railroad from Philadelphia, elsewhere noticed,) and extends along the eastern bank of the Susquehanna to Duncan's Island, where, crossing the river, it connects with the Juniata Division. From this place it follows the western bank of the Susquehanna to Northumberland, where the river separates into two branches, and the canal into two divisions. The ascent from Columbia is one hundred and eighty-two feet; distance, eighty-two miles; cost, $2,602,832. 3. The North Branch Division extends from this place to the Lackawanna, in Luzerne County. Ascent, one hundred and twelve feet; distance, seventy-three miles; cost, $1,491,894. This division is to be extended from this point to Bradford County, near the State line, so as to connect with the Chenango Canal in New York, and thus afford an outlet for the coal of the Lackawanna region. Over $3,000,000 have already been expended on this extension, which is now nearly completed, throughout its length, ninety miles, the ascent of which is nearly two hundred feet. 4. The West Branch Division extends from Northumberland to Panandsville, in Clinton County, an extensive region of bituminous coal. Ascent, one hundred and thirty-eight feet; distance, seventy-five miles; cost, $1,708,579. There are two small radiating branches in this division, one extending to Lewisburg, in Union County, a fertile agricultural district, not quite a mile in length; the other to Lock Haven, in Clinton County, nearly four miles in length. 5. The Juniata Division, a portion of the main line, and the most important division in the State, commences at Duncan's Island and extends to Hollidaysburg, in Blair County, situate at the foot of the Alleghany Mountains. Ascent of lockage, five hundred and seventy-six feet; distance, one hundred and thirty miles; cost, $3,437,334. 6. The Eastern and Western Divisions being here separated by the mountains, a railroad was constructed to connect them with each other. This is called the Portage Road. It is thirty-six miles in length, and cost $1,783,176. It ascends and descends the mountain with ten inclined planes, five on each side, which are now about to be avoided by a new route of railway already completed. The longest plane is three thousand one hundred and seventeen feet, overcoming an ascent of three hundred and eight feet, and the shortest is one thousand four hundred and eighty feet in length, with a rise of one hundred and thirty feet. The total elevation of the Alleghany summit, on this road, is about twenty-two hundred feet above tide-water. 7. This road, extending from Hollidaysburg to Johnstown, in Cambria County, meets the Western Division of the Pennsylvania Canal, at that place. The canal follows the valley of the Conewago and the Kiskiminitas to the Alleghany River, which it crosses, and then follows its western bank to Pittsburgh, where it again crosses, and meets the Monongahela. Descent, four hundred and seventy-one feet; distance, one hundred and five miles; cost, $2,964,882. 8. The Beaver Division extends up that river from the Ohio to the Shenango, and thence six miles beyond New Castle. Ascent, one hundred and thirty-two feet; length, thirty-one miles; cost, $756,000. This division, near New Castle, meets the Mahoning Canal, extending into the State of Ohio, which intersects the Ohio and Erie Canal of that State. 9. This division is a portion of a long line, originally intended to connect the Ohio River with Lake Erie, by way of Conneaut Lake. The Erie Extension is an unfinished line, divided into two branches, the Conneaut and the Shenango, which extends from the latter northward to Erie. The ascent from New Castle to Conneaut Lake is two hundred and eighty-five feet, from which point there is a descent to Lake Erie of five hundred and ten feet. The whole length of this extension is one hundred and five miles, and over $3,500,000 have been expended. 10. A navigable canal, called the French Creek Feeder, extends from Meadville, in Crawford County, to the Erie Extension at Conneaut Lake, twenty-seven miles. There is a branch from Meadville to the Alleghany River, at Franklin, twenty-two miles in length, with a descent of one hundred and twenty-eight feet. The cost of both works was nearly $1,000,000. After spending the above amount of money on these extensions, the State was finally induced to transfer them, to insure their completion, to incorporated companies, reserving to itself the right of controlling, to some extent, their policy and management. To the foregoing might be added some other lines, proposed or commenced, as the Gettysburg Railway, running from that place to the Baltimore and Ohio Railroad, in Frederick County, Md. This road, after over $700,000 had been expended upon it, was finally abandoned as perfectly useless. Lying in an extensive copper region, it may some day be found worth while to complete it, inasmuch as the Hanover Branch Railroad, connecting with the Pennsylvania Railroad at Columbia, extends within a few miles of Gettysburg. Such was the improvement system of Pennsylvania, for which a debt of over $40,000,000 has been incurred, and which has probably consumed, in losses by freshets and otherwise, in interest of capital invested, and in various other ways, of more than one hundred millions of dollars! Indeed, the *prime cost* of all the improvements constructed by the State, including subscriptions to turnpikes and bridges, would hardly fall short of this astounding sum.

STATE CAPITOL AT HARRISBURG.

Little more remains to be said of Harrisburg. The capitol stands on a handsome sloping elevation, rising in the north-east end of the town. It is sufficiently elevated to afford a fine view of the surrounding country, whose peculiar beauties we have already mentioned. The capitol grounds are enclosed with an iron-rail fence, and laid out in handsome gravel walks, shaded with numerous trees, which are still young and in vigorous growth. The main building is one hundred and eighty feet in length by eighty feet in width, and two stories in heighth. It is a plain but substantial brick building, sufficiently characteristic of our old commonwealth. A large circular portico, faced with six heavy stone columns, constitutes the front entrance to the building. In the interior is a large rotunda, with the high dome overarching, from which is entered the Senate Chamber on the left, and the Hall of the Representatives on the right. Both these halls are large and neatly furnished—but there is nothing extravagant about them. The chair occupied by the Speaker of the House is the identical one used by John Hancock whilst President of the Continental Congress, and during the consummation of the Declaration of Independence. It is a plain, but withal a very elegant chair. The wood, if we remember correctly, is black walnut. It is still

JOHN HANCOCK'S CHAIR.

in a tolerably good state of preservation, but time and constant use are beginning to attack its points. Such is its substantial build, however, that it will probably yet outlive a dozen other chairs of more modern manufacture. On the second story of this building are apartments for the State Library, the Canal Commissioners, and the Supreme Court—as well as for Committees of both houses of the Legislature. The State Library, we should judge, is in a bad condition, Its *written* catalogues, at least, are in a wretched state, showing neither ability of classification, orthography, or decent penmanship. Some of the large table-books—especially those of prints—are mutilated and soiled, which could entirely be avoided if suitable revolving platforms were constructed for their accommodation. One of these books, containing the series of valuable prints of Boydell's Shakspeare Gallery, cost eight hundred dollars, and if better care is not taken of it, it will not be in a condition to be seen in a few years more. This is especially important, because the original edition of this great work is very rare. Many books, too, are missing—no doubt *thousands.* What has become of them?

On both sides of the capitol two other buildings will be noticed, much smaller, but somewhat similar in their architectural features. That on the right is occupied by the Land Office of the State, and the Board of Property. The other, on the left, is occupied by the Treasury Office on the first floor, and by the Secretary of State's Office on the second floor. The Governor also has an apartment here for the transaction of his official business. Both these buildings are entirely too small, as well as miserably arranged and constructed. The truth is, the whole establishment reflects little credit upon the State, and we should not regret to see it displaced by a larger and more imposing edifice. A mansion is now about to be erected for the accommodation of the Governor, an appropriation for that purpose having been made during the last session of the Legislature. This is a step which should long since have been taken; and to make amends for the delay, the building should be one of substantial elegance, corresponding with the high functions of the Executive office of the State. Governor Bigler is not remarkable for his talents, but he is an honest man, and thus far has discharged his high duties with ability and general satisfaction to the people. We like the broad national ground he maintains in his political career, while we admire the energy and persevering industry with which he arose from a comparatively humble position in life to one of commanding influence and honor. Not many years ago he was editor of an obscure country paper, having previously regularly served the usual time as an apprentice to the printing business. Without the assistance of fortune or influential family connections, he has fought his way, and triumphed in every progressive step. That he fills his present position with the best motives for the general good, we have no doubt; but that he is liable to pursue the wrong, or misapprehend the true course in which the general and permanent good of the people may best be promoted, is probably the only fear which a political opponent may reasonably entertain. His Secretary of State is a man of some ability, which is supported by great energy of character. His past political history, mainly connected with the political movements of the day in Schuylkill county, where he has long resided, is by no means above reproach—because it too often exhibits a mere unscrupulous partizan. He was one of the daring spirits who, in 1844, undertook to prove Mr. Polk as good a tariff man as Mr. Clay, and for all practical purposes he was probably the most successful man of the tribe. He, however, appeared to have been disappointed in Mr. Polk's course, in this respect, and subsequently exerted himself to repair the damage which had been done; but in the meantime the people, among whom his statements had obtained currency, suffered severely the paralyzing effects of their credulity, and bitterly realized the deception under which they had previously labored. The Secretary of the Commonwealth, in addition to his other duties, is also Superintendent of Common Schools. The system of education, as now carried out in Pennsylvania, is one of the noblest features in the general character of the State—and yet, strangers passing through it seldom credit its citizens with the intelligence which they really possess. They see a large number of Germans, in some parts of the State, and at once set them down as unlettered. This is a great mistake. The fault is entirely in their vocal language, which is awkward, impure, and *inexpressive.* An educated youth, raised in a community where the leading characteristics of this peculiar Pennsylvania dialect prevail, will be sure to fall under its influence. Being a kind of familiar, every day dialect, every body adopts it. It is the most free-and-easy way of conducting business and social intercourse. The educated man is thus somewhat crippled in the most essential part—*fluency.* Feeling his utter incapacity to express himself effectively, he assumes none of the exterior trappings of learning, but is satisfied with the moderate and occasional use of it, which his business and momentous exigencies may require. A stranger, therefore, may stumble over a plain, industrious farmer, and be surprised to find a man of profound learning. The truth is, our growing population is very generally educated to a certain extent—that is, the essential points comprising a good English education are taught to all, with very few exceptions. But the isolated position of a large portion of our citizens exposes them to the influences of the mixed German-and-English dialect so peculiar to our agricultural districts, and thus the learning they acquire never receives the *colloquial finish* which intercourse with people speaking the English language alone would probably give it. Besides, our German population adhere, as far as practicable, to books and newspapers printed in their own language; but most of them can read understandingly English and German alike.

The present School law was adopted in 1836. It is an improvement on those previously existing, all of which flowed from an express provision of the constitution, requiring a system of general education under the auspices of the State Government. By its provisions, every man is required to pay a tax corresponding with his wealth, no matter whether he may have children to educate or not. All have to contribute for the support of government, and to provide for the common defence; so that all may be secure in the blessings of our institutions and the possession of the rights of person and property. To promote the general welfare—to prevent crime, immorality, and pauperism—education is necessary, and therefore it is properly reduced to a system, under the control and patronage of the government. Thus, the State is divided into some fourteen hundred districts, the citizens of which may determine by ballot, at stipulated periods, whether schools shall be supported in their midst or not. If they decide in their favor, the State contributes a certain amount in their aid—equal to about half their actual expenses. The other portion is collected from the citizens, according to their estates, to assess which directors are appointed, who also superintend the pecuniary and general affairs of the school, appoint teachers, &c. If the citizens decide adversely to schools, of course no benefit from the School-fund of the State can accrue to them. Of the fourteen hundred districts now comprising the School system, probably at least twelve hundred have accepted the law. The non-accepting districts, for the most part, are those where the people are all generally in "good circumstances," and prefer educating their children under their individual direction. Sometimes a district is under the control of rich old bachelors, or penurious, childless husbands—among whom, of course, such a law as this stands no chance of favor! During the sixteen years that the Common School system has been in operation, the people of this State have expended nearly *sixteen millions of dollars* in its support—exclusive of the large sums annually paid to sustain the numerous *private* academies, colleges, and seminaries, which are also giving their valuable aid to the cause of general education. The number of schools in the State, including Philadelphia, which comprises a separate division of our school system, is nearly twelve thousand. In 1836 there were but eight hundred and eight! The pupils now number half a million, and the annual cost is nearly *one and a half millions of dollars!* It is worthy of remark, that the ablest and most eloquent advocate this law ever had, either in or out of the Legislature, was Thaddeus Stevens, a bachelor!

There are several other matters which ought properly to be treated of in connection with Harrisburg; but it is time to pursue our journey, and thank fortune we have before us for the remaining two hundred and fifty-six miles, one of the smoothest and most substantial roads in the United States. So, then, all aboard!

About two miles from Harrisburg, on the right, is situated an imposing structure, not yet fully completed, which is exhibited in the accompanying figure. This is the Pennsylvania Lunatic Hospital, erected by the State, for which fifty thousand dollars were appropriated. It is placed in the charge of a board of trustees, who act without compensation. The insane from all parts of the State can be received here, at the expense of the counties to which they respectively belong; or, if able to pay themselves, at an ordinary cost of about $2.50 per week, including board and medical attendance. It is thus contemplated to make the institution pay its own expenses, without becoming a further charge to the State. It has accommodations for two hundred and fifty patients, and a farm of one hundred and thirty acres of land is attached. The institution was thrown open for the reception of patients in October, 1851, and there are now probably over one hundred enjoying its treatment.

The hospital building consists of a centre building, and wings extending in a linear direction on each side; each wing is so arranged

PENNSYLVANIA LUNATIC HOSPITAL.

that the second projection recedes twenty feet behind the first, and the third the same distance behind the second, so that the second and third projections of the wings on each side of the centre building are open at both ends, which renders them light and cheerful, and insures at all times a free natural ventilation. The centre building is of three stories above the basement or ground floor, has a large Tuscan portico, with a flight of twenty steps to the main entrance, and is surmounted by a large dome, from which a very extensive view of the surrounding country is obtained. The hospital is lighted throughout with gas brought from the works of the Harrisburg Gas Company. In the improvement of the grounds, and the cultivation of the garden, it is expected that much assistance will be derived from the patients themselves—out-door exercise of this character being part of the system of treatment resorted to whenever the patients can be persuaded thus to employ themselves, and generally attended with gratifying results. The institution is open to visitors under certain judicious restrictions.

A few miles above the hospital we cross the Susquehanna on one of the finest structures in the country, and amidst a scene of unrivalled picturesqueness and beauty. The Susquehanna has here forced its way through a range of abrupt mountains, which constitute the western termination of the great anthracite coal region of Schuylkill County. We have already devoted a separate chapter to the consideration of this wonderful region, touching upon all its leading features, and including a description of the processes of mining, &c. We can

RAILROAD BRIDGE OVER THE SUSQUEHANNA.

only, in this connection, refer the reader to the article alluded to, which comprises Part II. of "Off-hand Sketches." The coal veins, as they approach the Susquehanna, are flattened out, while the coal itself is soft and of a semi-bituminous character. The region, from its proximity to the Susquehanna, is favorably situated; but the position of the coal strata is such that mining operations can never be prosecuted with much success, or to any great extent. An immense amount of money has already been spent in gigantic improvements to develop a comparatively small and imperfect coal district. The other fork, called the Lykens Valley, is better developed in coal, and the strata lie in a position more favorable for mining purposes. A *soft coal never can be profitably mined,* because much of it is lost in the process of excavation, and much more in the subsequent handling and transportation. Having said thus much, we pass on "over the bridge and far away."

The railroad, after crossing the bridge, runs some ten miles along the narrow bank of the river. Forcing its way between it and the projecting mountain-spurs, it occasionally exhibits some "tall" specimens of side-cutting. This is particularly the case at the Cove, ten miles from Harrisburg. Here the wagon-road is forced into a tight place—unable to pass below, it ascends about eighty feet above the railroad, and winds around the frowning precipice in abject obedience to the "inevitable force of circumstances." The railroad, too, where it runs directly alongside the Susquehanna, is elevated from fifteen to twenty-five feet above it, and is protected from the ice and water freshets to which the river is subject, by strong stone embankments, sloping to the water's edge. Like all other mountain-streams, the Susquehanna is particularly known for its propensity to get "high." At such times it foams and rages terribly, though few would suspect it from its usual gentleness and docility. In the spring, when the snow melts, the swollen stream breaks up the lazy ice, which, charged with sand and pebble, and fragments of trees and timber, descends in huge masses. At

THE COVE.

some points below, the river becomes very narrow and correspondingly deep, being hemmed in by high walls of rock on both sides. The ice here frequently dams up, throwing back the water with its floating fields of ice, while these, rubbing against each other in fierce conflict, create other dams; and thus the river, for many miles, is often completely and emphatically *dam'd!* The water and ice, thus savagely arrayed against each other, adjourn their "muss" to the adjacent villages along the shores, where, seizing piles of lumber, boats, trees, small houses and stables, they bear them along, and finally dash them to pieces over some old villain of a rock. Sometimes they vent their furies upon the unoffending railroad, canals, and bridges; and it was thus that the noble structures at Clark's Ferry, above, and at Harrisburg, Columbia, and McCalls' Ferry, below, were at different times carried away or materially injured. The railroad here, however, is pretty safe; and so, too, is the bridge. The river is unquestionably becoming tamer every year, and by the time the Maine Law is put in force in this State, the Susquehanna will probably cease to "get high" altogether!

In the fall of the year the traveller will notice all along the Susquehanna, as well as the Juniata, the Lehigh, and other streams, a contrivance for catching fish, called a fish-basket, which is exhibited in the annexed figure. The thing, as will be seen, is very simple; but it plays "hob" among the cat-fish and eels, and perch, and other natives of these rivers. A stone wall is built in the stream (which, in the fall, outside of the channel, is usually not more than three feet deep) in the manner indicated in the engraving. At the neck of these walls, a wooden structure is placed, one end sunk under the water, and the other gradually rising some six feet above it, supported by strong wooden props. Upon this are placed five or six lath platforms, about four feet square, one projecting over the other. The lath are nearly an inch in width, and fixed half an inch apart, so as to allow the descending water to pass through. The fish, as they come idling down the stream, in friendly social glee with each other, suddenly fall over the platform of lath, and, no doubt much to their surprise, find themselves unable to swim back. They, however, appear to have a merry time of it, kicking and frisking around amongst each other, while new customers are constantly rolling in. Occasionally an old sucker is "sucked in," and makes a tremendous floundering among the assemblage, huddled together like democrats at a county meeting.

A FISH-BASKET IN THE SUSQUEHANNA.

But the day breaks forth through the foggy vapor, and the drowsy fisherman, emerging from his bed of straw in the cabin, begins leisurely to haul them up. They are all consigned to a large slimy bag; and thus ends the whole process, which, unhappily, forever seals the fate of the fish.

A few miles above the Cove Mountain is the village of DUNCANNON, well known in connection with its iron manufactures. Some twelve miles from this place are situated the Perry Warm Springs, with which there is a mail-stage connection. We know nothing of the merits of the spring, but we do know the proprietor, and cannot too warmly commend him to the notice of our rambling friends.

The iron works have lately been much enlarged, and include extensive rolling-mills, nail factories, and machine-shops, with a large furnace establishment a mile or two distant. There are upwards of two hundred persons employed in these establishments. The village stretches along the Susquehanna until it meets that of PETERSBURG, from which it is separated by the Little Juniata. Both these villages are pleasantly situated, both in respect to trade and surrounding scenery. A few miles above, Duncan's Island bridge stretches over the broad river, and the canal, as already noted, forms two divisions —one continuing along the Susquehanna, the other following the Juniata. Fourteen miles beyond Duncan's Island, on the Susquehanna, is LIVERPOOL, remarkable for the wild and picturesque beauty of its scenery, a tolerable specimen of whose general aspect is afforded at this and intermediate points. While nature has all along this river evidently done her utmost to give "enchantment to the view," man has mingled with it some of the noblest works of his enterprise and genius.

Having travelled some thirty miles along the Susquehanna, we obtain our last view of it near Duncan's Island, where the railroad gracefully leaves it, and follows the bright, the blue, the wild Juniata. Flowing, in one broad sheet, to the Conewago hills, it opened splendidly to our view, and only doubled its attractions as we rolled along the green banks and the soaring spurs of the Kittatinny; but here, nestled amongst a troop of mountain peaks and rocky cliffs, with a belt of green fields stretching to the Tuscarora, it fades serenely, but gloriously, in our departing gaze. The county we are now in, we should have mentioned before, is Perry, which lies between the Kittatinny and the Tuscarora Mountains. Entering this county from Franklin, adjoining it on the south-west, where the Tuscarora is known as Cove Mountain, while the Kittatinny is cut up into detached knobs, they keep close together for some distance, but finally separate and get far apart—the distance between them, at the Susquehanna, being at least twenty miles. Between these mountain ranges the prevailing scenery is that of a rich and thrifty farming district—agreeably diversified with an occasional sprightly village, a view of the tortuous river and canal, and the usual "thousand and one" concomitants of a wild, sloping, and beautiful valley. The land is well supplied with beds of limestone and iron ore, as well as numerous streams of water, calculated to furnish driving power for nearly every description of useful manufacture. It was long the abode of Indians, who perpetrated upon the early white settlers some of the bloodiest tragedies to be met in their eventful history. Duncan's Island, in the Susquehanna, was their favorite retreat—their summer resort; and in passing to and fro the valley resounded their frightful yells, and drank up the blood of their savage tomahawks. The early settlers consisted of Irish and Scotch Presbyterians, with a few English Quakers; the population now, however, is principally composed of Germans, who are invariably found wherever there is a good soil, and their patient industry and frugal habits always enable them to supplant every other class of people.

At NEWPORT, which is twenty-seven miles from Harrisburg, the Juniata makes a sharp curvature, apparently for the express purpose of meeting a little stream, called Buffalo Creek, which it gallantly carries off in its course. This ridiculous curve of the Juniata's put the railroad company to a considerable expense in constructing two elegant and substantial bridges over it, both of which would have been unnecessary had the stream kept in its straightforward course. We hate foolishness!

A short distance above Newport, which is a small village, two or three hundred yards from the railroad, the canal is transferred to the opposite side of the river. The water is deepened by a dam, and the boats towed over by a rope, in an ingenious manner. The rope,

MILLERSTOWN, ON THE JUNIATA.

winding around a wheel on each side of the river, is made to travel to and fro by water-power supplied by the canal. The boats are attached to this rope, and are thus towed over.

MILLERSTOWN, one hundred and forty miles from Philadelphia,

two hundred and twenty-three from Pittsburg, and thirty-two from Harrisburg, though a small village, is still the largest one in Perry County, containing a population of about five hundred. It is beautifully situated, and contains several well-built, spacious, and neat dwellings—but it looks best when seen at a "respectful" distance. The inhabitants are mostly German, and live as Germans do—quietly, decently, soberly—never flame out with sign-boards and gim-cracks over their stores and shops—never bustle about with quills stuck above their ears—never drive "fast" horses, or give wrong change—never have meals cooked in a hurry—never serve ham and eggs without placing around them, in a little circle, dishes of apple-butter, molasses, fruit, preserves, &c.; in brief, they live coldly, and accumulate wealth prudently, and "further this deponent saith not."

TUSCARORA STATION.

The railroad, a short distance above, enters the Tuscarora Mountain, and runs along its northern slope for many miles. The country, or certain portions of it, is not so well developed in fertility as it is below—but a glimpse is now and then obtained of a good farm, sloping to the river from the mountain sides. The gorge of the mountain passed, we enter the county of Juniata, which lies wholly between the Tuscarora and Shade Mountains, running in nearly parallel order, and only a few miles apart. The county is thus about five times as long as it is broad—but situated as it is, the land is, upon the whole, rather productive. Like Perry, it has its seams of iron ore and limestone, and its furnaces have long enjoyed good reputation for the quality of the metal they produce. Tuscarora Creek runs along the mountain of that name, and empties into the Juniata near MIFFLIN,

MIFFLIN.

which is the seat of justice, as well as the most populous town in the county. It is delightfully situated, like most other towns along the Juniata, on a sloping eminence, from which a view of the surrounding scenery is afforded. The canal passes under the bridge represented in the engraving, and has, of course, materially increased the trade of the town. Several little villages on the Juniata, between this place and Millerstown, are merely depots for the storage and shipment of the produce of the surrounding country, as Thompsontown, Mexico, Perryville, &c.

Opposite Mifflin the railroad company has erected several large buildings, including the hotel here exhibited, which is, in our opinion, one of the finest establishments of the kind in the United States, at the same time that it is extremely plain in its internal structure, and cost, comparatively, a trifling amount of money. Several brick hotels along this road, of about the same dimensions, but constructed by private capital, cost at least twice as much as this house, while they are palpably inferior to it in architectural beauty, convenience of arrangement, and general completeness of design and appearance. The truth is, nothing

THE PATTERSON HOUSE, ON THE PENNSYLVANIA RAILROAD.

has been more neglected in this part of the country, and generally speaking, throughout the entire State, than the subject of architecture. It is not known or recognized as an art at all; and the natural result is, the houses and buildings of every description are dull, stupid, and monotonous—destitute alike of real convenience, of harmonious proportion, and of true economic principle. They are totally unrelieved by a single meritorious feature, beyond the immediate one of protection from the elements. Ventillation—good taste—in short, everything is sacrificed to produce buildings as ridiculous and clumsy in appearance as they are uncomfortable for practical use. Good barns, it is true, are often seen; but even these are all constructed after the same antiquated plan—not one of them deviating so far as to exhibit a neat projecting border to relieve the blankness of the roof or gable-ends; nor is there any other ornament calculated to produce a sprightly effect to a building otherwise highly creditable.

The railroad company, in the construction of all its numerous buildings, has given a practical exposition of the laws of good taste and architecture, as founded on economical principles. There is not one of its buildings, even the little watering-stations, that does not rise far superior, in the spirit of the design, to any other on the route. And the force of our remarks will be appreciated when the above hotel is compared with other establishments characteristic of the country. It was erected with direct reference to the taste, comfort, and entertainment of the traveller—points which are too often overlooked by railway companies, when constructing their lines of improvement. They too often suppose that nothing more is expected of them beyond the mere transportation of the "flesh and blood," with the accompanying bag and baggage. No matter what inconveniences the weary traveller may be exposed to, it is no affair of theirs —it is not "found in the bond." This, however, is not the spirit that actuates the Pennsylvania Railroad Company. The Patterson House —named in compliment of a former President—is one that will tempt the traveller from the fatigues of his journey, and, being tempted, will obtain a hold upon the affections of the "in'ard man." The table is sumptuous—the dining-room cool and spacious—the servants black as charcoal, and polite as a Pennsylvania sun can make 'em. The proprietor, whoever he is, is evidently a learned man in his useful profession, and seems to act upon the lines of old Goldsmith—

Whoe'er has travelled life's dull round—
Where'er his journeyings may have been—
Must sigh to think he still hath found,
The *warmest welcome* at an inn!

This hotel is 156 miles from Philadelphia and 207 from Pittsburg. It is elevated 430 feet above tide-water, while the altitude of the surrounding mountains will average about 700 feet.

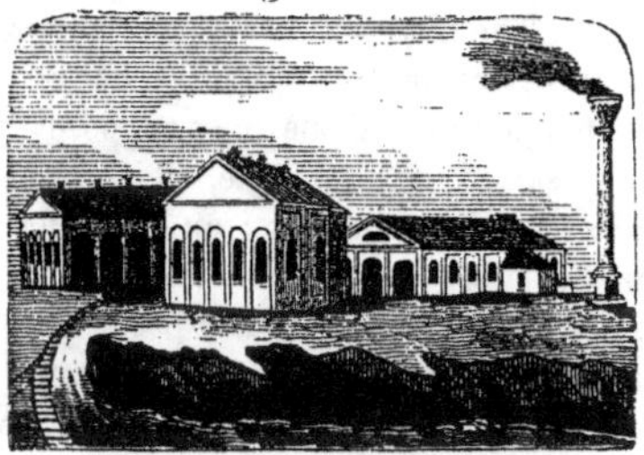

ENGINE-HOUSE AND WORKSHOPS.

In the rear of the Patterson House are situated the workshops and engine-house of the railroad, for the eastern section. The buildings are of brick, put up in the most substantial manner, with due regard to ornament. Whatever repairs of the running machinery of the road may be necessary, from time to time, will be attended to here.

Leaving this place, the railroad runs in a perfectly straight course to the north-western boundary of the county, where it enters that of Mifflin amidst the wildest and most beautiful scenery which the eye of man could desire. Passing through a narrow gorge, we have what is locally termed the Black-log Mountain on the left, and Shade Mountain on the right. Along the summits of both these mountains are huge rocky promontories, black and dismal, and broken into irregular peaks, with sharp intervening fissures and valleys of denudation. From these overhanging rocks immense avalanches have, from time to time, been detached, and precipitated in frightful confusion along the mountain sides, which are also very steep. These stones have no soil associated with them, but lying one on top of the other, they present a loose mass, apparently on the verge of sliding still further down. The whole stony assemblage is thus held in *statu quo* by some faithful tree or deeply-set rock. The amount of these detached rocks is altogether inconceivable—the mountains are literally covered with them, from top to bottom. Covered with moss, and of a dark and sombrous color, they give to the scene a stamp of positive wildness, the beauty of which is increased by the overhanging foliage, sloping to the rugged banks of the Juniata, which leaps over its rocky bed as if bewildered with the scene around. There are many other scenes in our mountain ranges similar to this; but there is none, in this broad continent, which surpasses it in picturesque outline. It is one of those things, too, that must be seen leisurely to be fully comprehended and enjoyed. The means by which these immense bodies of stone became detached, are perfectly plain—but are still none the less worthy our contemplation. Some may suppose that the mountains are occasionally seized with the ague, which sometimes prevails along the Juniata in the fall of the year; and shaking and trembling violently under its sickening spells, the rocks may thus have been detached and tumbled down from their lofty beds! But we are pretty sure the ague had nothing to do with it, and simply because those who are acclimated to the Juniata are not liable to take it; and of course mountains have resided in "these parts" long enough to be perfectly exempt from such ailments. No; the element which has been at work here, and sundry other places where similar effects are exhibited, is a simple but powerful one. It has evidently been busy throughout a vast space of time; nor has it abandoned its occupation. It is water. The Juniata, with the eccentric boldness which has always characterized it, commenced a violent onslaught to get through the mountains. Swelling indignantly at the obstacle before it, it finally broke through, in a wild resistless torrent, tearing the mountain savagely as it passed along. This splendid triumph achieved, it proceeded leisurely to clear out its course—one by one the rocks were split and hurled from their ancient positions, and gradually the incision in the mountain increased, until it reached its present level. In the meantime the waters of the river returned, time after time. Picked up by the wandering clouds, they rode back in swift-travelling gales, and again and again pounced upon the devoted mountains. First removing their coats of clay and sand, they seize a pebble here and another there, and roll them against some projecting rock. What can't be done in the regular way, must be accomplished by stratagem —therefore water enters into the state of snow and ice, and catching the loose particles laying on the surface, holds them in its freezing grasp, and carries them along when it resumes its annual spring campaign. By this means a gradual decomposition of rocks is produced, and sometimes the thawed earth lets loose immense avalanches. Thus, the substance of the mountains is daily diminishing, and water is the active agent that has charge of the whole business of transportation. Water, therefore, has been the leading and only agent in affecting the degradation of our mountain system; and to what extent its operations are still continued, may be estimated at every valley which receives the descending debris. Opposite Spruce Creek an example is afforded of the degrading effects of a single torrent. A deep incision is made in the side of the mountain—the surface scooped out all the way down, in some places to the depth of twenty feet or more. It will grow larger and deeper with every rain that falls, until, finally, the rocks of the mountain protrude, and detach themselves in the same frightful manner as in the Long Narrows, above described.

Mifflin County, of which Lewistown is the judicial seat, lies between two prominent mountain ranges, which run in a south-west and north-east direction. The southern boundary we have just passed in the long Narrows separating Black-log and Shade Mountains; the northern line runs between Jack's and Stone Mountains, adjoining Huntingdon and Centre, and meeting Union in the east. At this place the Juniata makes a sharp curvature to the south-west, passing through the greater portion of the county, when, making another sudden curve near Newton Hamilton, twenty-three miles distant, it cuts through Jack's Mountain, and again strikes to the north. Had the little thing only broke through the mountain, in a straight line between Petersburg and Lewistown, it would have saved more than

LEWISTOWN STATION.

half the distance it now travels. The Kishacoquillas, a large and beautiful stream, rising in the mountains north-east of Lewistown, joins the Juniata as it winds along the eastern slope of that borough. It furnishes a splendid water-power, which is extensively applied at various points on the route—at Lewistown constituting the driving power of two large and handsome flouring mills, the only branch of manufacture, we believe, which the town affords. This is surprising. Viewed from the railroad, LEWISTOWN has the appearance of a large and busy place; but it is not. The town is composed, for the most part, of inferior buildings, while the population (embracing about twenty-eight hundred) appear to be idle and without enterprise. The situation of the town is, in every respect, advantageous and pleasant. The entire county is one broad fertile valley, with rolling ridges dividing it into numerous subordinate ones, many of which contain a plentiful supply of iron ore and limestone, while the Juniata and its numerous tributaries supply any quantity of water-power at the same time that they thoroughly drain the land. It has all the natural beauty and many of the prominent outline features which have given such poetic and undying celebrity to Wyoming; but its citizens lack the spirit and stamina so characteristic of the people of that region

DEEP CUT NEAR LEWISTOWN.

throughout its extraordinary and romantic history. If the same kind of Connecticut Yankees had originally settled along the Juniata, or if they were induced to do so now, it would soon wear a different aspect. Their industry and practical enterprise would soon blaze forth in spirited rays, and the whole valley would resound with the clatter of machinery—the thunders of the hammer—the blaze of the furnace—the rattle of the loom.

Lewistown might become a more important place, even as a *summer resort.* Considered in this respect, it combines every requisite advantage, and all that is wanted is one or two spacious hotels, showy and well-conducted. There are two or three respectable inns in the centre of the town, but these are too much exposed to gossipping loafers to merit any considerable patronage from summer tourists. The hotels for the accommodation of travellers should be located nearer the railroad. A citizen of Lewistown, in a letter published some years ago, in one of the borough papers, says: "The scenery is the finest in the world; we breathe the pure mountain air. Our clear streams abound with fish, particularly trout. Our forests are filled with game of every description; and Milliken's Spring, on a farm adjoining the town, is

LEWISTOWN.

ascertained to possess all the medicinal qualities of the Bedford water, particularly in bilious complaints."

The celebrated Indian Chief, Logan, lived in this part of the valley when the whites first arrived. His cabin was near the wild gorge in Jack's mountain, and the history of the county of which this was formerly a part, is full of anecdotes relating to him. Logan was probably the most eloquent Indian orator of which we have any account. He was the Clay of the Indian people; and in natural dignity, independence of spirit, and loftiness of purpose, few whites have surpassed him.

Near the village of REEDVILLE, about six miles from Lewistown, is a beautiful spring, near which the orator often dwelt. The following anecdote, related by the late Judge Brown, one of the earliest settlers, is connected with this spot:

"The first time I ever saw that spring," said the old gentleman, "my brother, James Reed, and myself, had wandered out of the valley in search of land, and finding it very good, we were looking about for springs. About a mile from this we started a bear, and separated to get a shot at him. I was travelling along, looking about on the rising

LOGAN, THE INDIAN CHIEF.

ground for the bear, when I came suddenly upon the spring; and being dry, and more rejoiced to find so fine a spring than to have killed a dozen bears, I set my rifle against a bush, and rushed down the bank and laid down to drink. Upon putting my head down, I saw reflected in the water, on the opposite side, the shadow of a tall Indian. I sprang to my rifle, when the Indian gave a yell, whether for peace or war I was not just then sufficiently master of my faculties to determine; but upon my seizing my rifle and facing him, he knocked up the pan of his gun, threw out the priming, and extended his open palm toward me in token of friendship. After putting down our guns, we again met at the spring, and shook hands. This was Logan, the best specimen of humanity I ever met with, either *white* or *red*. He could speak a little English, and told me there was another white hunter a little way down the stream, and offered to guide me to his camp. There I first met your father. We remained together in the valley a week, looking for springs and selecting lands, and laid the foundation of a friendship which never has had the slightest interruption.

"We visited Logan at his camp, at Logan's Spring, and your father and he shot at a mark for a dollar a shot. Logan lost four or five rounds, and acknowledged himself beaten. When we were about to leave him, he went into his hut, and brought out as many deer-skins as he had lost dollars, and handed them to Mr. Maclay, who refused to take them, alleging that we had been his guests, and did not come to rob him—that the shooting had been only a trial of skill, and the bet merely nominal. Logan drew himself up with great dignity, and said, 'Me bet to make you shoot your best—me gentleman, and me take your dollar if me beat.' So he was obliged to take the skins or affront our friend, whose nice sense of honor would not permit him to receive even a horn of powder in return."

Logan was one of the most successful hunters ever known among the Indians, and supported his family entirely by killing deer, dresssing the skins and selling them to the whites. He once sold a large lot to a tailor, of the name of De Young, living somewhere in Ferguson's Valley—tailors, in those days, dealed extensively in buckskin breeches. Logan received his pay, according to agreement, in wheat. The wheat, however, on being taken to the mill, was found so worthless that the miller refused to grind it. Logan was much chagrined, and attempted in vain to obtain redress from the tailor. He then took the matter before his friend Brown, who was a magistrate; and on the judge's questioning him as to the character of the wheat, and what was in it, Logan sought in vain to find words to express the precise nature of the article with which the wheat was adulterated, but said that it resembled in appearance the wheat itself. "It must have been *cheat*," said the judge. "Yoh!" said Logan, "that very good name for him." A decision was awarded in Logan's favor, and a writ given to Logan to hand to the constable, which he was told would bring him the money for his skins. But the untutored Indian—too uncivilized to be dishonest—could not comprehend by what magic this little paper would force the tailor against his will to pay for the skins. The judge took down his own commission, with the arms of the king upon it, and explained to him the first principles and operations of civil law. "Law good," said Logan; "make rogues pay."

But how much more simple and efficient was the law which the Great Spirit had impressed upon his heart—*to do as he would be done by!*

Lo the poor Indian, whose untutored mind
Sees God in clouds, or hears him in the wind!
His soul proud science ne'er taught to stray
Far as the solar walk or milky way—
Yet simple nature to his hope hath given
Beyond the cloud-capped hills an humble heaven—
And thinks, admitted to that equal sky,
His faithful dog will bear him company!

Mr. Jefferson, in his Notes on Virginia, gives the following incident in the history of Logan, after leaving the Juniata:

In the spring of 1774, a robbery and murder were committed on an inhabitant of the frontiers of Virginia, by two Indians of the Shawnee tribe. The neighboring whites, according to their custom, undertook to punish this outrage in a summary manner. Colonel Cresap, a man infamous for the many murders he had committed on those much injured people, collected a party and proceeded down the Kanaway in quest of vengeance; unfortunately, a canoe with women and children, and one man only, was seen coming from the opposite shore unarmed, and unsuspecting an attack from the whites. Cresap and his party concealed themselves on the bank of the river, and the moment the canoe reached the shore, singled out their objects, and at one fire killed every person in it. This happened to be the family of Logan, who had long been distinguished as a friend to the whites. This unworthy return provoked his vengeance; he accordingly signalized himself in the war which ensued. In the autumn of the same year a decisive battle was fought at the mouth of the great Kanaway, in which the collected forces of the Shawnees, Mingoes, and Delawares were defeated by a detachment of the Virginia militia. The Indians sued for peace. Logan, however, disdained to be seen among the suppliants; but lest the sincerity of a treaty should be disturbed, from which so distinguished a chief abstracted himself, he sent, by a messenger, the following speech to be delivered to Lord Dunmore:

"I appeal to any white man if ever he entered Logan's cabin hungry, and he gave him not to eat; if ever he came cold, and he clothed him not. During the course of the last long and bloody war Logan remained idle in his cabin, an advocate for peace. Such was my love for the whites, that my countrymen pointed as they passed, and said, Logan is the friend of the white man. I have even thought to have lived with you, but for the injuries of one man. Colonel Cresap, the last spring, in cold blood, murdered all the relations of Logan, even my women and children.

"There runs not a drop of my blood in the veins of any living creature; this called on me for revenge. I have fought for it. I have killed many. I have fully glutted my vengeance. For my country I rejoice at the beams of peace: but do not harbor a thought that mine is the joy of fear. Logan never felt fear. He will not turn on his heel to save his life. Who is there to mourn for Logan? Not one!"

Such was the great Indian orator who once frequented these lofty mountains and rich rolling valleys. Such was the son of the great Cayuga Chief—Logan. The Juniata was the favorite retreat of the Indian, and no wonder! If it is wild and beautiful now, what must it have been in its primitive glory—when the Indian alone "was monarch of all he surveyed!"—when he pursued the roving deer amidst the solitude of the deep forest, or skimmed the blue waters of the stream in his light canoe! *Apropos*, a favorite song:

BRIGHT ALFARATA OF THE BLUE JUNIATA.

Wild roved an Indian girl,
  Bright Alfarata,
Where sweep the waters
  Of the blue Juniata!

Swift as an antelope
Through the forest going,
Loose were her jetty locks
In wavy tresses flowing.

Gay was the mountain song
  Of bright Alfarata—
Where sweep the waters
  Of the blue Juniata.
Strong and true my arrows are,
In my painted quiver—
Swift goes my light canoe,
A-down the rapid river.

Bold is my warrior good,
  The love of Alfarata,
Proud waves his snowy plume
  Along the Juniata.
Soft and low he speaks to me,
And then his war-cry sounding,
Rings his voice in thunder loud
From height to height resounding.

So sang the Indian girl,
  Bright Alfarata,
Where sweep the waters
  Of the blue Juniata.
Fleeting years have borne away
The voice of Alfarata,
Still sweeps the river on
  Blue Juniata.

Twelve miles above Lewistown, on the Juniata, and nearly equidistant from Philadelphia and Pittsburg, is McVEYTOWN, a handsome little village, doing some business on the canal. Hanawalt's Cave is near here, and has some interest. It often contains saltpetre, in a crude state, and numerous stalactitic curiosities peculiar to limestone formations. Ten miles further is NEWTON HAMILTON, another pleasant village, principally engaged in the trade of the canal. The great fertility and productiveness of this portion of the country renders this trade by no means insignificant—though much of it will ultimately be attracted to the railroad during that portion of the year when the canal is closed. An enormous quantity of pig iron, flour, grain, pickled meat and butter, &c., is forwarded from these places in the summer, the stock accumulating during *the winter*. A heavy capital

RAILROAD AND CANAL.

is thus *inactive* for several months, and the advantage of having a railroad, by which a favorable condition of the market may speedily be embraced, is of the highest importance. A much larger business may, in this way, be done with one-fourth the capital, which, together with the saving of time, will more than compensate the trifling difference, if any, in the cost of transportation.

At this place, winding around the gorge of Jack's Mountain, the Juniata follows the boundary line for a few miles, and then enters Huntingdon in a direction nearly north. We are now emphatically and unmistakably in the midst of mountains—bold, rugged, *thundering* mountains!—the most of which have a range nearly north and south, and cross the straight line of our course. No matter—we will pass 'em, and even mount their lofty summits. In Bedford County, which adjoins on the east, there are no less than twelve mountains, including those of the great Alleghany on the west, and Cove on the east, which constitutes its western and eastern boundaries. These mountains are each known by various local names, and are more or less broken and disrupted as they enter Huntingdon and Blair in the north. The celebrated Broad Top Mountain coal district lies in Bedford, some fifteen miles or more south of the village of Newton Hamilton. This splendid coal district is entirely isolated from the great Alleghany region, with which it was originally connected. The coal, too, is of a better quality, being semi-bituminous in character, and similar to that of the Dauphin coal field, as it approaches the Susquehanna. Being the only spot in a wide expanse of territory where coal is to be had at all, it must be regarded as of great value, and arrangements are now being made to extend a railroad so as to connect it with the lines of improvement on the Juniata. The isolated position of this coal mountain, with other connecting circumstances, go to prove that our anthracite and bituminous beds formed originally one entire and almost complete assemblage, and that they are both of cotemporaneous formation. The mountain ranges are higher and

HUNTINGDON, ON THE JUNIATA.

bolder toward the south-west, where the coal strata are prolonged; while toward the east they have been cut down and detached, and the coal washed away, leaving only a comparatively small amount behind, and that lying in a position low down, with the stratification generally in a semi-vertical dip. The whole Apalachian chain may, therefore, be regarded as one immense coal-bearing system—but such have been the destructive effects of time, during the countless millions of years it has been exposed, that a very small portion of the original amount of vegetable matter, constituting the coal measures, is now left behind.

Passing the unimportant stations of MOUNT UNION, MAPLETON, and MILL CREEK, we reach the borough of HUNTINGDON, two hundred and four miles from Philadelphia, one hundred and fifty-nine miles from Pittsburg, and ninety-seven miles from Harrisburg. This borough is elevated six hundred and ten feet above tide-water, while the average height of the mountains is about the same. The borough of Huntingdon is the seat of justice of the county of that name, which formerly included the adjoining one of Blair, and was laid out in 1770 by Dr. Smith, Provost of the University of Pennsylvania. The name was bestowed in compliment to the Countess of Huntingdon, who was a liberal contributor to the funds of that institution. The present population of the borough, we should judge, is about eighteen hundred, or more, and is now rapidly increasing under the stimulus of recent and forthcoming important public improvements. The population, too, is very intelligent, and embraces many wealthy families.

The situation of this place, as may be inferred from the sketch of it, is extremely pleasant. Some of the wildest scenery in the State may be found a few miles adjacent—among which may be mentioned the celebrated Pulpit Rocks, on Warrior Ridge, a view of which, from the turnpike, is afforded in the annexed figure, extracted from the work of Mr. Trego, on the Geography of Pennsylvania. The rocks appear equally bold from the railroad, which passes directly along the base of the ridge. These rocks are a coarse-grained cemented sandstone, varying in color, but generally of a yellowish-white, with particles of bright flint. They have attained their present curious appearance from the gradual effects of the atmosphere and rain, which, working out irregular fissures, have thus left standing lofty columns, that frown from their high summits upon the no less rugged and narrow valley below. The mountains of this county are nearly all cut up into bold sharp ridges similar to the above—though it is otherwise one of the most mountainous in the State. Jack's mountain presents a continuous range on the east, and Tussey's on the

PULPIT ROCKS ON WARRIOR RIDGE.

west; but the others are detached knobs and ridges, which, at one time, probably constituted an unbroken elevation. This broken and disruptured character of the county has been occasioned solely by water, and to substantiate this belief, we need only to point to its numerous streams, comprising the head waters of the Juniata. Thus, rising in Bedford County, and emptying into the Juniata a few miles from Huntingdon, is the snake-like Raystown Branch; rising in Blair, and traversing the middle of that county, is the Frankstown Branch, emptying into the Juniata near Alexandria; rising in the same county, further north, is the Little Juniata, which, at the place above named, forms the main stream, and afterwards receives five or six others—all in the County of Huntingdon. This county is, therefore, more liberally supplied with streams than any other in the State, and every stream, following the narrow gorges of the mountain ridges, presents favorable opportunities for applying its water-power. All the streams have their mills, forges, furnaces, and other iron works, while the whole county is well supplied with timber for the production of charcoal, as well as with beds of limestone and iron ore. Indeed, we are now in the most extensive iron region not only of the

United States, but of the *world!* It is really scarcely known to what an immense extent the iron manufacture of this State has reached. Pennsylvania now produces more iron annually than was manufactured in all Great Britain thirty years ago. Compared to the present manufacture of the article in France, that of Pennsylvania is at least equal—it is more than Russia and Sweden united, and exceeds that of all Germany. Pennsylvania may well be called the Iron State of the Union; and from these mineral treasures, she must build up a prosperity more splendid and permanent than if wrought from gold, for gold is the ultimate product of her iron.

The Juniata, from its mouth at the Susquehanna to its head waters on the Alleghany, forms one continuous and unbroken iron country, and the productions of its furnaces and forges, we have already stated, are amongst the best which enter the market.

UNION FURNACE.

As we are now in the "head quarters" of this stupendous iron district, we shall endeavor to point out some of the *prominent features* in its manufacture, and therefore commence here, which is the oldest establishment in this part of the country. It was erected nearly fifty years ago by Dorsey and Evans, and was worked successfully during the greater portion of the time until within two or three years past. It is a charcoal furnace, as are also all the others. Compared with more recent furnaces this affair is quite a curiosity. It is of small calibre, and its life was prolonged from time to time, with additions, supports, and patches, until finally it could stand it no longer, and yielded up its fiery breath. Mr. Ritts, to whose hospitality and that of his amiable wife we must make, as we do with grateful pleasure, our acknowledgments, is the present proprietor, and was the last to work it. Should the iron trade revive, it is probable it may again be set in operation, with such improvements and additions as its decayed condition may call for.

But we cannot enter into the spirit of iron manufactures until we shall have seen BIRMINGHAM—*our* Birmingham. Though not so large as its great namesake in England, it still occupies *high ground*, and is at least in the midst of a tremendous iron country. It is a village of four hundred population, more or less, and is romantically situated, if nothing more can be said of it. The scenery all around it is varied, but wild beyond measure. Speaking of Birmingham, suppose we run over to *Ironville*—a strong name, to be sure, for a small village; yet, standing on an iron foundation, it is properly an iron village. Bridges! bridges!—is there to be no stop to these bridges? This is the most

rascally little river we have yet met with; having crossed during the last five miles at least a dozen of these elegant iron structures, there is yet no end to 'em, for here we are again perched seventy feet in air, over the same stream, looking down at Ironville, nestling there on the hill! We dash on, and wind around through a deep cut, when—confound the stream—here it is again with another bridge! But hold—here is Blair County, and before us one of the prettiest villages we have yet seen. This is TYRONE, two hundred and twenty-six miles from Philadelphia, one hundred and thirty-nine from Pittsburg, and eight hundred and eighty-six feet above tide water. So, sir, you see we are "getting up in the world," as the saying is. We already breathe the fresh air of the Alleghany, and it will not be long ere we find ourselves on the top of its loftiest summit. Here then, within sight of old iron-bound Huntingdon, let us indulge some observations in reference to the manufacture of iron, of which this place is in many respects the principal theatre. These extensive works include a furnace at Bald Eagle, a few miles distant, a forge a few hundred yards above, at the water-station, the forge below the railway, (indicated in the picture,) and some other works scattered over the land connected with them, embracing several thousand acres. The name of the firm is Lyon, Shorb & Co., and the business is conducted here by J. T. Matthias, Esq., a son-in-law of Mr. Shorb. This is, in our opinion, a model iron establishment. There are upwards of two hun-

TYRONE.

dred hands regularly employed, averaging ten dollars per week, each; and a more cheerful set of men we never saw—every one, too, we should judge, a sincere admirer of Mr. Matthias. We are probably diverging from the straight line of our object in alluding thus, *en passent*, to a gentleman occupying a position purely private. But it is so unusual a circumstance to hear overseers complimented by their workmen, that for the novelty of the thing we must beg to be excused for mentioning it. When we see a village of cheerful and contented people—neat, intelligent, industrious and orderly; when we see the manager in his store, "busy as a bee" in attending to the wants of his hard-working customers, and only leaving it to dispense the sweet little courtesies of hospitality in his dwelling; when we see all this, nothing could prevent us from believing that *everything is just as it should be!* And if we are not greatly mistaken, the success of these works is as much the result of their careful and judicious management as anything else that could be suggested; while it is notorious that for the want of such management, many of the best establishments are often rendered unproductive and comparatively worthless. While upon iron, we should allude to

THE CORNWALL ORE BANKS.—This celebrated deposit of iron ore, situated in South Lebanon township, Lebanon County, the largest in the State of Pennsylvania, or perhaps in the United States, has been worked for upwards of a century, and has long been celebrated for the quality of the iron produced from it. Mr. Richard C. Taylor, in a report made in 1851, thus speaks of them:

"I need scarcely mention here the well known fact, that the ore banks of Cornwall have acquired no slight celebrity in times past by reason of the peculiar physical features which they exhibit, and on account of the immense quantity of black magnetic iron ore which they contain, and which, for a long series of years, they have furnished to the adjacent furnaces, and even now present unmistakable evidence of a far greater supply as yet untouched, above the ordinary level of the surrounding country."

The largeness of this deposit, and the cheapness with which it is mined, for it requires no underground work, but merely to be quarried, makes it the most valuable mine of iron ore in the State. The average per centage is fifty, though there is much of it that will yield sixty-five or seventy per cent., being the pure magnetic oxide of iron.

LEBANON FURNACES, situated on the summit of the Union Canal, in

North Lebanon township, Lebanon County, are owned by G. D. Coleman. They consist of two blast furnaces, capable of making ten thousand tons of pig iron per annum, and a large foundry for the manufacture of cast iron pipe. These works, when in full operation, consume about twenty-five thousand tons of anthracite coal, and twenty thousand tons of iron ore; the latter is obtained from the

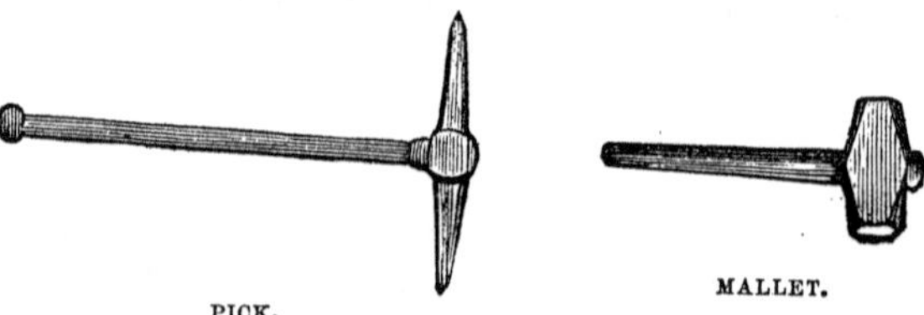

PICK. MALLET.

Cornwall Mines. The coal used is from the Pine Grove region. This region of country, on account of the cheapness and richness of the ore of the Cornwall Mines, and its great abundance, must become

WEDGE. SLEDGE.

one of the great centres for the manufacture of iron, as it presents facilities unequalled in any part of the State.

*The instruments or tools for mining* are here annexed. The *pick*,

SHOVEL.

made, according to circumstances, of various forms; but one point is generally edged, and the other pointed. The *mallet* is used for driving wedges, and striking the hand-drill. The *wedge* is driven into crevices,

HAND-DRILL. TAMPING-BAR.

or small openings, made with the pick to detach pieces from the rock or mine. The *sledge* is a mallet of from five to six pounds weight, and

LEBANON IRON WORKS.

is used to break larger pieces of rock or mine. A miner's *shovel* is pointed, so as to penetrate the coarse and hard fragments of minerals and rocks. All these tools should be well steeled and tempered, and kept in good repair.

Besides these, the miner requires the following *blasting* tools: a *hand-drill*, which is a bar of iron or steel, edged at one end and

FRONT VIEW OF A PENNSYLVANIA BLAST FURNACE.

headed at the other—both well hardened and tempered; the *scraper*, a small iron rod with a hook on one end, to take the bore-meal out of the hole ; and a *copper needle*, which is a simple wire one-fourth of an inch thick, somewhat tapered at one end. The *tamping-bar* is a bar of round iron, with a groove to fit the needle.

The erection of a furnace is a very complicated and hazardous task.

VERTICAL SECTION OF A MODERN CHARCOAL FURNACE.

The stack is always a piece of splendid masonry, requiring the most compact and heat-resisting stones. The engraving on page 111 exhibits a front view of a furnace stack, as they are ordinarily built—there being little difference in their external appearance, between charcoal and anthracite furnaces; this figure exhibits the prominent features of both.

Charcoal furnaces are built upon one general principle, but vary materially in size and appearance, as well as in their interior structure, according to the kind of ore and fuel, and similar circumstances governing their operations. The interior of the furnace-stack is lined with a wall of fire-brick, or else with fire-grained white sandstone, both of which are well adapted to resist the extraordinary heat to which it is exposed. The lining is situated a few inches from the main stack, the space between being filled in with fragments of stone, sand, and occasionally coarse mortar. This serves to protect the stack from the decomposing effects of the heat. The furnace stack is, moreover, secured from expansion by strong iron girders imbedded in it, as indicated in the engravings. The stack is generally surmounted with an iron or wooden-railing. The height of the furnace, of which the engraving on page 112 is a sketch, is thirty-five feet. The hearth measures five and a half feet from the base to the boshes; its width at the bottom is twenty-four inches, and at the top thirty-six inches. The boshes are nine feet and a half in diameter, and measure from the top of the crucible four feet, thus giving a slope of about 60°. The tryeres are twenty feet above the base of the hearth. The blast is conducted through iron pipes, laid below the bottom-stone of the hearth, into the tryeres. There is little difference, either in the interior or outward structure, between charcoal and anthracite furnaces; but to render our treatise as complete as possible, we append a view of the cross-section of the latter—that of Dr. Eckert, situated near Reading. (The Doctor is one of the most experienced, intelligent, and practical men connected with the iron-trade of this State. He formerly represented the fourteenth district in Congress—the largest and most important one in the Union. He is at present Director of the United States Mint, in Philadelphia, and we know of no man more thoroughly versed in all the practical intricacies and political economy of the coal and iron trade of Pennsylvania.) The height of this furnace is thirty-seven and a half feet; the top six feet in diameter; hearth, five feet high; tryeres twenty-two inches above its bottom; hearth, five feet square at the base, and six feet at the top; boshes inclined 67½°, or six inches to the foot, and measure fourteen feet at their largest diameter. Many of the anthracite furnaces receive their charges of ore and fuel by a very ingenious contrivance, which was first introduced at the Crane works, near Easton, and is applied at Phœnixville, Safe Harbor, and other places. A reservoir of water is put upon the trunnel-head bridge, where it is kept filled by means of force-pumps from the blast-engine. An iron chain suspended over a pulley carries one or two buckets of sheet iron, sufficiently heavy, when filled, to balance a charge of ore or coal. When

CROSS-SECTION OF AN ANTHRACITE FURNACE.

either of these is loaded below, the filler turns a stop-cock, and fills the water bucket or barrel, which descends and lifts up the charge. A valve in the botton of the water-cask, which is opened by a simple arrangement, permits the water, when it arrives at the proper place, to escape. The platform containing the ore or coal, relieved from its burthen, is charged with empty boxes or barrows, after which it descends, and the water barrel again rises. This arrangement is peculiarly advantageous when the furnace has no rear elevation by which to approach the trunnel-head, as is the case at the points mentioned, as well as the furnace of Ex-Governor Porter, situated along side of the railroad, at Harrisburg.

The opposite figure exhibits the interior of a Pennsylvania charcoal furnace of the usual dimensions—width of boshes 9½ feet; hearth 5 feet high, two feet in width at bottom, and two and a quarter at the top. Two tons and a half of ore generally produce one ton of metal. For each ton about 180 bushels of charcoal are consumed—varying more or less, according to the quality or kind of wood charred. In charcoal furnaces the fuel, of course, forms one of the leading features. West of the Alleghany, some iron establishments, and a few furnaces, are supplied with coke. To extract the tar, or bituminous matter from the coal, (as also the *sulphur*, which is injurious in the furnace, and disqualifies the coal for use,) the coals are piled on heaps, or in an oven, and heated to a certain extent, when the atmosphere is shut out by covering them with a coat of earth. At many places, however, ovens are

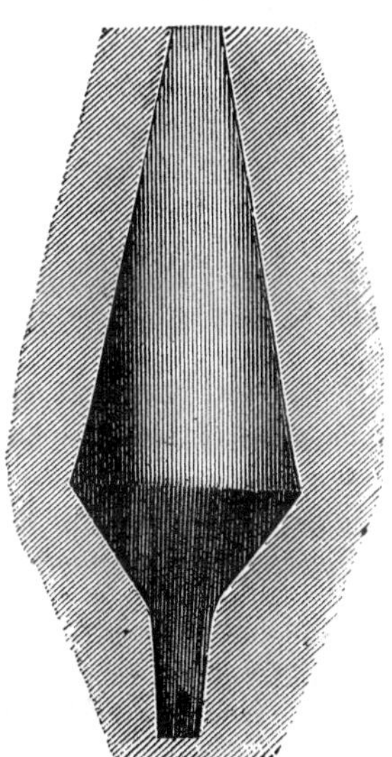

INTERIOR OF A CHARCOAL FURNACE.

COKE OVEN.

built at considerable expense, of which the annexed figure affords an illustration. The oven is erected on the side of a hill, so as to

COKE OVEN.

allow the coal to be hauled and thrown in the top with facility. The interior of the ovens is nearly circular, so as to prevent the matter hanging on the walls, as well, also, to facilitate the process of charring. Two or three tons of bituminous coal are placed in the furnace, when the fire is kindled, after which the doors, *b b* and *a*, are closed, and the bituminous matter separates from the coal, and leaves behind a spongy substance of about the same quality as charred wood. The process is very little different from that of charring wood, and various

SETTING THE WOOD FOR CHARRING.

plans are adopted to effect the same object. As there is neither anthracite nor bituminous coal in any of the counties drained by the

Juniata, (except the great Broad Top Mountain district, lately alluded to, which is consequently highly valuable,) charcoal necessarily constitutes the sole fuel of the iron works. A great difference exists in the value of wood, and charcoal has all the variety and quality of character of fossil coal—the more compact and fine-grained wood yielding the best coal, and principally because it contains less water or sap than other kinds. Tough oak, therefore, for the production of charcoal, is worth one dollar per cord; when common pine is worth about sixty cents. The Alleghanies, particularly on their western slope, are abundantly supplied with oak timber, as well as nearly every other variety generally peculiar to the State, as oaks, poplar, beech, sugar-maple, birch, pines, hickory, &c. The most common mode of burning charcoal, in this State, is in heaps, as represented in the engraving on page 116. The sticks of wood are set close together, in a nearly vertical position. In the centre of the heap, (which is about fifty feet in diameter, or less,) the largest sticks are placed sufficiently

THE PROCESS OF CHARRING.

wide apart to form a chimney, *a*. After the wood is thus carefully arranged, brush-wood and loose earth are thrown over the pile, so as to smother the flame, and prevent its bursting out from the mass of wood. The fire is applied to the wood soon after this covering is effected, which is increased from time to time, as circumstances seem to require. For the purpose of attracting the fire all around the heap, holes are made in the sides to create draft, through which the watery elements of the wood are expelled, by the heat of the hydrogen, oxygen and carbon, which is, in turn, held in check by the exclusion of atmospheric air. Were the air allowed to circulate, the entire mass of wood would be reduced to ashes. The whole process is extremely intricate, depending for complete success on the state of the weather, as well as the skill and watchfulness of the colliers. The burning lasts two or three days and nights, according to the nature of the wood, and the success attending the operation.

OVEN FOR ROASTING ORE.

The fuel being thus prepared, the next point is in regard to ore. These are of various qualities, and need not be enumerated here. Nearly all ores have to be cleaned, that is, the earthy matter associated with them is removed, either by washing or pounding. After this many ores are roasted, to effect which kilns are often constructed, somewhat similar to those for burning lime.

In the engraving on page 118, *a* is the shaft hearth, where ore and fuel are thrown in; *b b* are the grate-bars, which can be removed to let down the roasted ore; *c c* are side arches, which permit access to the draft holes; *d d d d* are four arches, including the work arch. To start operations in such an oven, the grate bars are covered with wood; upon this is placed either charcoal or coal; then a layer of coal and ore alternately, until the oven is filled, after which the fire is kindled. When the lower strata of ore are sufficiently roasted, they are taken out at the grate bars. The air-holes, *d d d d* are designed to admit air when necessary, as well as to observe the progress of the work. As the top of the ore sinks, it is replaced by fresh layers. This kind of furnace is used only for the hydrates, carburets, and other easily worked ores, but will not answer for carbonates, sulphurets, or even magnetic ores—for they are too soon smelted. They are generally roasted in heaps in the open air.

COAL BARROW.

Putting a furnace into blast is a very delicate and responsible task—requiring great prudence, watchfulness, and activity. To supply the furnace with fuel, a barrow, similar to the above, is employed. A new furnace requires firing for two or three weeks before the regular charges of ore can be thrown in. After the stack and hearth-stone are sufficiently dry, the charges of ore are introduced in small quantities, and are afterwards gradually increased.

The furnace is always exposed, at the outlet, to the liability of chilling; that is, the iron clinkers in the interior, and suddenly cools near the mouth or top-hole, impairing the draft, and not unfrequently entirely choking it up. in which case the whole interior work has sometimes to be taken out and rebuilt. Under these circumstances, introduction of the charges of ore and fuel becomes an important matter. A measure often used, similar to the above, is constructed of two half-inch round iron-bars, so connected at one end that one bar sinks into the furnace, while the other serves as a handle; *b*

CHARGE MEASURE OF FURNACES.

forming the handle; *c* the measure, and the iron plate, *a*, prevents the sinking of the rod into the materials. There are various other devices to effect the same object, but these are the most common and simple.

The philosophical principles of the blast furnace, or the causes which separate the several substances with which it is charged and precipitate the metals of the ore, are by no means easily disclosed. While the effects produced are generally well understood, the nature of those chemical and mechanical changes and combinations, formed by the decomposition of the combustible material in the blast, during the various stages of ignition through which it passes, is not easily explained. The engraving on page 121 exhibits Mr. Overman's theory of the blast, according to which it appears that, at *a*, (the points where the blast is received,) the first operation of ignition commences. Here the chemical composition of the material in the furnace undergoes an important change, the immediate result of which is a new combination of fiery matter, which is borne along with great force by the ascending draft. The iron ore, being less combustible, is thrown against the walls of the furnace, where it is liable to form benches or projecting arms of clinker. The fiery draft, by its peculiar chemical qualities, penetrates the pores of the iron ore, and *uniting with its combustible parts*, precipitates the metal in a fluid state. The metal, as it sinks, still bears off a certain quantity of the gases, as well as more or less of the earthy matter originally conglomerated with it—hence the lava which always floats upon the surface of the pure metal. To thoroughly perform this operation in the furnace requires a due mixture of fuel and fluxes with the ore, while the ore itself has often to be mixed with other ores, combining different chemical or mineral qualities, to secure safe and easy working in the furnace, as well as pure metal. What is known, therefore, as *rich* ore, (or ore which yields a large per centage of metal,) is really not rich when estimated in reference to economical working in the furnace, for such ores are necessarily heavy, compact, and hard to penetrate by the blast, thereby consuming more fuel, and exposing the furnace to irregularities and other dangers. In short. rich ores have generally to be mixed with poor ores, and, in point of economy for smelting, one is scarcely more valuable than the other. The furnace is usually tapped at intervals of twelve hours. The cinders or lava is first allowed to escape, after which the metal flows out, and travels through iron troughs, or canals made in damp sand, and reaching the beds prepared for it, is cut off into pieces of about two feet in length, and probably eight inches in circumference. Here the metal is imbedded in moulds, and becomes cool in a short time. This is what is called *pig iron*, and here ends the whole process of smelting the ore in the furnace.

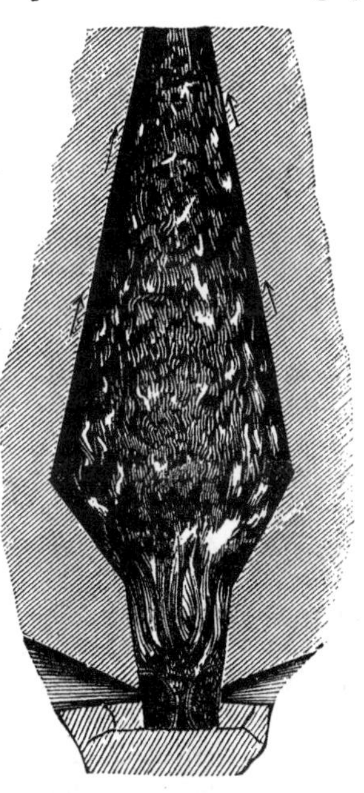

THEORY OF THE BLAST FURNACE.

One of the most interesting and beautiful sights which the varied arts of civilized man can afford, is the operation of tapping the metal from the furnace. The metal wildly issues forth in a red translucent liquid, leaping along the little banks and curvatures of the canals, as if right glad to escape from the prison where

*Black* spirits and *white*,
*Blue* spirits and *gray*,
Mingle, mingle, mingle,
They that mingle may!

Thus released from the clay and rock with which it was lately associated in the "bowels of the earth," it comes forward, crackling and sparkling, to play a very important part in the progressive and varied scenes of man. To see the stubborn, cold, hard rocks thus turned into threads of molten iron, leaping and creeping submissively at our feet, is, indeed, a scene interesting alike to our sight and pride, and gratifying to contemplate. Such scenes undoubtedly raise the human family still higher in the scale of moral grandeur—for it is here where civilized man forges his thunderbolts against ignorance and oppression;—it is here where he asserts the majesty of mind and the glory of labour;—it is here where, by the metals raised from the caverns of mountains, "he obtains strength for his hands, and subjects all nature to his use and pleasure."

After the metal leaves the furnace, it is subjected to various manipulations, depending on the use and form to which it is to be applied. The immediate object of these manipulations, is to render the metal purer and stronger, as well as to put it in size and shape for still further working. Pig iron is always converted into blooms, and this is done either in the forge-fire or the puddling furnace. The forge-fire is the oldest made, as well as the most simple—there being no material difference between it and the common fire of the blacksmith, except in the heavier character of the former. A mass of metal is thus melted and carried to the forge-hammer, which is moved by machinery, and gives a tremendous blow. Whatever impurities are in the metal, will be pretty well hammered out under this huge hammer, and the metal, as it cools, is formed into rounded pieces, about a foot in length, which are called *blooms*. In the annexed figure, p. 123, *a* is the hammer, weighing from one hundred to four hundred pounds. It is strongly wedged to the helve, *b*, which is moved by the projecting teeth *h*, of the cylinder *k*. This cylinder is made to revolve by the water or fly-wheel, *m*. The hot metal is laid under the hammer, upon the platform *d*. The metal is temporarily connected with an

FORGE HAMMER.

iron handle, so as to guide it under the hammer, and is cut off after the bloom is perfected.

After leaving the forge-hammer, the bloom is ready for conversion into rolled or bar iron of every description, preparatory to which it undergoes some additional working in the puddling or heating furnaces—especially the pig metal of the anthracite furnaces of the eastern counties of Pennsylvania, which is much more impure than charcoal iron. This iron, in fact, is not forged at all; but after being puddled is taken to the squeezer, formed into blooms, and is then ready, after re-heating, for the rollers. The puddling-furnaces are always erected in the interior of rolling-mills, and their tall chimneys are seen projecting all around the building. They are built singly and doubly, of various dimensions, but on one general principle. By their aid iron, otherwise valueless, can be made perfectly good, which cannot be done in the forge. Besides this, iron can be produced of any desired quality. The hearth is most generally composed of fire-brick, over which is a coating of furnace cinders, broken into small pieces. Upon this the metal is thrown, and heated until the whole is melted, when the puddler stirs up the mass, and rolls it into balls of sizes to correspond with the capacity of the rollers, or the size and nature of the pieces of bar iron to be produced. The engraving on page 125 exhibits the interior of an anthracite puddling furnace, from which the process of working the metal is indicated. The interior of these furnaces, as may well be supposed, is intensely hot, and great skill is required to work the metal and keep the furnace in proper working condition. The chimney tops of these furnaces, as we have just remarked, may be seen ranging along the roofs of all rolling-mills, and when they are in full operation, always present an aspect of great activity and industry. The large railroad mill at Safe Harbor, when all the doors are opened in the summer time, affords, in the evening, one of the most picturesque scenes that could be imagined. The fire of the numerous puddling and heating furnaces—the red glare of the blooms, as they are borne along to the squeezer—the pale translucent heat of the flat plates, as they are run through the rollers—the rattle and movements of the stupendous and complicated machinery—the peculiar buzz and extraordinary evolutions of the large fly-wheel—the hasty and determined movements of upwards of three hundred athletic artisans—all convey an idea of industry and enterprise perfectly magnificent to contemplate. It might be supposed that such a place, at such a *time*, would be almost as hot as the puddling furnaces themselves—but

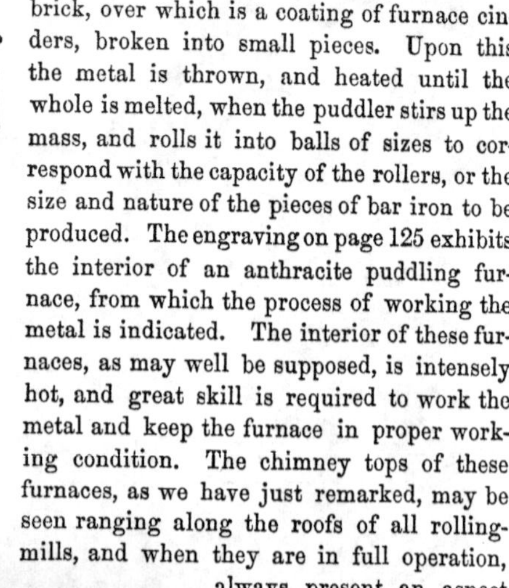

VIEW OF A PUDDLING FURNACE.

such is the ventillation of these large establishments, that they are by no means uncomfortable, notwithstanding the great heat of the fires, in the hottest weather of the season.

VIEW OF THE INTERIOR OF A PUDDLING FURNACE.

When the metal is sufficiently boiled and worked in the puddling furnace, it is rolled into as compact a ball as possible, and then with all convenient despatch is borne in iron pincers to the squeezer.

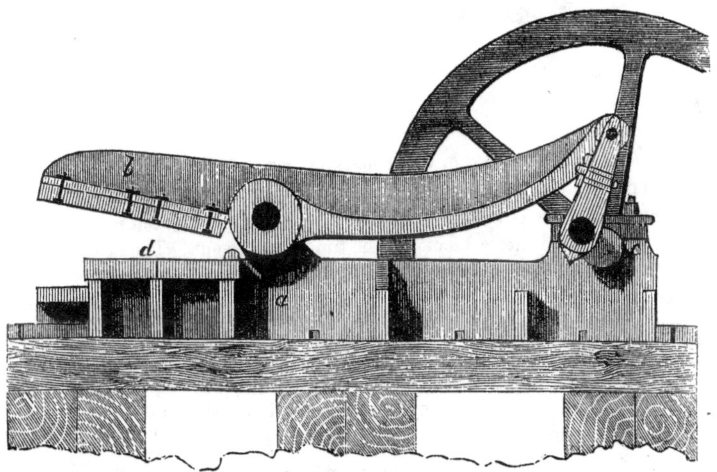


SQUEEZER FOR BLOOMS.

The object of the squeezer is indicated by the *name*. The red hot ball is placed into its iron jaws at *d*, and is thus pressed at every evolution of the wheel which drives it—the bloom being still held by tongs, and turned round as occasion requires. Whatever impurity may be in the metals is thus worked out by the squeezer, at the same time that the bloom is made perfectly solid and compact. The rotary squeezer is probably a much better machine for this purpose than any other now in use, inasmuch as it saves labor, and performs the work in a very brief space of time. The stationary part of

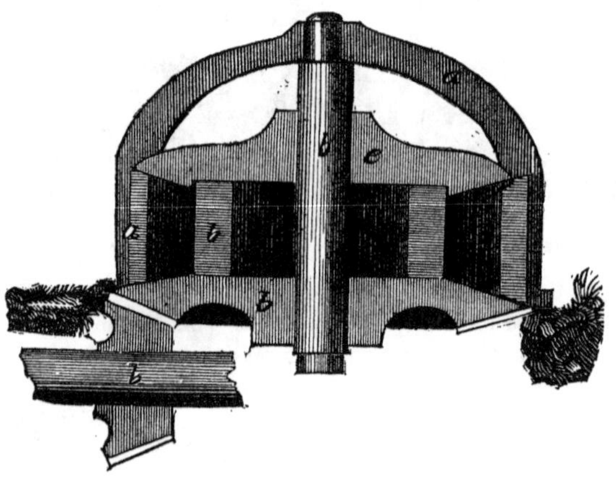


ROTARY SQUEEZER.

the apparatus is marked *a a*, and consists chiefly of a cast iron cloak, which encloses the movable parts, *b b b*. An eccentric space between the two main parts is thus left, in which the ball is placed, and is thus rolled and pressed into a bloom by the time it comes out. The heated ball makes a rumbling noise in its passage through the rotary squeezer, accompanied with one or two very loud reports or explosions. The squeezer, however, is very compactly built, and is so constructed that it cannot well be choked up or broken by too large a charge.

The blooms are generally conveyed directly from the squeezer to the roughing rollers, where they are thinned and considerably elongated. The engraving on page 127 represents a series of flat rollers, from which the gradual transition of the metal in a round to a long and

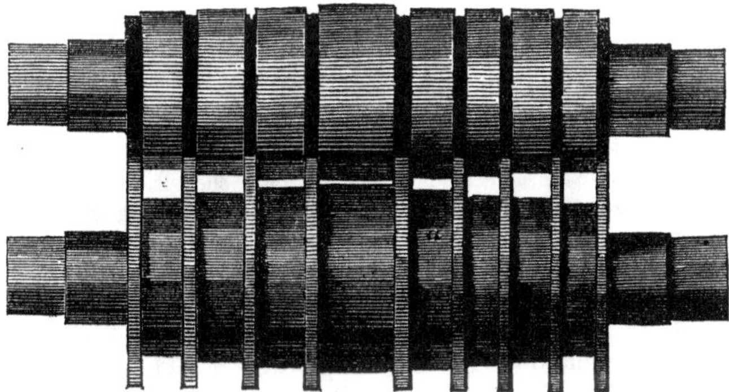

FLAT ROLLERS.

flat form will be indicated—entering the square space on the right and proceeding to the groove-rollers in the middle. The leading features of all rollers are very nearly similar—differing only in their form,

ROLLERS FOR SHEET-IRON.

strength, and dimensions. Roughing rollers are, therefore, merely adapted to the rough form of the bloom, which they elongate by pressure, and render nearly square. After leaving the roughing rollers, the iron is taken to such rollers as will reduce it to the desired shape—if it is to be run into square bars it will pass through the grooves of the flat rollers—if in broad sheets for sheet-iron, it will pass through rollers like those indicated on page 127, or if in small round or square bars, like those of the figure below. For sheet-iron and wire, charcoal iron is always best. In ancient times sheet-iron and other flat iron was hammered out from the blooms by forge-hammers, and then flattened, and the surface smoothed by smaller hammers over the anvil. This method is still pursued in some portions of Europe, where labor is not of as much consideration and value as it is in this country. For this reason we are compelled to resort to machinery whenever it can be done, and hence the proverbial ingenuity of our countrymen, as evinced in every department of the useful arts. The iron to be wrought into broad sheets must previously have been run

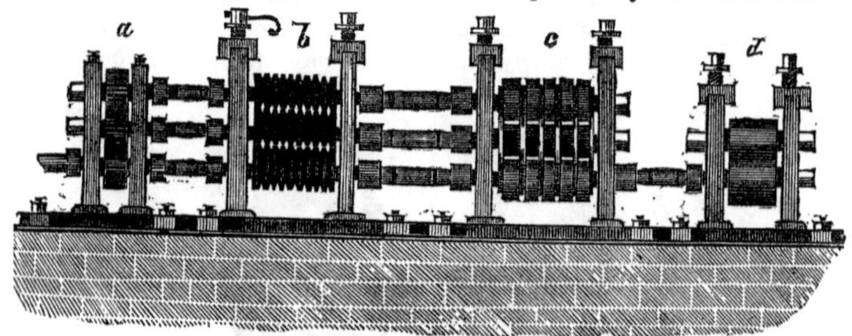


ROLLERS FOR SMALL BARS.

into flat bars. It should be a clear, white, and fibrous iron, and adapted to the progressive capacity of the rollers. The wrenches on the top screws of the rollers in the figure above, form a cross, so as to expose a handle to the workmen, by means of which they are enabled to regulate *the thickness* of the sheets, as the iron passes between the rollers. The sheet, which soon resembles sheet-iron in point of thinness, is then reheated, and again passed through the rollers, after which two sheets are rolled together. Sheet-iron is thus made of any required thickness, from the strong tenacious boiler-iron to the thin wafery sheet.

The iron for small round and square bars is run through rollers similar to the above—first passing the flat grooves at *a* and *d*, then *c*, and finally at *b*, where it is cut off into long thin rods, similar to those used by blacksmiths and machinists. The process is simple. In rolling railroad bars, the *ne plus ultra* of the art is achieved. The blooms are, of course, very heavy, and the whole process of rolling is on a scale much more stupendous than any other kind of work. The annexed figure shows the gradual transformation of the square billet, when introduced here. It is received at No. 1, and after passing through, is run through numbers 2 and 3. Number 4 presses the bottom and top smooth, and works the bottom flanch down to its

GROOVES FOR RAILROAD IRON.

proper thickness, and somewhat broader. Nos. 5 and 6 are almost of equal form and size, giving the finish to the rail. The decrement of the grooves is very limited, and there is no difficulty whatever in making a straight rail, even with one groove less. To run the heavy rails through the rollers is, as may well be supposed, a herculean task—but machinery is brought to bear in this, as in every other department of the business. Chains are suspended from sliding pulleys fixed in frame-work over the heads of the operatives, to which huge tongs are attached. These are guided by the workmen, and the long red hot rail is seized and conducted to the groove, where the revolving rollers grasp it. As soon as it appears on the other side, another set of men, with tongs in readiness, grasp it, and immediately return it through another groove. And thus, after five or six passages through the rollers, the rail appears with its peculiar form, and now only needs to be cut off smoothly at both ends, and, if crooked, straightened out, to complete it. The sawing machine, for cutting the ends of the rail,

R

is exhibited in the engraving below. The saws are circular, and are put in motion by the straps at *a*. But one end of the rail is cut at the same time—that done, the rail is drawn under the saw at the other end, and cut off in like manner. Equal lengths are not generally demanded by Railroad companies, hence they are sometimes a little longer and sometimes shorter than the uniform length sought. After the ends are thus cut off, the rail is subjected to a few simple processes to render it perfectly straight, after which the whole work is completed.

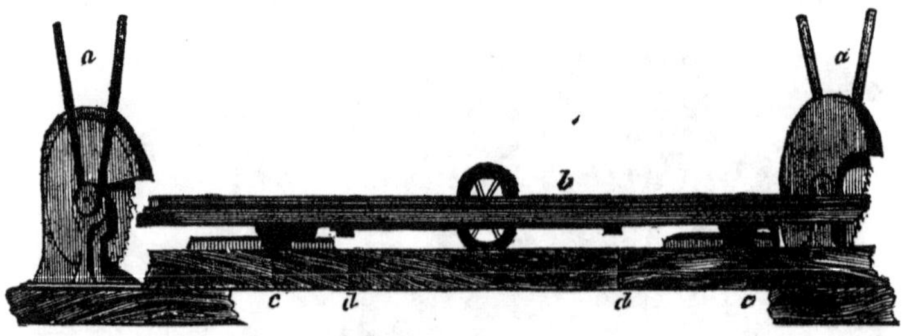

SAWING APPARATUS FOR CUTTING RAILROAD IRON.

The whole number of charcoal furnaces supposed to be in operation in Pennsylvania, is about two hundred and eighty, yielding an annual product of two hundred and fifty thousand tons. The number of anthracite furnaces we estimate at fifty, yielding one hundred thousand tons, making the total of furnaces in the State three hundred and thirty, and the aggregate yield three hundred and fifty thousand tons. Under the tariff act of 1842, the number of furnaces, especially anthracite, increased at a most extraordinary rate—having almost doubled the entire number during the few years it remained in force. The present law, however, has all along operated unfavorably, and while many works have been suspended very few new ones have been put up. There is at this time, however, probably not less than fifteen millions of dollars invested in the production of iron, exclusive of about $6,000,000 invested in rolling-mills, and similar works for the conversion of the metal into forms for use, making the aggregate sum about $21,000,000. This, we think, is a moderate estimate, based on practical data. The number of persons employed in mining the anthracite and iron ore, is about five thousand; in making the charcoal, fifteen thousand; total, twenty thousand. The number of persons directly dependent on this description of labor may be stated at seventy thousand; of those supported by their labor in the conversion of pig iron, ninety thousand; and of the population connected with the production of iron, one hundred thousand—making the total number of persons directly and indirectly concerned in iron manufactures, in Pennsylvania, including miners and colliers, two hundred and eighty thousand. Besides this a large number are employed in the manufactories of machinery; in the transportation, sale, shipment, and other branches of the trade.

We have thus endeavored to present the outline features of several of the more prominent points of inquiry upon iron manufactures. To describe particularly the whole process would require a volume thrice the extent of these pages. To those desiring more elaborate information, and scientific and practical data, we would commend the able work of Mr. Overman, already alluded to, which combines all on this subject that the most practical and curious could desire. Nearly all the foregoing illustrations are copied, by permission of the publisher, from this work: H. C. Baird, Philadelphia.

A short distance from Tyrone station is the celebrated SINKING SPRING, situated in a limestone formation, in the valley bearing the same name. It is an object of great interest, and well deserves a visit from the passing traveller. The spring, where the water emerges, is called Arch Spring, because it rises in a large limestone rock, with a high arch overhanging. As the stream runs along amidst the wildest scenery, it receives additions from smaller springs, when finally the whole volume of water disappears in a large cavern, and again enters the bowels of the earth. In the inside of this rocky cavern the stream continues from eighteen to twenty feet wide. The roof declines as you advance, and a ledge of loose rugged rock keeps in tolerable order upon one side, affording means to scramble along. In the midst of this cave are large quantities of brush, fragments of trees and branches, and such like matter, lodged quite up to the roof, thus indicating that the water, during freshets, is swelled up to the full capacity of its rocky jaws. This opening continues several hundred yards, when the cavern opens into a spacious room, at the bottom of which is a great vortex, into which the water is precipitated, and whirls round with amazing force. The stream is supposed to pass several miles under Brush and Cove Mountains, and to re-appear by two branches, which empty into the Frankstown branch of the Juniata.

Opposite the station at Tyrone, on the left, we have Tussey's Mountain, and on the right the bold ridge constituting the Bald Eagle Mountain, celebrated for its deposits of iron ore. The whole country here, however, is filled with iron ore and limestone, with some thin layers of the carbonate of lead, which, many years ago, excited a great deal of attention. A few hundred yards above this station is TYRONE CITY, a sprightly little village, sailing with flying colors under a prosperous breeze. All it wants to become a city in reality, is fair play and plenty of time. It already has the name, and makes a decent exhibit of several brick houses, among which is a hotel, which looks large enough to accommodate a crowd of hungry summer tourists. We say *hungry*, because we happen to know something about it. Turn a city "pale-face" loose among these mountains, let him ramble boldly amongst the foxes, and snakes, and bears that frequent them, and if he never before knew *Appetite*, he will soon become familiar with him.

Passing the unimportant stations of TIPTON'S RUN, FASTORIA, BELL'S MILLS, and BLAIR FURNACE, we arrive at the intersection of the branch and main line of the railway at ALTOONA. The branch road extends six miles to Hollidaysburg, where it meets the Alleghany Portage road, already mentioned as crossing the mountain by means of inclined-planes and stationary steam-engines. The main branch of the Central Railroad, between Altoona on the eastern, and a point a few miles from Conemaugh station on the western slope of the mountain, is still in an unfinished state, but will probably be ready for use in the course of the next ensuing eight months. This road boldly climbs the mountain without the assistance of inclined-planes. The ascent is accomplished in twelve and a half miles, by a maximum gradiate of eighty-four and a half feet on straight lines, reduced on curvatures, according to their diameter, to seventy-five feet upon those of minimum radii. To reduce the elevation to be overcome, from the foot to the summit of the mountain, a tunnel has been driven through it at the highest elevation of the road, which is over eleven hundred yards in length. The elevation of this tunnel above tide-water is nearly twenty-two hundred feet! The whole distance across the mountain from Altoona to CONEMAUGH station is about thirty-five miles, or about the same distance as the present Portage Railroad with its *ten inclined planes!*

A brief description of this road, or some of its most prominent

ALTOONA.

12

characteristics, may not prove uninteresting in this place. It is, without doubt, all things considered, the most complete, the most substantial, the most interesting railroad improvement yet constructed in the United States. It was commenced in 1847, and will be completed throughout, with single track and sidings, in the ensuing year. It extends from Harrisburg to Pittsburg, connecting the Ohio River with the capitol of the State, and by means of the railroads already finished, on the east, with Philadelphia, the second city in point of population on the Western continent, and first in the natural resources of the country tributary to it. The length of the Pennsylvania Railroad is two hundred and forty-eight miles, of which about two hundred and twelve miles are now in successful operation—while the whole of the remaining portion is under contract, and rapidly advancing towards completion. The route of this road, although it intersects in its course all of the mountain ranges of the State, is highly favorable.

The Alleghany Mountain is the only one not severed to its base by either the Susquehanna, Juniata, or Conemaugh Rivers, the valleys of which are followed by the railroad through the great gateways nature has opened for its accommodation. The distance from Harrisburg to Altoona, at the foot of the eastern slope of the mountains, is one hundred and thirty one miles, and the ascent overcome is eight hundred and fifty-eight feet. The steepest ascending gradient on this part of the road, passing eastwardly, is ten and a half feet per mile, and westwardly twenty-one feet per mile. From Altoona to Pittsburg the steepest gradient is fifty-two and eight-tenths feet per mile, with the exception of nearly twelve miles of the eastern slope of the mountain already referred to, where a maximum gradient of eighty-four and a half feet on straight lines, reduced to seventy-five feet by means of curvatures, is encountered, upon which extra locomotive power may be employed, locomotive stations being located on both sides of the mountain, at Altoona and Conemaugh, near Johnstown.

The Baltimore and Ohio Railroad overcomes this mountain by a maximum gradient fifteen miles in length, of one hundred and sixteen feet per mile, and descends upon the west side, at the same rate, for eight miles. It also overcomes Laurel Hill, which is avoided on our route by a tunnel four thousand two hundred feet long, with gradients on each side of one hundred and five feet per mile.

The Pennsylvania Road is graded for a double track in all the tunnels and rock cuttings, and much of the earth work; the masonry in all cases is constructed for a double track. Thirty-nine miles of the mountain division, and thirty-four immediately east of it, it is proposed to lay with a double track at once; and on the rest of the line there will be sidings every five miles.

The superstructure is of the most substantial character. The cross ties of white-oak, eight by eight inches, eight and a half feet long, placed two and a half feet apart, are imbedded in ballasts of broken stone, twenty inches in *depth*. This is one of the finest features of the road, for while it gives it a substantial and solid basis, it prevents the *accumulation of dust*, so annoying to passengers on every other railroad with which we are acquainted. Indeed, summer travelling on many railroads, by reason of the dust, is rendered a source not of pleasure, but of downright suffering and fatigue. Another splendid feature, which has already been alluded to in connection with the Safe Harbor and Phœnixville Railroad mills, is the heavy and substantial character of the railroad iron. The rails weigh sixty-four pounds to the yard, except on the steep grades of the Alleghany mountain, where their weight is increased to *seventy-six pounds to the yard!* These rails are all of American manufacture, and no one need to be told of their great superiority over similar iron imported from England. One-half the accidents occurring on railroads are to be attributed to the inferior quality and lightness of the railroad iron. The buildings and bridges, we have also observed before, are of the most approved, elegant, and substantial character; and the examples afforded by our illustrations will abundantly establish their superiority over structures of the same class on other railroad lines. In the words of the Ethiopian song:

> We've travelled East,
> And we've travelled West,
> And we've been to Alabama;

but in all our travels we never saw a more complete, systematic, and interesting railroad line than this, the pride of the Keystone State. While very little has been expended in unnecessary ornament, no expense has been spared which was required to secure substantial excellence. From Altoona to Pittsburg, one hundred and seventeen miles, there are only two wooden bridges, each of about one hundred feet span, all the others being constructed of stone or iron.

Bituminous coal abounds on the western part of the road from Pittsburg to the summit of the Alleghany Mountain, a distance of one hundred and five miles, the road passing in this distance through numerous veins varying from four to thirteen feet in thickness. The extensive coal field at Broad Top Mountain is within fifteen miles of the road, at a point one hundred and fifty-five miles east of Pittsburg and ninety west of Harrisburg, while, in the valley of the Susquehanna, the road is in the immediate vicinity of the anthracite coal region.

At Harrisburg commences the Harrisburg and Lancaster Railroad, thirty-six miles long, now leased and worked by the Pennsylvania Railroad Company. This road intersects the Columbia Railroad at Lancaster, completing the railroad communication to the city of Philadelphia. The distance from Harrisburg to Philadelphia is one hundred and six miles, but improvements are now in progress upon the Columbia Road which will save about four miles, making the whole distance, from Pittsburg to Philadelphia, three hundred and fifty miles. A railroad, running from Harrisburg *via* Cornwall, Ephrata, and Phœnixville, to Philadelphia, is now being surveyed, and will probably be completed in a short time. This will be a shorter route than the present State Road, and may possibly be used hereafter by this line for the transportation of its passengers.

At Harrisburg the line of railroads leading to Baltimore and Washington also commences. The distance from Harrisburg to Baltimore is eighty-five miles, and from Pittsburg to Baltimore, by this route, three hundred and thirty-three miles. The above eastern and southern connections are completed. Pittsburg, the western terminus of the Pennsylvania Railroad, is a most important manufacturing city, and has been appropriately styled the "Birmingham of America." Its population, including that of the suburban towns, is about one hundred thousand. The position of the city, on the Ohio, at the head of navigation for first-class steamers, connects it, through the Mississippi and its tributaries, with the south and west by several thousand miles of continuous steamboat navigation, which alone will draw to it sufficient business for transportation to and from the seaboard to insure the success of the Pennsylvania Railroad. But as this navigation is subject to interruptions from low water, regularity of intercourse, as well as a direct connection with the interior and the lakes, seemed to demand railroad facilities to secure the control of the travel and carriage of valuable freights to this route.

The railroads and canals hitherto constructed to accommodate the population of Ohio and Indiana have generally a north and south direction, connecting the fertile central region with the Ohio River and Lake Erie. Within a few years the importance of a more direct eastern communication with the seaboard has been appreciated, and several great leading lines have been projected and commenced to secure this object. That which has made most progress is the Ohio and Pennsylvania Railroad, commencing at Pittsburg, and extending westwardly through the most fertile and populous part of Ohio, to the new town called Crestline, on the Columbus and Cleveland Railroad, a distance of one hundred and eighty miles. From Crestline a railroad is completed to the city of Cincinnati, on the Ohio River; another has been commenced to Fort Wayne, in Indiana, which will be extended to Chicago, on Lake Michigan; another will be completed in the coming twelve months, from Crestline, through Bellefontaine and Indianapolis, to Terre Haute, on the western boundary of the State of Indiana, a distance of two hundred and seventy-five miles. The extension of this to the city of St. Louis, on the Mississippi, one hundred and seventy miles further, has been commenced. Of the completion of this entire direct continuous railroad from Philadelphia to St. Louis, a distance of nine hundred and seventy-six miles, within two years, there can be no doubt. The region traversed by this route is equal in fertility to any portion of the globe, and is inhabited by a people who have the sagacity and enterprise to improve and draw from it all that the bountiful hand of the Creator has designed for it.

The Ohio and Pennsylvania Railroad is now completed to Massillon, one hundred and four miles. At Alliance, eighty-four miles from Pittsburg, it intersects the Cleveland and Pittsburg Railroad, which is completed from that point to Cleveland, making a direct railroad communication between Pittsburg and Lake Erie, one hundred and forty miles long. From Cleveland to the City of New York, by way of the Pennsylvania Railroad, the distance is now forty miles shorter than by the New York and Erie Railroad; and must consequently command the travel from the Western States to that commercial emporium.

The Pittsburg and Steubenville Railroad will connect this line with the Steubenville and Indiana Railroad, and accommodate the centre of Ohio; while the Hempfield Railroad from Greensburg, thirty miles east of Pittsburg, to Wheeling, will connect it with the southern portions of that State, through the Marietta, Chilicothe and Cincinnati road, upon which line an unbroken guage of track may be secured to St. Louis. These form the leading eastern communications in Ohio, already alluded to; and to these, and especially the Ohio and Pennsylvania road, all the north and south lines from Cleveland, Sandusky, Lexington, Louisville, Evansville, &c., will become tributaries, concentrating the trade and travel of the great Mississippi basin, and pouring it over the Pennsylvania Railroad and the main trunk connecting the commercial and manufacturing interest of the East with the rich agricultural regions of the West.

Calculations of the amount of transportation and travel that will pass over this great highway appear, in view of these facts, to be superfluous. All its rivals are inferior in character, more expensive to work, and encumbered by a disproportionate debt. It has therefore nothing to fear from rivalry, either on the north or the south; and its business will only be limited by the capacity of a first class double track railroad.

In its present incomplete condition it yields a net revenue of more than *eight per cent.* upon the capital expended in its construction, and has attained a tonage, ere it has reached its western terminus, nearly as great as can be carried with regularity upon a single track road.

The entire estimated cost of the road, finished with a single track and sidings, and equipments, including freight and passenger stations at Philadelphia, is $12,300,000. The whole amount of subscriptions, thus far, exceed ten millions of dollars, and the work of the Company has been prosecuted without incurring a dollar of debt. The remaining amount to complete and equip the road is now being subscribed, and presents a splendid inducement for the investment of the capital.

The following statement exhibits the receipts and expenditures of the road for the year ending 1851:

| | |
|---|---|
| From Passengers, Mails, Express, &c. on Pennsylvania Railroad, - - - - - - | $315,145 33 |
| From Lancaster, Columbia, and Portage Railroads, | 371,164 54 |
| Total Receipts from Passengers, Mails, &c., - - | 686,309 78 |
| Total Receipts from Freight, - - - | 353,255 72 |
| Total Receipts, - - | 1,039,565 59 |

EXPENSES.

| | | |
|---|---|---|
| Maintenance of Way, - - | $51,547 66 | |
| Maintenance of Cars, - - | 20,611 28 | |
| Motive Power, - - - - | 78,173 68 | |
| Conducting Transportation, including tolls and expenses on State and Harrisburg & Lancaster Railroads, | 556,307 57 | |
| | | 706,640 19 |
| Balance, or Total Net Receipts, - - - | | $332,925 40 |

The following statement will exhibit the receipts of the Company since the commencement of the present fiscal year.

| | |
|---|---|
| For January, - - - - - - | $ 93,772 50 |
| February, - - - - - | 157,251 13 |
| March, - - - - - - | 245,373 71 |
| April, - - - - - - - | 206,408 94 |
| May, - - - - - - - | 163,183 92 |
| June, - - - - - - - | 123,752 83 |
| July, - - - - - - - | 114,479 52 |
| August, 1st week, - - - - - | 28,040 04 |
| Total, - - - | $1,132,262 59 |
| Same months last year, | 584,995 92 |
| Difference, or increase, - | $547,266 67 |

ALTOONA, (two hundred and thirty-eight miles from Philadelphia, one hundred and twenty-five from Pittsburg, and elevated eleven hundred and sixty-eight feet above tide-water,) will ultimately become one of the most important places on this route. And it is a source of satisfaction to perceive that there is plenty of room, and that admirably situated, for a large and flourishing town. The surrounding country, being the rich slope of the Alleghany, is highly cultivable, and only needs an industrious farming population to clothe it in the lively colors of growing crops. Altoona contains the machine-shops and engine-houses for the western section of the road, and the hands employed in them together with the agents of the road stationed here, will be quite sufficient to people a village of more than ordinary pretensions. The Railroad Company has already erected several handsome buildings, besides the machine shops referred to, (which will soon be enlarged to twice their present capacity) and after the piece of road overcoming the mountain is finished, a large and splendid hotel will be added, with numerous other buildings for private residences. If lots can be purchased, we know of no place along the line where money could be invested with a safer prospect of future profit.

Six miles further, following the branch road, we reach the foot of the Alleghany mountain, our steam-horse bringing us directly in front of a spacious and showy hotel, very properly called *the Mountain House.* HOLLIDAYSBURG, the termination of the eastern division of the Pennsylvania Canal, lies about one mile south of us. It is a busy place, suddenly called into existence and principally supported by the trade of the canal, and contains a population of about twenty-five hun-

THE MOUNTAIN HOUSE.

dred. It is otherwise without interest, and will probably lose some of the trade it now enjoys after this portion of the road is avoided by the Central Railroad.

The Mountain House stands on an elevation, above tide-water, of about twelve hundred feet. The ascent from this place to the summit of the mountain is nearly fourteen hundred feet—consequently, on reaching it, we will find ourselves in the region of clouds, upwards of twenty-six hundred feet above the Delaware River at Philadelphia!

Our course, therefore, is still onward and *upward*. Hitched to a little old rickety locomotive, of the extravagant days of Governor Ritner, —he, of whom it was sung, in Pennsylvania Dutch—

> "Der Joseph Ritner is der mon
> As unser Staat rigeren kon!"

we are tugged, two or three miles, over a steep ascending grade, to the foot of the first inclined-plane. Here the cars are attached to an endless wire-rope, winding round large iron wheels, placed horizontally, at each end of the plane. When all is ready, a signal is given to the engineer at the head of the plane, who immediately sets the stationary steam-engine in motion, and the rope begins its accustomed travel. It is prevented from touching or dragging the ground by numerous little wooden wheels, which revolve rapidly whenever the rope falls low enough to touch them. The ascent is soon made, and the same process is repeated at each of the other planes. At first the novelty of the operation excites some interest, but this gradually wears off, and the slow progress in travelling produces a feeling of impatience. The scenery is sometimes very inspiring; but, for the most part, the Alleghany has little of striking effect. The ascent, too, is so gradual, and the distance to the summit so considerable, that the actual height is never fully realized. The summit of the mountain forms the boundary line between Cambria and Blair Counties, and the village on its top *ought* to be the best place in Pennsylvania, because it is unquestionably the nearest to heaven.

Making maple-sugar, in this region of country, is one of the characteristic employments of the people—though it is carried on to a much greater extent in the adjoining county of Somerset. The quantity of this sugar, raised in this State in 1850, as appears by the census, was two million two hundred and eighteen thousand, six hundred and forty-four pounds, from which it will be seen that it is by no means an inconsiderable item of our domestic products. Indeed, we have no doubt but that this amount, large as it seems, might readily be trebled and quadrupled with profit, were the matter reduced to the common basis of a regular and systematic business. Immense districts, otherwise unproductive, might be timbered with these sugar-bearing trees, and large sums annually realized from their productions without in the least depreciating the value of the trees for timber. If we are not greatly deceived, this sugar-maple business

BOILING MAPLE-SUGAR ON THE ALLEGHANY.

will ultimately become important—that is, it will enter the market in such quantity as to offer a determined competition to the products of the sugar-cane.

The sugar-maple is a beautiful tree, reaching the height of seventy or eighty feet, the body straight, for a long distance free from limbs, and three or four feet in diameter at the base. It grows in older climates, between latitude forty-two and forty-eight, and on the Alleghanies to their southern termination, extending westward beyond Lake Superior. The wood is nearly equal to hickory for fuel, and is used for building, for ships, and various manufactures. When tapped as the winter gives place to spring, a tree, in a few weeks, will produce five or six pailfuls of sap, which is sweet and pleasant as a drink, and when boilded down will make about half as many pounds of sugar. The manufacturer, selecting a spot central among his trees, erects a temporary shelter, suspends his kettles over a smart fire, and at the close of a day or two will have fifty or a hundred pounds of sugar, which is equal to the common West India sugar, and when refined equals the finest in flavor and in beauty. "When the sap has been boiled to a syrup and is turning to molasses, then to candy, and then graining into sugar, its flavor is delightful, especially when the candy is cooled on the snow." The person in the engraving is represented as blowing the wax or candy to ascertain how far the boiling has advanced. Maple-sugar boiling is also a favorite amusement with the Yankees down in "Varmeount." A correspondent of the Boston *Atlas*, writing from Burlington, in that State, gives us the following particulars, which are equally applicable to this quarter: "At this season, the spring, sugar orchards become places of much resort, especially for those who love the sweet things of life. In this village parties are frequently formed, to take a trip to some sugar orchard in an adjacent town, and there regale their palates with maple molasses. The maple-sugar manufactories are generally located in romantic spots—in some beautiful valley, or on some delightful hill-side, where the air is pure and invigorating, and the landscape views enchanting and picturesque. Vermont contains thousands of such delightful retreats; and at this season of the year, when the crystal waters of the brooks are released from their frozen bands, and come leaping down the mountain sides, waking the beautiful trout from his winter's sleep, and filling the valleys and groves with sweet music, it is pleasant to visit these sugar orchards, drink sap, lap maple molasses, and make love. Make love! Ah! thereby hangs a tale. Let the Vermont ladies beware; for in such places they may fall in love, while they would not dream of such a thing in their quiet homes. The delicious saccharine qualities of maple molasses, presented to the swelling lips of a beautiful lass by the hand of a smiling swain, has a wonderfully softening effect upon the head, and creates a pleasant dreamy sensation through all the nervous system, especially when it is powerfully aided by romantic woodland scenes, and the music of a thousand cascades. And young gentlemen, too, may need a word of caution on such occasions, and under the pressure of such peculiar circumstances. An able English writer said, many years ago, when human nature was just what it is now, that it was dangerous for a Benedict to select a wife in a ball-room, when her disposition was sweetened by the music of the violin. But what are the streaming notes of the fiddle, in sweetening the female heart, when compared with the luxury of maple molasses? But a word to the wise is sufficient; we will not follow out the comparison." Indeed, there are no more joyous seasons than these festive scenes that serve as occasions to bring together the "guide country folks and the lads and lasses," while the cool bracing air gives zest to the labor of preparing the delicious sweets.

A few miles from the summit, in descending the western slope of the mountain, we avoid the last inclined-plane, as well as the long level of the Portage road, and again strike that of the Central Railroad. The first object here that meets our attention is a singularly

ROCK CUTTING AND CURVATURE OF THE CONEMAUGH.

wild scene on the Conemaugh. A steep and narrow ridge of rocks, projecting from the western slope of the valley, drives the little stream nearly a mile out of its course, where it suddenly wheels round and passes along within a few feet of the place of deviation, on the opposite side of the ridge. The above figure exhibits a view of the scene on the eastern side, and the passage of the railroad through the ridge, whose position to the river is somewhat like the blade of a sword—the massive handle representing the mountain, while the projecting blade shows the nature of the rocky ridge around which the stream is forced to travel. In re-appearing on the western side of the ridge, the river has descended some twenty feet or more, and is crossed by a stone viaduct. This is one of the finest specimens of massive architecture to be found in this country—consisting of a single arch, with a span of eighty feet, and nearly seventy-five feet above the water of the stream. While it can scarcely be surpassed in the neatness and symmetrical proportions of the design, it is as durable as the eternal foundation upon which it rests. The Conemaugh, passing through the

THE BIG VIADUCT, AND THE SHARP RIDGE ON THE WESTERN SIDE.

Big Viaduct in a north-eastern course, soon winds round again, and descends towards the west. A short distance from the viaduct, therefore, it again crosses the railroad, or rather, the railroad crosses it, with a splendid iron and wooden bridge, having an elevation of seventy-three feet. The bridge is approached through a cut over one hundred feet

BRIDGE BELOW THE VIADUCT.

in depth—the elevation thus penetrated consisting of a drift-shale rock. This excavation is the deepest to be found on the road—probably the deepest of any other railroad in the country. The scene here, like the others just mentioned, is perfectly wild—deeply, gloriously, sublimely wild! Stones are scattered along the mountain sides in dire disorder, among which shoot up tall and stately pine trees, with an occasional oak and maple. Below rushes the torrent as if mad with the opposing obstacles which have lately impeded its course, and as the fruits of its savage impetuosity, large rounded stones are scattered along its banks, which have been borne down from the stony ridges obtruding in its bed.

We are now entering—nay, we are already within, the great Alleghany bituminous coal region—that vast and extraordinary assemblage of vegetable matter which, in an economical and moral view is worth fifty Californias, were each California a lump of massive gold! This, with the adjacent fields east and west of the Missouri River, contains at least twice the amount of workable coal of all the rest of the world *combined;* and lying, as they all do, within the valleys drained by the Mississippi and its numerous tributaries, their future value in dollars and cents is perfectly incalculable—entirely beyond the scope of human computation. The whole country, from the Gulf of Mexico to the Gulf of the St. Lawrence and Newfoundland, at one period of the earth's history—with here and there a probable barren spot—comprised *one vast coal region,* the most extensive remaining portion being the field we are now in, extending over a large portion of the area drained by the Ohio River and its tributaries. Detached portions of this once unbroken coal region are scattered along the Rio Grande and Chihuahua Rivers in Mexico, as well as their branches rising in the interior of Texas; other beds are found, separated only by a few miles, along the Red and Arkansas Rivers, flowing into the Mississippi; while further north lies the great Missouri coal field, separated from that of the great Illinois field only by the Missouri and Mississippi Rivers, which pass through both. The south-eastern point of this field is pierced by the Ohio River, and it approaches within some fifty miles of the great Alleghany field, which, lying in a north and south position, runs parallel, with the Atlantic coast, nearly seven hundred and fifty miles, at an average distance from it, in a straight line, of about two hundred miles. Thus it originally traversed portions of the State of New York and the New England States, where deposits of anthracite are still found; but the encroachments of the sea, in its northern course, have overflowed the beds, and left those of Nova Scotia and Newfoundland exposed, literally emerging, as they do, from the watery depths encircling them.

The single State of Pennsylvania has a greater area of coal than all Great Britain, Ireland, Scotland, Wales, Spain, France and Belgium, *united!* It is only exceeded by the British provinces of New Brunswick, Nova Scotia and Newfoundland, which, as before stated, are but continuations of our own great region, and which contain a net area of some eighteen thousand square miles—that of Pennsylvania being not quite sixteen thousand square miles, or one-third of its entire physical area. The coal strata in the British provinces are not as productive as our own—the seams are "faulty," and the coal full of earthly impurities. Large as this body of coal appears, some other States, (with a larger territory, however,) have a still greater amount; or, rather, a greater amount of coal *area*—for in regard to *quality*, Pennsylvania, from the *inclined dip of* her *anthracite coal veins*, still has more coal than those States where the veins lie in a horizontal position, no matter what their *area* of coal land. The following table exhibits the quality of coal land of each State where it is found:

| States. | Area of the States. | Coal areas. |
|---|---|---|
| | SQUARE MILES. | SQ. MILES. |
| 1. Alabama, | 50,875 | 3,400 |
| 2. Georgia, | 58,200 | 150 |
| 3. Tennessee, | 44,720 | 4,300 |
| 4. Kentucky, | 39,015 | 13,500 |
| 5. Virginia, | 64,000 | 21,195 |
| 6. Maryland, | 10,829 | 550 |
| 7. Ohio, | 38,850 | 11,900 |
| 8. Indiana, | 34,800 | 7,700 |
| 9. Illinois, | 59,130 | 44,000 |
| 10. Pennsylvania, | 43,960 | 15,687 |
| 11. Michigan, | 60,520 | 5,000 |
| 12. Missouri, | 60,384 | 6,000 |
| | 565,283 | 133,382 |

Reduced to acres, we have, in the twelve States above mentioned alone, something like three hundred and sixty-one millions, eight hundred and eighty-one thousand one hundred and fifty-one acres of coal land! Of this Pennsylvania has over twenty-eight millions two hundred thousand acres, embracing (which the other States do not,) anthracite, bituminous, and semi-bituminous coal. The greatest length of the Alleghany coal-field is seven hundred and fifty-six miles; its greatest breadth one hundred and seventy-three, and its average breadth about eighty-five miles. There are some thirty or forty points where coal is regularly mined, the average thickness of the veins worked being about ten feet—some of them being twenty-two, while those along the Conemaugh and the Portage Railroad, near us, are from ten to fifteen feet thick At Frostburg there is a vein twenty-six feet in thickness. The annual amount of coal shipped to the eastward, by way of the Juniata and the north branch of the Susquehanna, will not exceed sixty thousand tons. Of course the greatest amount raised is consumed in and around Pittsburg, and sent down the Ohio River, the actual amount of which in tons we can make no estimate, there being no record or statistics by which it can be ascertained satisfactorily. The annual consumption, at a single estimate, (taking in view all the outlets, including that to Lake Erie,) we should compute at one *million tons* —most probably more; certainly not less. The anthracite coal region we have discussed at sufficient length in Off-hand Sketches, Part II., another part of this work, to which we beg leave to refer our readers. Our remarks upon that region are illustrated with cuts, nearly all of which would be applicable to this region, there being little difference between the two districts except in the process of mining, and a material dissimilarity in the structure and position of the strata.

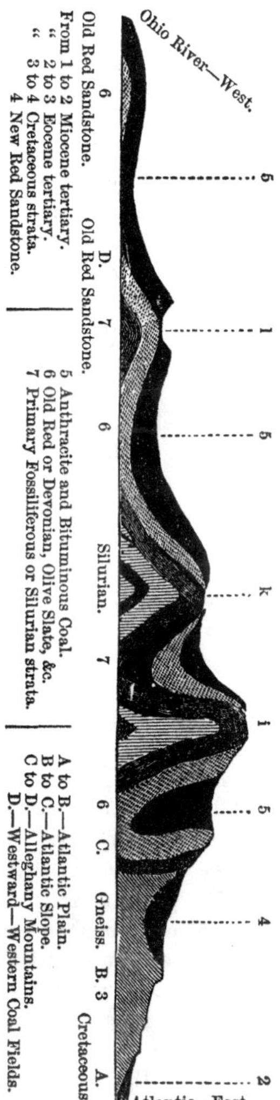


STRUCTURE AND ORIGIN OF THE ALLEGHANIES.—The within geological section (extending from near the Atlantic some four hundred miles westward) exhibits the general order of structure, and relative elevation of the wonderful region we are traversing. Starting in the east, we first cross a low plain, extending to the letter A, which was called the alluvial plain by the first geographers. A portion of this plain has been omitted in the engraving, for the want of room. It is occupied by tertiary and cretaceous strata nearly horizontal, and in general contains no hard and solid rocks, and is usually not more than from fifty to one hundred feet high, at any point along the coast, from New Jersey to Virginia. In these States this zone is not many leagues in breadth, but it acquires a breadth of one hundred and one hundred and fifty miles in the southern States, and a height of several hundred feet towards its western limits. This is called the *Atlantic plain.*

The next belt, running from B to C, consists of granite rocks, chiefly gneiss and mica-schist, covered occasionally with *unconformable* red sandstone (marked figure 4). It is from twenty-five to thirty miles in width, and traversed by the Conewago Hills, or South Mountain range. This is called the *Atlantic slope,* because it brings us to the first mountain range belonging to the Alleghany series, the Kittatinny; and contains itself numerous elevations, cut up by streams and otherwise much worn away, ranging in height from one to three hundred feet. On either side of this sandstone district are extensive layers of limestone, which we have elsewhere spoken of at length.

From C to D we have all the mountain elevation constituting the great Alleghanies. Leaving the new red sandstone formation, (which is distinguished from the older formation by its occupying a horizontal position, or rather by its not laying in regular order with the older strata) we cross a narrow projecting belt of the primary fossiliferous group, constituting what is called the Silurian system of rocks. This group of strata, it will be seen, occupies the most elevated position of the mountain region. The next formation above it is that of the old red sandstone, which, it will be seen, forms the bed of the coal throughout the whole extent of country, whenever it appears, extending westward from D some five hundred miles, with the exception of the district previously referred to, separating the Alleghany from the Illinois coal-field, and where the Silurian Rocks again rise to the surface. But from D westward the coal invariably lies in a horizontal or flat position, with a few local exceptions, where it has been raised to a slight elevation, not exceeding one hundred feet in any case.

A few days' observation of the identity of the fossil plants, and the relative position of the anthracite, satisfied me that it was of the same age as the bituminous coal which I had seen at Blossberg. This opinion was, I believe, first promulgated by Mr. Featherstonehaugh, in 1831, at a time when many geologists were disposed to assign a higher antiquity to the anthracite than to the bituminous coal measures of the United States. The recent surveys have now established this fact beyond all question, and hence it becomes a subject of great interest to inquire how these two kinds of coal, originating, as they did, from precisely the same species of plants, and formed at the same period, should have become so very different in their chemical composition. In the first place, I may mention that the anthracite coal measures, occurring in the eastern or most disturbed part of the Alleghany chain, are fragments or outlayers of the great continuous coal-field of Pennsylvania, Virginia, and Ohio, which occurs about forty miles to the westward. This coal-field is remarkable for its area, for it is described as extending continuously from north-east to south-west for a distance of seven hundred and fifty miles, its greatest width being about one hundred and eighty miles—extending from the northern border of Pennsylvania as far south as near Huntsville, in Alabama. It must have measured originally, before reduced by denudation, at least nine hundred miles in length, and in some places more than two hundred miles in breadth. By reference to the section it will be seen that the strata of coal are horizontal to the westward of the mountain region, and become more and more inclined and folded as we proceed eastward, from D to C. Now it is invariably found that the coal is most *bituminous* towards its western limit, where it remains level and unbroken, and that it becomes progressively debituminized as we travel south-eastward towards the more bent and distorted rocks. Thus, on the Ohio, the proportion of hydrogen, oxygen, and other volatile matters, ranges from forty to fifty per cent. Eastward of this line, on the Monongahela, it still approaches forty per cent., where the strata begin to experience some gentle flexures. On entering the Alleghany Mountains, where the distinct anticlinal axes begin to show themselves, but before the dislocations are considerable, the volatile matter is generally in the proportion of eighteen or twenty per cent. At length, when we arrive beyond *i*, associated with the boldest flexures of the Appalachian chain, where the strata have been actually turned over, as near Pottsville, we find the coal to contain only from six to twelve per cent. of bitumen, thus becoming a *genuine anthracite.—Trans. of Ass. of Amer. Geol., p.* 470.

It appears from the researches of Liebig and other eminent chemists, that when wood and vegetable matter are buried in the earth, exposed to moisture, and partially or entirely excluded from the air, they decompose slowly and evolve carbonic acid gas, thus parting with a portion of their original oxygen. By this means they become gradually converted into lignite or wood-coal, which contains a larger proportion of hydrogen than wood does. A continuance of decomposition changes this lignite into common or bituminous coal, chiefly by the discharge of carburetted hydrogen, or the gas by which we illumine our streets and houses. According to Bischoff, the inflammable gases which are always escaping from mineral coal, and are so often the cause of fatal accidents in mines, always contain carbonic acid, carburetted hydrogen, nitrogen, and olifiant gas. The disengagement of all these gradually transforms ordinary or bituminous coal *into anthracite,* to which the various names of splint coal, glance coal, calm, and many others have been given.

We have thus seen that, in the Alleghany coal-field, there is an intimate connection between the *extent* to which the coal has parted with its gaseous contents, and the *amount of disturbance* which the strata have undergone. The coincidence of these phenomena may be attributed partly to the greater facility afforded for the escape of volatile matter, where the fracturing of the rocks had produced an infinite number of cracks and crevices, and also to the heat of the gases and water penetrating these cracks, when the great movements took place, which have rent and folded the mountain strata. It is well known that, at the present period, thermal waters and hot vapors burst out from the earth during earthquakes, and these would not fail to promote the disengagement of volatile matter from the carboniferous rocks.

To the elaborate and faithful surveys of Prof. Rogers, and the late R. C. Taylor, we owe the discovery of the clue to the general law of structure prevailing throughout this important range of mountains; which, however simple it may appear when once made out and clearly explained, might long have been overlooked, amidst so great a mass of complicated details. It appears that the bending and fracture of the beds is greatest on the south-eastern or Atlantic side of the chain, and the strata become less and less disturbed as we go westward, until at length they regain their original or horizontal position. By reference to the section, it will be seen that on the eastern side, or on the ridges and troughs nearest the Atlantic, the south-eastern dips predominate, in consequence of the beds having been folded back upon themselves, as in *i*, those on the north-western side of the arch having been inverted. The next set of arches (such as *k*) are more open, each having its western side steepest; the next (*l*) opens out still more widely, and so it continues on until the strata become low and level. In nature, or in a true section, the *number* of bendings or parallel folds is so much greater, that they could not be expressed in a diagram without confusion. It is also clear that large quantities of rock have been removed by aqueous action or denudation, or will appear by the cuts or fissures at *l*, *k*, and *i*, as well as hundreds of other places which could not be indicated in the section.

The movements which imparted so uniform an order of arrangement to this vast system of rocks, must have been contemporaneous, or belonging to one and the same series, depending on some common cause. Their geological date is unusually well defined. We may declare them to have taken place *after* the deposition of the carboniferous strata (5), and *before* the formation of the red sandstone, figure 4. The greatest disturbing and denuding forces have evidently been exerted on the south-eastern side of the chain, and it is here that *igneous* or *plutonic rocks* are observed to have invaded the strata, forming dykes, some of which run for miles in lines parallel to the main direction of the mountains, or N. N. East and S. S. West.

According to the theory of Professor Rogers, the wave-like flexures, above alluded to, are explained by supposing the strata, when in a plastic state, to have rested on a widely extended surface of fluid lava, and elastic vapors and gases. The billowy movement of this subterranean sea of melted matter imparted its undulations to the elastic overlying crust, which was enabled to retain the new shapes thus given to it by the consolidation of the liquid matter injected into fissures.*

For my own part, I cannot imagine any real connection between the great parallel undulations of the rocks and the real waves of a subjacent ocean of liquid matter, on which the bent and broken crust

* Trans. of Ass. of Amer. Geology, 1840—2, p. 515.

may once have rested. That there were great lakes, or seas of lava, retained by volcanic heat for ages in a liquid state, beneath the Alleghanies, is highly probable, for the simultaneous eruptions of distant vents in the Andes leave no doubt of the wide subterranean areas permanently occupied by sheets of fluid lava in our own times. It is also consistent with what we know of the laws governing volcanic action to assume that the force operated in a linear direction, for we see trains of volcanic vents breaking out for hundreds of miles along a straight line, and we behold long parallel fissures, often filled with trap or consolidated lava, holding a straight course for great distances through rocks of all ages. The causes of this peculiar mode of development are as yet obscure and unexplained; but the existence of long narrow ranges of mountains, and of great faults and vertical shifts in the strata, prolonged for great distances in certain directions, may all be results of the same kind of action. It also accords well with established facts, to assume that the solid crust overlying a region where the subterranean heat is increasing in intensity, becomes gradually upheaved, fractured, and distended—the lower part of the newly opened fissures becoming filled with fused matter, which soon consolidates and chrystallizes. These uplifting movements may be propagated along narrow belts, placed side by side, and may have been in progress simultaneously, or in succession, in one narrow zone after another.

When the expansive force has been locally in operation for a long period, in a given district, there is a tendency in the subterranean heat to diminish—the volcanic energy is spent, and its position is transferred to some new region. Subsidence then begins, in consequence of the cooling and shrinking of subterranean seas of lava and gaseous matter; and the solid strata collapse in obedience to gravity. If this contraction takes place along narrow and parallel zones of country, the incumbent flexible strata would be forced, in proportion as they were let down, to pack themselves into a smaller space, as they conformed to the circumference of a smaller arc. The manner in which undulations may be gradually produced in pliant strata, by subsidence, is illustrated on a small scale by the *creeps* in coal-mines; there both the overlying and underlying shales and clays sink down from the ceiling, or rise up from the floor, and fill the galleries which have been left vacant by the abstraction of the fuel. In like manner the failure of support arising from subterranean causes may enable the force of gravity, though originally exerted vertically, to bend and squeeze the rocks as if they had been subjected to lateral pressure.

> "Earthquakes have raised to heaven the humble vale,
> And gulfs the mountain's mighty mass entombed,
> And where the Atlantic rolls, wide continents have bloomed."

In applying these lines to the physical revolutions of the territory at present under consideration, we must remember that the continent which bloomed to the eastward, or where the Atlantic now rolls its waves, was anterior to the origin of the carboniferous strata which were derived from its ruins; whereas the elevation and subsidence supposed to have given rise to the Appalachian ridges was subsequent to the deposition of the coal-measures. But all these great movements of oscillation were again distinct from the last upheaval which brought up the whole region above the level of the sea, laying dry the horizontal new red sandstone, (No. 4,) as well as a great part of, if not all, the Appalachian chain.

The largest amount of denudation is found, as might have been expected, on the south-eastern side of the chain, where the force of expansion and contraction, of elevation and subsidence, has been greatest. The first set of denuding operations may have taken place when the strata, including the carboniferous, were first raised above the sea; a second, when they sank again; a third, when the red sandstone (No. 4.), after it had been thrown down on the truncated edges of the older strata, participated in the waste. The great extent of solid materials thus removed, must add, in no small degree, to the difficulty of restoring in imagination the successive changes which have occurred, and of recounting in a satisfactory manner for the origin of this mountain chain.*

Hollo! Here's CONEMAUGH STATION! We are down the mountain —we are west of the Alleghany, to-be-sure! Ah! we are down to twelve hundred and twenty-six feet again *above* tide-water; two hundred and seventy-six miles from Philadelphia, and only eighty-seven from Pittsburg. Here is a neat brick engine-house and machine-shop for the accommodation of the iron nags who tug us over the western divisions of the road, and a characteristic water-station to

* Abridged from Sir Charles Lyell's Travels in America.

CONEMAUGH STATION.

refresh them when they are dry. Running along the northwestern bank of the Conemaugh, we reach JOHNSTOWN, two miles below. At

JOHNSTOWN.

this place the western division of the Pennsylvania canal commences, and the miserable Portage Railroad, with its short splintery rails and curvatures, its stationary steam-engines and abominable inclined plains, terminates. The traveller, who has crossed the mountain over it, will not regret to leave it, but will thank the stars that a better road will soon supersede it. The friendly Conemaugh, as it passes this place, shakes hands with the Stony Creek, and the two proceed together, in a nearly northern course, around Laurel hill, where they strike due west. Johnstown lies on a level flat, surrounded by steep hills. It is pleasantly situated, but is without the least interior attraction. The buildings are small and without ornament, and the population, consisting of about two thousand, embraces a conglomerate of character, of which the most part follow the "raging canawl," or business appertaining to the trade of which it is the distributor. The original settlers of the county were Welsh, Scotch, and Irish, and the latter appear to have the "megority" here.

EBENSBURG, a few miles distant, is the seat of justice. It is situated on one of the high ridges of the Alleghany, commanding a fine view of the surrounding country. The *Mountaineer*, published in this borough in 1840, contained the following obituary notice of one of the most remarkable men that ever resided in this part of the country:

"Died, on the 6th instant, at Loretto, the Rev. Demetrius Augustine Gallitzin, who for forty-two years exercised pastoral functions in Cambria County. The venerable deceased was born in 1770, at Munster, in Germany. His father, Prince de Gallitzin, ranked among the highest nobility in Russia. His mother was the daughter of Field Marshal General de Schmeltan, a celebrated officer under Frederick the Great. Her brother fell at the battle of Jena. The deceased held a high commission in the Russian army from his infancy. Europe in the early part of his life was desolated by war—the French revolution burst like a volcano upon that convulsed continent; it offered no facilities or attractions for travel, and it was determined that the young Prince de Gallitzin should visit America. He landed in Baltimore in August, 1782, in company with Rev. Mr. Brosius. By a train of circumstances, in which the hand of Providence was strikingly visible, his mind was directed to the ecclesiastical state, and he renounced forever his brilliant prospects. Already endowed with a splendid education, he was the more prepared to pursue his ecclesiastical studies, under the venerable Bishop Carroll, at Baltimore, with facility and success. Having completed his theological course, he spent some time on the mission in Maryland.

"In the year 1789, he directed his course to the Alleghany mountain, and found that portion of it which now constitutes Cambria County a perfect wilderness, almost without inhabitants or habitations. After incredible labor and privations, and expending a princely fortune, he succeeded in making 'the wilderness blossom as the rose.' His untiring zeal has collected about Loretto, his late residence, a Catholic population of three or four thousand. He not only extended the church by his missionary toils, but also illustrated and defended the truth by several highly useful publications. His 'Defence of Catholic principles' has gained merited celebrity both here and in Europe.

"In this extraordinary man we have not only to admire his renunciation of the brightest hopes and prospects, and his indefatigable zeal, but something greater and rarer—*his wonderful humility.* No one could ever learn from him or his mode of life what he had been, or what he exchanged for privation and poverty.

"To intimate to him that you were aware of his condition, would be sure to pain and displease him. He who might have revelled in the princely halls of his ancestors was content to spend thirty years in a rude log-cabin, almost denying himself the common comforts of life, that he might be able to clothe the naked members of Jesus Christ, the poor and distressed. Few have left behind them such examples of charity and benevolence. On the head of no one have been invoked so many blessings from the mouths of widows and orphans. It may be literally said of him, 'if his heart had been made of gold he would have disposed of it all in charity to the poor.'"

Cambria County contains, what probably no other county in Pennsylvania will *boast*, a *deserted village*. It is called *Beulah*, and was laid out with the prospect of becoming the seat of justice; but this having subsequently been established at Ebensburg, four or five miles distant, the entire village soon became depopulated, and is now in ruins.

Following the course of the canal, after leaving Johnstown, the railroad runs parallel with it for several miles. We pass the unimportant stations of MAGILL'S FURNACE, NINEVEH and NEW FLORENCE, and pause a moment for breath at LOCKPORT. The canal here comes over to our side of the river, crossing the stream in one of the finest stone aqueducts we have yet seen. It does not leak a drop, and the little clusters of creeping weeds and water-plants holding on to its sides give it quite a romantic and poetic appearance.

We are now in the great county of Westmoreland—great alike in its agricultural fertility, its numerous streams, and its extraordinary mineral resources. The Conemaugh separates it from Indiana and Armstrong on the north-east, and Laurel Hill from Cambria and Somerset on the east and north-east—Fayette lying south and Washington and Alleghany Counties west of it. Like Huntingdon, it is full of small streams, which, for the most part, have their rise in its midst. Between Laurel Hill and Chesnut Ridge is the celebrated Ligonier Valley, remarkable for the extent of its timber as well as the fertility of the soil. This beautiful valley has a north-east and south-west course, and consists of a very long but narrow belt of land lying between the parallel elevations mentioned. It commences in Virginia, traverses the eastern portion of Fayette and Westmoreland, and finally spreads out into the adjoining county of Indiana. During the revolution, and for many years prior to it, this extensive valley was the scene of many interesting and stirring events, some of which we may

PACK-SADDLE ROCKS IN CHESNUT RIDGE GAP.

probably allude to hereafter. For the present we must keep an eye on the Conemaugh, which is very busy in receiving the little streams pouring in upon it, and sending them along, under its auspices, to the Kiskiminetas, one of the principal forwarding agents of the Alleghany. The Kiskiminetas (which means, kiss-me-in-a-minit, if we are not wrong!) is but a continuation of the Conemaugh; but as this stream is *united* with the former, of course the propriety of the change of name is not to be questioned. The Kissme—— ah, ha! This is what we have been looking out for! "'Ere," as Captain Cuttle would remark, after making a close "hobserwation,"—"'ere is a scene as is a scene!" Chesnut Ridge, as it is the last lofty and prominent elevation properly belonging to the Appalachian chain, determined to end the mountain series with a flourish of trumpets—a *coup d'etat.* The Conemaugh, too—grown amazingly since we first saw him, a mere untamed urchin, wandering down the Alleghany—but a powerful hand in scooping out the gap, which is at once high, bold, precipitous, and rugged. Along this frightful bluff runs the railroad, elevated nearly a hundred feet above the glassy river; and the idea of danger—a mere passing shadow—only gives spice to the grandeur of the scene. Winding hurriedly around a curvature, we catch a glimpse of the overhanging pack-saddle, so called from its resemblance to that time-honored travelling appendage. Travelling three or four miles thus, we suddenly leave the Conemaugh, and rein in our fiery steed at the branch of the road leading to BLAIRSVILLE, six miles distant. The canal also goes that way, and passes through some magnificent scenery on the route. A few miles below Blairsville, it passes through a tunnel nine hundred feet in length, and emerging from it, is carried over the Conemaugh on a stupendous stone aqueduct, amidst the wildest scenery imaginable.

On leaving the Conemaugh, and winding around the gap of Chesnut ridge, the railroad has nothing more to do with the Alleghany mountain, and accordingly directs its course due west, very nearly through the centre of Westmoreland County. We pass three or four unimportant stations, and find ourselves near the Loyalhanna River, emptying into the Kiskiminetas, which, in turn, runs into the Alleghany, then into the Ohio at Pittsburg, and so on. Latrobe is three hundred and twenty miles from Philadelphia, two hundred and seven from Harrisburg, and forty-three from Pittsburg. Its elevation above tide water is one thousand and four feet. This place has sprung up within the last year or two, and bids fair for future importance. A railroad from it to Uniontown, via Mount Pleasant, and thence through Kanawha to the Big Sandy, in Virginia, where it would connect some other railway routes, is projected. In this age of steam and railroading, it is not improbable that this contemplated road may ultimately be undertaken. A large portion of the stock, we understand, is already subscribed for. The line would cross the northwestern railroad leading to Parkersburg, in some point in Harrison County, and the main line of the Baltimore and Ohio Railroad, leading to Wheeling, at some point in the county of Taylor or Marion, in Virginia. The people of that section of Virginia, (which is exceedingly rich in its mineral resources,) would thus be brought within several hours ride of Wheeling, Baltimore, and Philadelphia, while they are now comparatively shut out from each of these places.

LATROBE.

The great interior region of Westmoreland comprises the rich table lands of the Alleghany—the country being sufficiently rolling to adapt it admirably for all the purposes of agriculture, including that of sheep husbandry, which is extensively pursued in the adjoining county of Washington, and others adjacent to the Monongahela, in this State and Virginia. East of Latrobe, in the adjacent counties of Indiana, Jefferson, Armstrong, and Clarion, is one of the finest and most extensive lumber regions in the State, to penetrate which it is sometimes proposed to extend the Blairsville branch of the railroad, at least so far as Indiana, the seat of justice of that county. A portion of all these counties is drained by the Clarion River, emptying into the Alleghany; a portion of Armstrong and Indiana by the

Crooked Creek and its forks; and the remaining portion of both, with Westmoreland on the west, by the Conemaugh, the Kiskiminetas, and the Loyalhanna, emptying into the former. The country is thus well supplied with streams, which are sufficiently large, when swollen in the spring by the melting snow, to bear the lumber to the Alleghany, where it is formed into rafts, and floated down the Ohio River. All the towns and cities on this noble river, (including Pittsburg, Cincinnati, and Louisville,) have been built almost entirely from the descending lumber of this great region; while the Monongahela has furnished the oaks and heavier and finer-fibred timber for purposes of steamboat, ship and bridge building, as well as cabinet-ware.

The lumbermen are essentially original characters. A more devil-may-care set of fellows never handled an axe or swung an oar; good-natured, robust, and hard-working, they have an inexhaustible fund of humor and forest-adventure, which does much to smooth down their exterior roughness. In the fall of the year they repair to the depths of the forest, and commence the preliminaries of the winter campaign.

Wide around their woodland quarters
  Sad-voiced autumn grieves;
Thickly down the swelling waters
  Float his fallen leaves;
Through the tall and naked timber,
  Column-like and old,
Gleam the sunsets of November
  With their skies of gold.

As soon as the timber district to be cleared is fixed upon, the woodmen commence operations, and one after the other fall the stately inhabitants of the forest—leaving behind, as monuments of their past glory, their bright stumps gleaming through the dismembered branches. The ox-teams are busy tugging off the saw-logs, and the saw-mill hard by tears savagely through their woody fibres.

Be it starlight—be it moonlight*
  In these vales below,
When the earliest beams of sunlight
  Streaks the mountain's snow,

* John G. Whittier.

FELLING TIMBER — WINTER.

Crisps the hoar frost keen and early
  To our hurrying feet,
And the forest echoes clearly
  All our blows repeat.

* * * * * *

Make we here our camp of winter,
  And through sleet and snow,
Pitch-knot and beechen splinter,
  On our hearth shall glow;
Here, with mirth to lighten duty,
  We shall lack alone,
Woman, with her smile of beauty,
  And her gentle tone.
But her hearth is brighter burning
  For our work to-day,
And her welcome at returning
  Shall our loss repay.
Strike, then, comrades! Trade is waiting
  On our rugged toil,
For ships are waiting for the freighting
  Of our woodland spoil!
Cheerily on the axe of labor
  Let the sunbeam dance,
Better than the flash of sabre,
  Or the gleam of lance!
Strike! With every blow is given
  Freer sun and sky,
And the long-hid earth to heaven
  Looks with wondering eye.
Loud behind us grow the murmurs
  Of the age to come!
Clang of smiths and tread of farmers
  Bearing harvest-home!
Here her virgin lap with treasures
  Shall the green earth fill—
Waving wheat and golden maize-ears
  Crown each beechen hill!

In the spring, when the snows melt, the sawed lumber is drifted down the mountain streams to the Alleghany, where long rafts are formed, and then piloted down the river. Each raft contains from four to eight hands, who lodge in a little board cabin erected in the centre of it. Here they live in merry social glee, consuming the night in song and revelry, interspersed with anecdotes and tales of their forest life—of "hair breadth 'scapes" from wolves and bears—of feats of strength and agility—of travels and fortune—of places, men, and women!

When the white frost gilds the valleys, the cold congeals the flood;
When many men have naught to do to earn their families bread;
When the swollen streams are frozen and the hills are clad in snow,
Oh! we'll range the wild woods o'er, and a-lumbering we'll go!
 And a-lumbering we'll go, so a-lumbering we'll go—
 Oh, we'll range the wild woods over, and a-lumbering we'll go!

When you pass through the proud city, and pity all you meet,
To hear their teeth a-chattering as they hurry through the street—
In the frost-proof flannel we're encased from top down to the toe,
While we range the wild woods o'er, as a-lumbering we go.
 And a-lumbering we'll go, so a-lumbering we'll go,
 Oh, we'll range the wild wood o'er, while a-lumbering we go!

You may boast of your gay parties, your pleasures, and your plays,
And pity us poor lumbermen while dashing in your sleighs;
We want no better pastime than to chase the buck and doe,
As we range the wild woods over, and a-lumbering we go!
 And a-lumbering will go, so a-lumbering we'll go,
 Oh! we'll range the wild woods over while a-lumbering we go!

The music of our burnished axe shall make the woods resound,
And many a lofty ancient pine will tumble to the ground;
At night, round our good camp-fire, we'll sing while rude winds blow;
Oh! we'll range the wild woods over while a-lumbering we go!
 So a-lumbering we'll go, and a-lumbering we'll go,
 Oh! we'll range the wild woods over, and a lumbering we'll go!

When winter's snows are melted, and the ice-bound streams are free,
We'll run our rafts to market, then haste our friends to see;
How kindly true hearts welcome us—our wives and children too—
We will spend with these the summer, and again a-lumbering go!
 And a-lumbering we'll go, and a-lumbering we'll go—
 We'll spend with these the summer, and again a-lumbering go!

When our forest days are ended, and we cease from winter toils,
And each through the summer warm will till the virgin soil—
Enough to eat, and drink, and wear—content through life to go—
Then we'll sing our adventures o'er, and no more a-lumbering go!
 And no more a-lumbering go, no more a-lumbering go—
 Oh! we'll tell our adventures o'er, but no more a-lumbering go!

Pennsylvania has long been one of the most productive lumber States in the Union, and while it is true that her splendid forests are annually disappearing or growing smaller, yet she can never become entirely destitute of this necessary article. The extensive use of coal, if nothing else, will materially contribute to avert this evil. As the pines disappear, they are succeeded by oaks, equally valuable and necessary; while, again, the oaks are generally succeeded by pines. There would seem to be a law of succession in our forests, as well as in our grains—one crop of vegetable matter, drawing from the earth certain of its constituent elements, prepares it for another growth, which draws some other particular substance from it, and in turn deposits that which was previously extracted. This natural law, backed with the obvious *interest* of man, will always secure us large regions of timber, in those localities where the situation and quality of the soil are unfavorable to profitable culture.

THE LUMBER TRADE OF THE SUSQUEHANNA.—The principal lumber points of the Susquehanna River are Harrisburg, Middletown, Marietta, Columbia and Wrightsville, in Pennsylvania, and Port Deposit, in Maryland. About two million feet of the manufactured article, on an average, pass down the Susquehanna to these points every year, besides a large quantity which comes via the canals. Of this quantity it is estimated that about seventy million feet will arrive at Baltimore during the current year. It is not easy to get at the exact amount of business done in this rapidly increasing article of trade, in consequence of the imperfect data required by law to be kept, but it is generally conceded that the receipts of the year, ending on the 20th of April last, reached very nearly two hundred and fifty million feet, which is but little more than one-half the total number of feet which arrived at Albany, New York, (one of the greatest lumber markets in the United States,) during the year 1851, which amounted to not less than four hundred and sixty million feet, a large portion of which, however, came from Canada, whilst all that is brought down the Susquehanna is from Southern New York and Pennsylvania.

At Harrisburg the lumber is purchased to supply that city and the adjacent country, embracing the Cumberland Valley. A large portion is also purchased at the other points mentioned. The towns of Columbia and Wrightsville, opposite, are depots for purchasing and piling lumber, to season for the Baltimore and Philadelphia markets, as well as for the supply of all the manufacturing towns along the lines of railroads thence to both cities; whilst Port Deposit, being at the head of tide-water, affords facilities for shipping to all the markets southward, as well as those on the Delaware; and many buyers and manufacturers meeting here, a large quantity of lumber consequently changes hands at this point. The following is an approximation to the quantity *sold* in each market which we have enumerated:—Harrisburg, five million feet; Middletown, twenty-five million feet; Marietta, ten million feet; Columbia, fifty million feet; Wrightsville, ten million feet; Port Deposit, fifty million feet—total, one hundred and fifty million feet. Besides which, fifty million feet are shipped from Columbia and Port Deposit for Baltimore and Philadelphia. All this amount is exclusive of an average of twelve hundred rafts of square timber, the greater part of which goes to Philadelphia and New York.

We have thus briefly stated the comparative importance of the principal lumber points on the Susquehanna, from whose business some millions of dollars change hands yearly. But it is limited in extent, compared to what it promises to be in a very few years. Ten years ago the lumber trade of Baltimore was hardly worthy of mention, whilst at this period something like one million of dollars worth is sold in this market, and the trade is only in its infancy. Improvements are continually going forward in the timber regions for getting that article to the mills to be manufactured, and vast tracts of country, now abounding in a wild primeval growth of timber, must, ere long, yield to the stroke of the woodman's axe and the magic influence of the lightning saw, to be sent down the "big stream," to the markets of consumption, and planted again in smiling towns and villages, not in the rude fashion of native wildness, but in the improved shape of thousands of human tenements.

This quarter of the State is scarcely less remarkable for its coal, iron, and lumber, than for its excellent *salt springs*. The saline properties of the water, underlying the upper stratum of the soil, were noticed at an early period; but no efforts were made to extract the salt until 1813, at which time, owing to the war, it became exceedingly scarce and correspondingly high in price. The water, in many places along the Kiskiminetas and the Alleghany, oozed out of the ground, and those spots were much frequented by deer, who would stand around them licking up the water with great satisfaction. This fact finally led to experiments, which have since resulted in extensive mining operations. The water is pumped up, in immense quantities, generally by means of small steam-engines. The water thus raised is boiled until it attains a consistency approaching chrystallization; after which it is transferred to cisterns, in which the sediment is deposited, and thence, purified, it is put into large kettles, in which it soon becomes chrystallized without further trouble. The process of chrystallization is both interesting and curious. Spear after spear, of the most delicate structure and fantastic shape, bounces into existence as the water is absorbed, and soon the whole kettle appears white with the salt. Thirty gallons of water are usually evaporated for every bushel of salt. In all these works, coal, of course, constitutes the sole fuel. The pumps, we should add, are sunk to various depths, from three to eight hundred feet.

GREENSBURG.

Greensburg, laid out in 1783, was named in honor of Gen. Greene, of the Revolutionary War. It is the seat of justice of Westmoreland County, and contains a population of about twelve hundred, having recently somewhat increased. The town lies on an eminence overlooking the surrounding country. The late Judge Coulter, one of the Judges of the Supreme Court of Pennsylvania, resided here. He was one of the soundest lawyers and purest men in the State. This place is three hundred and twenty-four miles from Philadelphia, two hundred and seventeen from Harrisburg, and twenty-four from Pittsburg. It is ten hundred and ninety-one feet above tide-water. The railroad at this place, not quite finished as yet, will be in operation a few months ensuing. A railroad is also being constructed from Greensburg to Wheeling, which will probably intercept a large amount of the trade of the Ohio, during the season of low water above that point, as well as a portion of the travel now enjoyed by the Baltimore and Ohio Railroad. With these splendid improvements, concentrating at Greensburg, it will probably assume an active business aspect in striking contrast to its past career.

The remains of Gen. Arthur Saint Clair are buried in the Presbyterian church-yard of this place, over which a handsome pyramidal monument has been erected by his brethren of the Masonic fraternity. Gen. St. Clair, in addition to his connection with the Revolutionary struggle, bore a conspicuous part in the political movements of this State, having been a member of Congress, a member of the Convention to form the Constitution of the State, a candidate for Governor, &c., &c. His military career, on several occasions, exposed him to censure; but he was a true patriot, and withal a useful and valuable man. The latter portion of his life was embittered with pecuniary misfortunes, and he lived for some time in comparative seclusion amongst the hills of the Chesnut Ridge.

Twenty-one miles from Greensburg, and ten from Pittsburg, we pass the scene of Gen. Braddock's battle with the French and Indians, which took place in 1755. The entire region of country watered by the Ohio and its tributaries had long been claimed by the French, upon the strength of the original discoveries of La Salle. They accordingly built forts at various points along the Ohio, the Alleghany and Monongahela, and established themselves in the friendship of the Indians then living along those rivers. In the meantime, the authorities of Virginia and Pennsylvania conceiving that the country rightfully belonged to the British Crown, and constituted a portion of their respective colonies, took measures to oppose the further occupancy of the French. In the prosecution of this work, the fort at Pittsburg was commenced in 1754, under the authority of the Governor of Virginia; but, before its completion, the French captured it, and held it under the name of Fort *Du Quesne*, until 1758, when it

BATTLE OF BRADDOCK'S FIELD.

was abandoned to Gen. Forbes. Soon after it was enlarged and improved by Gen. Stanwix, and named Fort Pitt, in honor of the distinguished British statesman, and from which Pittsburg subsequently took its name. It thus remained in possession of the English until the commencement of the Revolutionary war, when it was seized and ever after held by the Americans—including, also, the whole surrounding country—the claim of Virginia in the meantime having been amicably arranged.

It was in view of the incursions of the French, that General Washington, in 1753, then a mere youth, was dispatched by the Governor of Virginia on an expedition to the head waters of the Ohio, to ascertain and report the state of the country. This was one of the most perilous journeys that could have been undertaken—his path laid through immense wildernesses, covered with the snows of a severe winter, and beset, at every turn, with hostile Indians. One of the most memorable incidents of this journey, (which exhibits the remarkable boldness and indefatigable energy of him who afterwards became the embodiment of our revolutionary struggle,) was the passage of the Alleghany River, about two miles above the present city of Pittsburg. The man who could thus push his way across a wild,

WASHINGTON CROSSING THE ALLEGHANY.

icy stream, was well qualified, subsequently, to cross the Delaware with the American army under his directing eye.

"I took my papers," said General Washington, in describing this perilous enterprise, "pulled off my clothes, and tied myself up in a watch-coat. Then, with gun in hand, and pack on my back, in which were my papers and provisions, I set out with Mr. Gist, fitted in the same manner, on Wednesday the 26th of December. The day following, just after we had passed a place called Murdering Town, (where we intended to quit the path and steer across the country for Shannopin's Town,) we fell in with a party of French Indians, who had lain in wait for us. One of them fired at Mr. Gist or me, not fifteen steps off, but fortunately missed. We took this fellow into custody, and kept him until about nine o'clock at night, then let him go, and walked all the remaining part of the night without making any stop, that we might get the start so far as to be out of the reach of their pursuit the next day, since we were well assured they would follow our track as soon as it was light. The next day we continued travelling until quite dark, and got to the river about two miles above

Shannopin's. We expected to have found the river frozen, but it was not, only about fifty yards from each shore. The ice, I suppose, had broken up above, for it was driving in vast quantities.

There was no way for getting over but on a raft, which we set about with but one poor hatchet, and finished just after sun-setting. This was a whole day's work; we next got it launched, then went on board of it, and set off; but before we were half way over, we were jammed in the ice in such a manner that we expected every moment our raft to sink, and ourselves to perish. I put out my setting-pole to try to stop the raft, that the ice might pass by, when the rapidity of the stream threw it with so much violence against the pole, that it jerked me out into ten feet water; but I fortunately saved myself by catching hold of one of the raft-logs. Notwithstanding all our efforts, we could not get to either shore, but were obliged, as we were near an island, to quit our raft and make to it. (Now called Washington's Island.)

"The cold was so extremely severe, that Mr. Gist had all his fingers and some of his toes frozen; and the water was shut up so hard, that we found no difficulty in getting off the island on the ice in the morning, and went to Mr. Frazier's."

A short time after his return to Virginia, the expedition of General Braddock was undertaken. He had arrived in this country in 1755, with the 44th and 45th regiments of royal troops, which were subsequently joined by others, with numerous wagons and horses, obtained through the exertions of Dr. Franklin, in the eastern portion of Pennsylvania. The army moved on the 9th of June, but its progress through the wilderness was much retarded by the wagon trains, which were finally left behind at the suggestion of Washington, acting as aid-de-camp to the commanding General. The General with thirteen hundred men moved forward, leaving Col. Dunbar to follow with the baggage and the remaining troops. The army crossed to the left bank of the Monongahela, below the mouth of the Youghiogany, being prevented by the rugged hills from continuing along the right bank to the fort at Pittsburg, the object of attack. At noon, however, they recrossed to the right bank of the river, near Turtle Creek, some eight miles from Pittsburg. Here occurred the battle. The sloping hills were covered with a dense forest to the water's edge, and their sides were worn with deep ravines and gulleys, rendering the spot a favorite one for the Indian. Captain Orme, an aid of Braddock, says:

"As soon as the whole had got on the fort side of the Monongahela, we heard a very heavy and quick fire in our front. We immediately advanced in order to sustain them, but the detachment of the five hundred men gave way, and fell back upon us, which caused such confusion and struck so great a panic among our men, that afterwards no military expedient could be made use of that had any effect upon them. The men were so extremely deaf to the exhortation of the officers, that they fired away in the most irregular manner all their ammunition, and then ran off, leaving to the enemy the artillery, ammunition, provisions and baggage; nor could they be persuaded to stop till they got as far as Gist's plantation, while many of them proceeded as far as Col. Dunbar's party, some six miles distant. The officers were absolutely sacrificed by their unparalleled good behaviour—advancing sometimes in bodies and sometimes separately—hoping, by such example, to engage the soldiers to follow them; but to no purpose. The whole number of killed and wounded amounted to between six and seven hundred, among which were many officers, including Gen. Braddock himself, with two of his aids. Gen. Morris, speaking of this action, says:—"The defeat of our troops appears to be owing to the want of care and caution in the leaders, who have been too secure, and held in great contempt the Indian manner of fighting. Even by Captain Orme's account they were not aware of the attack. And there are others that say that the French and Indians lined the way on each side, and in the front and behind ravines, that we knew nothing of till they fired upon us." Gen. Washington, nine days after the battle, wrote to his mother, as follows: "When we came there we were attacked by a party of French and Indians, whose number did not probably exceed five hundred men, while ours consisted of about thirteen hundred well-armed troops, chiefly regular soldiers who were struck with such a panic that they behaved with more cowardice than it is possible to conceive. The officers behaved gallantly in order to encourage their men, for which they suffered greatly, there being near sixty killed and wounded—a large proportion of the number we had. The Virginia troops showed a good deal of bravery, and were nearly all killed; for I believe out of three companies that were there, scarcely thirty men are left alive. Capt. Peyrouny and all his officers, down to a corporal, were killed. Capt. Polson had nearly as hard a fate, for only one of his was left. In short, the dastardly behavior of those they call regulars exposed all others that were inclined to do their duty, to almost certain death; and at last, in despite of all the efforts of the officers to the contrary, they ran, as sheep pursued by dogs, and it was impossible to rally them.

"The General was wounded, of which he died three days after. Sir Peter Halkett was killed in the field, where died many other brave officers. I luckily escaped without a wound, though I had four bullets through my coat, and two horses shot under me. Capts. Orme and Morris, two of the aids-de-camp, were wounded early in the engagement, which rendered the duty hard upon me, as I was the only person then left to distribute the General's orders; which I was scarcely able to do, as I was not half recovered from a violent illness, that had confined me to my bed and a wagon for above ten days. I am still in a weak and feeble condition, which induces me to halt here two or three days, in the hope of recovering a little strength to enable me to proceed homeward."

And to his brother John he writes at the same time: "As I have heard, since my arrival at this place, a circumstantial account of my death and dying speech, I take this early opportunity of contradicting the first, and of assuring you that I have not yet composed the latter. But, by the all-powerful dispensations of Providence, I have been protected beyond all human probability or expectation; for I had four bullets through my coat,* and two horses shot under me, yet escaped unhurt, although death was levelling my companions on every side of me!"

Arrived at our journey's end, we have little more to say. Pittsburg lies in a triangular position, between the Alleghany and the Monongahela—the first flowing in from a north-east, and the other from a south-east direction. Passing on each side of the city, they unite their waters at its western point, and thus form, and thus

* When Washington went to the Ohio, in 1770, to explore wild lands near the mouth of the Kenawha river, he met an aged Indian chief, who told him, through an interpreter, that during the battle of Braddock's field he had singled him out as a conspicuous object, fired his rifle at him many times, and directed his young warriors to do the same; but none of his balls took effect. He was then persuaded that the young hero was under the special guardianship of the Great Spirit, and ceased firing at him. He had now come a long way to pay homage to the man who was the particular favorite of heaven, and who could never die in battle.

stretches forth the broad and beautiful Ohio. In this respect Pittsburg is situated somewhat similarly to New York, and it will not be many years before it will bear comparison with that great commercial emporium in many other respects. The present population is about fifty thousand, including the manufacturing villages adjacent, and Alleghany city opposite, which properly constitutes a portion of the main city. The land upon which it stands was originally owned by the Penn family, under whose auspices the town was surveyed and laid off in lots, in 1765, at which time it contained but a few log houses, hurriedly thrown up by the Indian traders and other adventurers. Its progress since has been rapid and extraordinary; and its future prospects are brighter than ever.

As a manufacturing and distributing point, Pittsburg must ultimately become, if it is not now, the most important interior city in this country. The transportation of nearly every article of manufacture intended for western consumption, from the Atlantic sea-board to the western waters is, and always must be, an item of serious expense, which the consumer naturally desires to avoid. This is particularly the case in reference to heavy bodies, as iron, wood, glass, earthen and other wares, and it is in the production of these articles that Pittsburg is most extensively engaged. With inexhaustible beds of coal and iron, with abundance of salt, minerals, lumber, wool, and an endless variety of agricultural facilities—standing at the head of the longest continuous river in the world, with uninterrupted navigation throughout, as well as to points radiating from it, comprising thousands of miles of uninterrupted water navigation, penetrating every point of the great, grand, and glorious West—to say nothing of the canal and railway system forming a net-work of more interior intercourse—Pittsburg is, and ever must be, the principal theatre of the productive greatness of this vast continent! Nature has so ordained it—she has fixed her stamp of greatness upon it, and her right to enjoy it there is none to dispute.

The annexed view of Pittsburg is afforded from the hill above Sligo, nearly opposite its western point. The editor of the Wheeling *Times*, (which has always been a rival city,) in speaking of the visit of a Board of Inquiry appointed to select a site for the United States Marine Hospital, some years ago, used the following eloquent language:

"This Board found Pittsburg a much larger place than Wheeling;

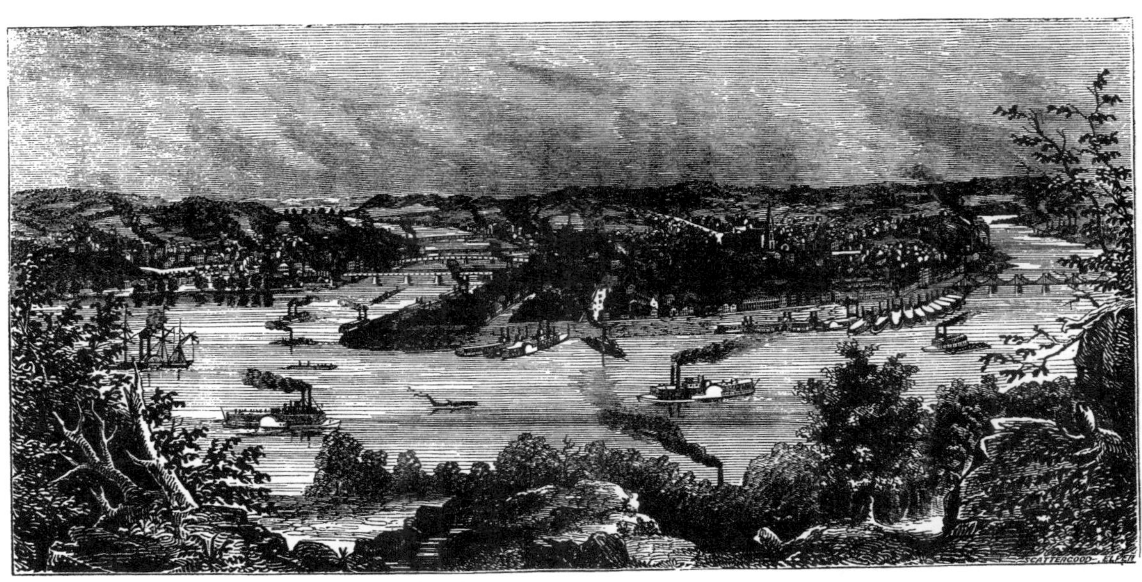

PITTSBURG.

they found it a thriving place, with numerous engines, furnaces, and machinery; they found it with a rich and industrious population—a people that would work, and would therefore prosper—at the same time they found them an hospitable, gentlemanly class of beings, possessed of intelligence and willing to impart it. They doubtless took an early excursion upon the hills that environ the city. They looked down, and a sea of smoke lay like the clouds upon Chimborazo's base. No breath of air moved its surface; but a sound rose from its depths like the roar of Niagara's waters, or the warring of the spirits in the cavern of storms. They looked around them, and saw no signs of life or human habitation. They looked above them, and the summer sun, like a haughty warrior, was driving his coursers up the eastern sky. Then from the sea of smoke a vapor rose—another and another cloud rode away, and a speck of silvery sheen glittered in the sunbeams.

"Again, a spire came into view, pointing heavenward its long slim finger; then a roof—a house-top—a street; and lo! a city lay like a map spread out by magic hand, and ten thousand busy mortals were seen in the pursuit of wealth, of fame, of love, of fashion. On the left, a noble river came heaving onward from the wilderness of the north, bearing on its bosom the treasures of the forest. On the right, an unassuming but no less useful current quietly yielded to the vessel's prow that bore from a more genial soil the products of the earth. They looked again, and extending downward through fertile and cultivated vales, checkered with gently swelling hills, they saw the giant trunk formed by the union of these noble branches. Ruffling its mirrored surface, they saw the noble steamer leaping like the panting courser, bearing a rich burden from the far sunny south; another, gathering strength and rolling onward to commence its long journey past fertile fields, high hills, rich and flourishing cities, and forests wide and drear, bearing the hand-work of her artisans to Mississippi, Texas, Mexico, the groves of India, and the hills of Pernambuco—nay, to every land to which the sun in its daily course gives light. Such they saw Pittsburg; and as such, as a citizen of the West, we are proud of her."

And as such, a citizen of Pennsylvania, *we* are proud of her. But, alas! for the smoke! There is too much of *that* here—our anthracital experience never could be overcome sufficiently to allow our bituminization. You cannot walk the streets with a clean face or a white collar—pshaw! Pittsburg is full of soot and smoke, emitted from a thousand tremendous coal fires, and the whole aspect of the place is as black as a negro's skin. This makes a residence in it unpleasant—kills taste for dress—prevents the ladies from promenading—destroys their complexions, and plants cabbages where only roses should grow. The sulphurous air, however, prevents eruptions of the skin, and people suffering with these disagreeable diseases should forthwith eschew medicine, and take up a residence for a short time in Pittsburg. It will cure 'em.

Many of the extensive manufactories spoken of as being situated at Pittsburg, are not within the limits of the city proper, but are scattered around within a circle of five miles radius from the courthouse. Within this compass are the cities of Pittsburg and Alleghany, the latter already a large place of some twenty thousand inhabitants, containing many extensive manufactories, particularly of cotton, iron, and white-lead, and doing a large proportion of the lumber business of the district, the boroughs of Birmingham and Lawrenceville, and the towns and villages of Manchester, Stewartstown, Sharpsburg, East Liberty, Wilkinsburg, Croghansville, Minersville, Arthursville, Riceville, Oakland, Kensington, Sligo, Cuddysville, Temperance Village, Tarentum, Millersville, and New Troy. The manufacturing establishments located in these surrounding villages, have their warehouses, owners, or agents, within the city, and so far as general business interests are concerned, may be considered a part of the city itself, that being the centre where the greater part of the business is transacted.

Judge Breckenridge, in giving some of his recollections of Pittsburg in its infancy, says:

"At the time to which I allude, the plain was entirely unincumbered by buildings or enclosures, excepting the Dutch church, which stood aloof from the haunts of men, unless at those times when it was forced to become the centre of the hippodrome. And the races, shall we say nothing of that obsolete recreation? It was then an affair of all-engrossing interest, and every business or pursuit was neglected during their continuance. The whole town was daily poured forth to witness the Olympian games, many of all ages and sexes as spectators, and many more, directly or indirectly, interested in a hundred different ways. The plain within the course, and near it, was filled with booths as at a fair—where everything was said, and done, and sold, and eaten or drunk—where every fifteen or twenty minutes there was a rush to some part, to witness a *fisticuff*—where dogs barked and bit, and horses trod on men's toes, and booths fell down on people's heads! There was Crowder with his fiddle and his votaries, making the dust fly with a four-handed, or rather four-footed reel; and a little further on was Dennis Loughy, the blind poet, like Homer casting his pearls before swine, chanting his master-piece in a tone part nasal and part guttural:

> "Come, gentlemen, gentlemen all,
> Genral Sincleer shall rem'ber'd be,
> For he lost thirteen hundred men all
> In the Western Tari-to-ree."

"All at once the cry, to horse! to horse! suspended every other business or amusement as effectually as the summons of the faithful. There was a rush towards the starting post, while many betook themselves to the station best fitted for the enjoyment of the animating sight. On a scaffold, elevated above the heads of the people, were placed the *patres patriæ*, as judges of the race, and —— but I am not about to describe the races: my object was merely to call to mind the spot where they were formerly executed; yet my pen on this occasion was near running away with me, like the dull cart-horse on the course, who feels a new fire kindled under his ribs, and from seeing others scamper, is seized with a desire of trying his heels also. The Dutch church, after some time spent in searching was found by me; but as for the race field, it is now covered with three-story brick buildings, canal basins, and great warehouses—instead of temporary booths, erected with forks, and covered with boughs just cut from the woods.

"It will be the business of the annalist, or of the historian, to trace the gradual progress of increase, or the various changes which the city has undergone. Who would imagine, on beholding the concourse of country merchants from all quarters, laying in their supplies of merchandise for the purpose of retail, that but a few years ago, the business was done in small shops, part cash and part country produce, that is, for skins, tallow, beeswax, and maple sugar? Who would imagine that the arrival and encampment of Cornplanter Indians on the banks of the Alleghany would make a great stir among the merchants? It was quite a cheering sight, and one which made brisk times, to see the squaws coming in with their packs on their backs, and to whom the business of selling as high, and buying as cheap as possible was entrusted. Now an Indian is not to be seen, unless it be some one caught in the woods a thousand miles off, and sent to Washington in a cage to make a treaty for the sale of lands.

"I can still remember when the mountains were crossed by pack-horses only, and they might be seen in long files, arriving and departing with their burdens swung on pack-saddles. Wagons and wagon-roads were used in the slow progress of things, and then the wonder of the West, a turnpike, was made over the big hills; and now canals and railways bring us as near to Philadelphia and Baltimore, as the Susquehanna was in those times. The western insurrection is not so much a matter of wonder, and there is no trifling excuse for the dissatisfaction of the West, when we reflect on their situation at that period. The two essentials of civilized and half-civilized life, iron and salt, were almost the only articles they could procure. And how could they procure them? There was no sale for their grain down the Ohio and Mississippi, on account of the Indian war, and the possession of New Orleans by the Spaniards. There was no possibility of transporting their produce across the mountains, for sale or barter. There was but one article by means of which they could contrive to obtain their supplies, and that was whiskey! A few kegs were placed on each side of a horse, transported several hundred miles, and a little salt and iron brought back in their place. Is it any wonder that the excise, in addition to the expense of transportation, almost cut them off even from this miserable resource?"

From 1790 to 1800, says the editor of Harris' Directory, the business of Pittsburg and the West was small, but gradually improving; the fur trade of the West was very important, and Messrs. Peter Maynard and William Morrison were engaged largely in it, and from 1790 to 1796 received considerable supplies of goods, through Mr. Guy Bryan, a wealthy merchant in Philadelphia, and the goods were taken to Kaskaskia in a barge, which annually returned to Pittsburg, laden with bear, buffalo, and deer skins, and furs and peltries of all kinds, which were sent to Mr. Bryan, and the barge returned laden with goods. At that period there was no regular drayman in Pittsburg, and the goods were generally hauled from the boats with a three horse wagon—until in 1797 a Mr. James Rattle, an Englishman, settled in this city, and was encouraged to take up the business, and drayed and stored goods, until a box of drygoods was stolen from his yard and shed, for then we had no warehouse, or regular commission merchant in Pittsburg—and this broke the poor man up, and he died broken-hearted and unhappy.

A French gentleman, Louis Anastasius Tarascon, emigrated in 1794 from France, and established himself in Philadelphia as a merchant. He was a large importer of silks, and all kinds of French and German goods. Being very wealthy and enterprising, in 1799 he sent two of his clerks, Clarles Brugiere and James Berthoud, to examine the course of the Ohio and Mississippi rivers, from Pittsburg to New Orleans, and ascertain the practicability of sending ships and clearing them from this port, ready rigged, to the West Indies and Europe. Those two gentlemen returned to Philadelphia, reported favorably, and Mr. Tarascon associated them and his brother, John Anthony, with himself, under the firm of "John A. Tarascon, brothers, James Berthoud, & Co.," and immediately established a large wholesale and retail store and warehouse, a shipyard, a rigging and sail loft, an anchor smith shop, a block manufactory, and in short everything necessary to complete vessels for sea. The first year, 1801, they built the schooner Amity, of one hundred and twenty tons, and the ship Pittsburg, of two hundred and fifty tons—and sent the former, loaded with flour, to St. Thomas, and the other, also with flour, to Philadelphia—from whence they sent them to Bordeaux, and brought back a cargo of wine, brandy, and other French goods, part of which they sent here in wagons at a carriage of from six to eight cents per pound. In 1802 they built the brig Nanino, of two hundred and fifty tons; in 1803 the ship Louisiana, of three hundred tons; and in 1804, the ship Western Trader, of four hundred tons.

A curious incident connected with this subject was mentioned by Mr. Clay on the floor of Congress. "To illustrate the commercial habits and enterprise of the American people, (he said) he would relate an anecdote of a vessel built and cleared out at Pittsburg for Leghorn. When she arrived at her place of destination, the master presented his papers to the custom-house officer—who could not credit him, and said to him, 'Sir, your papers are forged; there is no such port as Pittsburg in the world: your vessel must be confiscated.' The trembling Captain laid before the officer the map of the United States, directed him to the Gulf of Mexico, pointed out the mouth of the Mississippi, led him a thousand miles up it to the mouth of the Ohio, and thence another thousand up it to Pittsburg: 'There, sir, is the port from whence my vessel cleared out.' The astonished officer, before he had seen the map, would as readily have believed this vessei had been navigated from the moon."

In or about the year 1796, three of the royal princes of Orleans came to Pittsburg, and stopped at a hotel situated on the bank of the Monongahela, where Jno. D. Davis's warehouse now stands. They were very affable and conversant, and remained for some time in the city: at length they procured a large skiff, part of which was covered with tow linen, laid in a supply of provisions, and (having procured two men to row the skiff) proceeded on to New Orleans. One of these princes was Louis Phillippe, the late king of France—who, in his exile, visited our city, and spent his time very agreeably with Gen. Neville, Gen. James O'Hara, and several other respectable families who then lived on the bank of the Monongahela river. Louis Phillippe and his companions had previously descended the Susquehanna, to Harrisburg, where they embarked up the Juniata, and over the mountains to Pittsburg.

We remember well during the Embargo times and last war, when the internal trade and commerce of Pittsburg, by the Ohio, Western, and Southern rivers, brought us comparatively nigh to Wheeling, Cincinnati, Louisville, Nashville, St. Louis, Natchez, and New Orleans; but the slow process of keel-boats and barges was such that it consumed almost a whole summer for a trip down and up—when all was done by the hardy boatmen, with the pole or by warping; and when a barge arrived, with furs from St. Louis, cotton from Natchez, hemp, tobacco, and saltpetre from Maysville, or sugar and cotton from New Orleans and Natchez, it was a wonder to the many, and drew vast crowds to see and rejoice over it. And then the internal commerce during the war allied us closely with Richmond, Baltimore, Philadelphia, and New York—these cities getting much of their sugar, saltpetre, &c., by boats and wagons, through Pittsburg—which then did an immense carrying trade for the United States.

The first steamboat on the western waters was built in Pittsburg, in 1811, and was called the New Orleans. There were but six or seven built previously to 1817. From that period the number has rapidly been increasing, as well as improving in character, model, and workmanship. As late as 1816, the practicability of navigating the Ohio with steamboats was esteemed doubtful—none but the most sanguine regarded it favorably. In 1817, however, Capt. Shreve, an enterprising man, made a trip from New Orleans to Louisville in twenty-five days. The event was celebrated by rejoicing, and by a public dinner to the daring individual who had achieved the miracle. Previous to that period, the ordinary passages by barges, propelled by oars and sails, was three months. A revolution in western commerce was at once effected. Every article of merchandise began to ascend the Mississippi, until we have seen a package delivered at the wharf at Cincinnati, from Philadelphia, *via* New Orleans, at one cent per pound. From the period of Capt. Shreve's celebrated voyage till 1827, the time necessary for the trip has been gradually diminishing. During that year the Tecumseh entered the port of Louisville from New Orleans in eight days and two hours from port to port! * *

We cannot better illustrate the magnitude of the change in everything connected with western commerce and navigation, than by contrasting the foregoing statement with the situation of things at the time of the adoption of steam transportation, say in 1817. About twenty barges, averaging one hundred tons each, comprised the whole of the commercial facilities for transporting merchandise from New Orleans to the "upper country." Each of these performed one trip down and up again to Louisville and Cincinnati, within the year. The number of keel-boats employed on the upper Ohio cannot be ascertained, but it is presumed that one hundred and fifty is a sufficiently large calculation to embrace the whole number. These averaged thirty tons each, and employed one month to make the voyage from Louisville to Pittsburg; while the more noble and dignified barge of the Mississippi made her trip in the space of one hundred days, if no extraordinary accident happened to check her progress. Not a dollar was expended for wood in a space of two thousand miles, and the squatter on the banks of the Ohio thought himself lucky if the reckless boatman would give the smallest trifle for the eggs and chickens which formed almost the only saleable articles on a soil whose only fault was its too great fertility. Such was the case only a few years ago—what is it now? What is it *not?* Of the 16,674 miles of steamboat navigation of the Mississippi and its branches, there was employed in 1851 an estimated aggregate steamboat capacity of 300,000 tons; 4,500 boats not propelled by steam, of 75 tons average each; making 337,500 tons; which, with the numerous descending flat-boats, making two or three trips per year, with an estimated tonnage of 700,000, gives the extraordinary total of 1,337,500 tons! The value of the merchandize thus annually transported, exceeds $100,000,000!

And this is but a commencement—a mere foretaste of what is to be done in the future; for we have just shown that only a few years since the first steamboat was built, and the trade of the rivers was then comparatively insignificant.

If Pittsburg enjoyed a clear pure atmosphere—that is, if the disagreeable consequences flowing from the use of its coal could be conveniently obviated, it would attract to a greater extent than it now does men of fortune and enterprise within its bounds. As it is, it will still increase and prosper; but we doubt if it will ever prove an interesting place for those who *have* it, to enjoy their wealth in its midst. Various attempts have been made to kill the effect of the smoke and soot, but they have thus far proved ineffectual. We repeat, this is an insufferable evil, which a stranger, accustomed to a clear atmosphere, cannot overcome.

Let us, in conclusion, accompany our kind reader to the river, where the tall and stately floating palaces are arranged along the shore. Here are generally lying to, from twenty to forty large and elegant steamboats, and the greatest rivalry exists amongst them for the conveyance of passengers, which even seizes the negro cooks and waiters, and firemen. Here is a specimen of their advertisements, set off in rhyme in good Ethiopian spirit:

"Come all ob you passengers
What want to ride fast,
Come on de Corneelyah,
You will nebbe be passed;
De Corneelyah is a good boat,
She knows how to move,
But what will she do
When her engines get smoothe?
CHORUS—O, go it Corneelyah,
She is de boat, I reckon.

"She is a fast boat,
She never comes in late:
Leaves Pittsburg at five,
And Cairo at eight;
But when she is comin',
De ladies dey will say,
'Behold it is Corneelyah,
She has come before day.
O, go it Corneelyah,
She is de boat, I reckon.

"Come all ob you passengers,
  Dat want to know your fare,
Jus' walk up to de office,
  You'll find our clerk dare:
Our clerk he is a good man,
  One ob de bery best,
He treats all ob de passengers
  To de honest good jest.
    O, go it Corneelyah,
    She is de boat, I reckon."

Turning around the projecting angle of the city, we see the glories of the broad Alleghany on both sides, strewn with arks, rafts, and flat-boats—some loaded with lumber, some with coal, and others with iron and miscellaneous freights. These boatmen are merry fellows, and have their own sport going down the river:

  Heigh-ho! boatmen row,
  A-floating down de riber de Ohio!
De boatman dance—de boatmen sing—
De boatmen 're up to eb'ry ting—
Dance, boatmen dance—dance boatmen dance!
We'll dance all night till broad daylight,
And go home wid the gals in the morning!
  Heigh-ho, boatmen row!
  A-floating down de riber de Ohio!

## PHILADELPHIA AND THE LAKE TRADE.

The Sunbury and Erie Railroad Company, which was projected to open out the northern counties of the State, and develope their rich mineral and agricultural resources, as well as to pour into the lap of Philadelphia the vast trade of the great lakes, was first chartered in 1837. The charter authorizes the company to construct a Railroad from Sunbury to Erie, 286½ miles. They propose, however, first to build that portion of the road between Williamsport and Erie, a distance of 240 miles; there will then be, in connection with the "Cattawissa and Williamsport," "Little Schuylkill," and Reading railroads, a continuous chain of railroads from Philadelphia to Erie, without transhipment, the entire route of 425 miles lying within the State. The highest grade will be 52 8-10 feet to the mile, and that for only eight miles. The cost of the road, graded and bridged for a double track, with a single track laid, and with sixty miles of siding, completed ready for the locomotive, it is estimated will not exceed six millions, and that it can be completed in two years.

The advantages possessed by this road are very great; the harbor of Erie, its terminus, is by far the best and safest on the lake, having an average depth of about 20 feet, an extent of six square miles, and is free from obstruction by ice considerably earlier in the spring and later in the fall. It is the point at which the competition between the Atlantic cities must take place for the passenger and freight trade of all the railroads running westward from it, by reason of the law of our Legislature, compelling the break of guage between the Eastern and Western railroads to occur there. At Erie, also, it will connect with some 3000 miles of railroad penetrating in every direction through the Western and North-western States. The country traversed by the road is rich in every source of intrinsic wealth; vast quantities of timber of the best kind for building or shipping purposes, as well as an inexhaustible supply of anthracite and bituminous coals and iron ore, together with the produce of a rich farming district, will seek a market over this road.

The great bituminous coal basin is situated in Clinton County, and is extensively worked. It furnishes the State with coal for locomotive use on the Columbia Railroad. The coal is free from sulphur, will make gas, and is good for blacksmiths' use.

It is only within a few years past that these northern counties of the State, such as Clearfield, Potter, Forest, McKean, Tioga, Lycoming, Warren, Elk, Jefferson, Clinton, Erie, &c., have excited the attention of settlers in any numbers. But they are rapidly filling up with settlers of the best class, such as Danes, Swedes, Norwegians, and New Englanders. Large companies from Vermont have settled there, and are writing every day for their friends. Ole Bull has purchased large tracts of land, and established a Norwegian colony in Potter County, which is thriving and flourishing rapidly. He has called his town Oleona; it was christened with much ceremony on the 8th of September, 1852. Here, if his town grows and is as prosperous as he can but wish it to be, is a glorious opportunity to render his name as lastingly famous as it now is from his violin celebrity. Many of the foresighted capitalists of Philadelphia make summer trips to this region and are so well pleased with it that they make large purchases. The best route is to Williamsport via the Pennsylvania Central Road to the Junction, about fifteen miles above Harrisburg, thence by packet up the Susquehanna; or, by the Reading Railroad to Pottsville, thence by stage via Danville to Williamsport. Jno. F. Cowan, Esq., of Williamsport, a gentleman who owns large tracts of land, we feel confident, would give any information that might be desired. This gentleman, by judicious purchases, and by re-selling his land to actual settlers, on easy terms, has amassed a large fortune within comparatively a short period.

The construction of the proposed road must inevitably induce a very extensive improvement in each of those branches of industry, and develope an incalculable amount of wealth now comparatively unproductive for want of such an outlet—while in time these improvements will add largely and rapidly to the business and profits of the road and the revenues of the State.

The great Western Lake *Country*, with which the proposed road will unite Philadelphia, contained in 1810 a *population* of only 272,000. The five States which have been formed out of the Northwest Territory, bordering on the lakes, now number a population of four millions and a half—being fifty per cent. more than the entire population of the United States at the time of the Declaration of Independence.

The territory embraced between the Ohio River and the Lakes, from the western boundary of Pennsylvania to the Upper Mississippi, containing about 180,000,000 acres of *arable* land, measures 280,000 square miles, being nearly twice as large as France, and about six times as large as the whole of England.

The trade of the lakes, with which Philadelphia will have a direct connection, in the aggregate of foreign and domestic imports and exports at the several ports, largely exceeds $200,000,000 per annum.

For passengers who arrive at Erie it will be the shortest route to a good market, as will be seen by the following:

| | |
|---|---|
| From Erie to New York via Dunkirk, | 505 miles. |
| From Erie to New York via Buffalo, | 568 miles. |
| From Erie to Philadelphia, | 425 miles. |

In connection with this road, and the products of the counties which it will open, we should perhaps allude more particularly to the lumber trade of Northern Pennsylvania.

When enumerating the sources of Philadelphia's prosperity, we speak of her manufactories, her commerce, her iron trade, and her vast and still increasing coal trade. Has it ever occurred to Philadelphians that the lumber trade of Northern Pennsylvania, which they nearly monopolize, is infinitely greater and more important than they have the least conception of? Not having the necessary data before us, we cannot make use of the exact figures; but, still, we have not the slightest hesitation in asserting that between three and six millions of dollars worth of lumber are now yearly manufactured in Northern Pennsylvania, and that for some years to come this amount will be increased in place of being diminished. The annual yield of anthracite coal in all parts of the State, is about four millions of tons, which is worth about six millions of dollars at the mines; and yet many Philadelphians are inclined to believe that the principal source of the prosperity of the city is in the enjoyment of this trade.

At a very considerable outlay of money, a Boom has been constructed in the Susquehanna some three or four miles above Williamsport, for the purpose of catching and retaining saw logs during the floating season. This is the mechanical object of the Boom. The commercial or speculative object is to bring distant lumber regions into market, and to bring the logs to the saw mill, instead of the saw mill to the logs. In other words, a man owning timber lands, say at the head waters of Pine Creek, or Cedar Creek, or Kettle Creek, instead of erecting a saw mill some sixty or eighty miles from the head of canal navigation, erects it in the neighborhood of Williamsport, within sight of the canal. During the winter, he gets out his logs, and as soon as the spring freshet takes place, boldly launches them into the rushing stream. Under the guidance of the hardy, red shirt "drivers," they are hurried down the creeks until they reach the "big river," and down the "big river" they are "driven" until they come within the clutches of the Boom.

Under the expectation that the Boom would accomplish all that its projectors promised for it, a number of large and very expensive steam saw mills have been erected in the immediate vicinity of Williamsport; and more mills, equally as large and equally as expensive, will be built. It may be inferred from this that great confidence was reposed in the successful operations of the Boom; and that unless that confidence was well grounded, immense pecuniary losses would be the consequence. The strength and capacity of the Boom was tested to its fullest extent this spring; and the experiment was made for the purpose of ascertaining whether lumber could be manufactured at a profit in the immediate neighborhood of Williamsport, and timber lands of distant regions could be made equally as available as the timber lands of what was once regarded more favorable locations. The experiment was far more successful than the most sanguine had the temerity to hope for. In the place of catching fifteen or twenty thousand logs, the Williamsport Boom Company caught and retained, in ten days, from fifty to eighty thousand logs—and demonstrated to the entire satisfaction of the community, that if the West Branch was dotted with saw mills from the mouth of the Lycoming Creek to the mouth of Loyalsock Creek, there would be no difficulty in procuring a sufficient amount of stock.

The company will be able to catch and retain upwards of two hundred thousand logs, or about four hundred thousand logs during the spring and fall freshets. Saw logs, when manufactured into boards, will average three hundred feet per log; consequently, the manufactured product of four hundred thousand logs would be one hundred and twenty million feet of boards, which at $16 per thousand, in the Philadelphia market, (a low figure) would bring $1,920,000. There is now in the vicinity ten large steam saw mills, with capacity sufficient to manufacture 40,000,000 feet of lumber per annum. If our figures are correct—and the Boom Company will make them correct, so far as the catching, retaining, and delivering the logs are concerned—it will require thirty large water or steam saw mills, in the neighborhood of Williamsport, to manufacture all the logs which can be, and which eventually will be delivered by the Williamsport Boom Company. The manufacture of lumber, directly on the West Branch River, is now in its infancy. It must rapidly increase from month to month, and from year to year, adding millions to the wealth and thousands to the population of the Susquehanna region. The start which Williamsport, Jersey Shore, Lock Haven, and Montoursville have taken is solely to be attributed to this trade. Williamsport, from her location, and from her large investment of capital, is and must continue to be, the soul and centre of the West Branch lumber region. Now that she has undertaken to manufacture lumber on a scale never before attempted by any one town in the United States, it may safely be assumed that her population will be doubled within the next ten years.

THE END.

# THE EPHRATA MOUNTAIN SPRINGS.

The Ephrata Mountain Springs opens for visiters on the 1st of June. This watering place is situated on the Ephrata Ridge, Lancaster Co., Penna., 13 miles N. E. of Lancaster, 18 miles S. W. of Reading, and 40 miles E. of Harrisburg.

The buildings being located on the most elevated spot in the county, renders the atmosphere as light and as pure in the night as during the day. Here the south mountain is the dividing ridge, the waters on the south side run into the Chesapeake, and on the north to the Delaware.

Persons desirous of exercising in the shade will find here graded walks through a dense forest to the different Springs and mountain dousche, (where the water falls twenty-five feet,) and to the Observatory, from which is presented the richest, most extensive, and the greatest variety of landscape scenery to be met with any where. The city of Lancaster, and numerous other towns and villages, are in full view. Points in five other counties, and in the State of Maryland, are visible to the naked eye; some at a distance of forty miles.

The water is pure, soft sand stone and slate, consequently is very superior for drinking and bathing; as a recommendation of its quality to restore to health and vigor diseased and frail constitutions, we would refer only to those who have availed themselves of its use, without any further comment. The temperature of the different Springs is, some very cold, others more moderate.

Since last season extensive additions have been made; there are now comfortable accommodations for three hundred and fifty persons. Additional Baths, cold and warm, a superior Ten-pin alley and a Billiard Saloon, and Stables, have been erected.

Three daily Coaches (except Sunday) leave Lancaster on the arrival of the trains of cars from the east, south, and west. A daily Coach (except Sunday) will leave the Bird-in-Hand (after the 1st of July,) on the arrival of the train leaving Philadelphia at eight o'clock, A. M. Families preferring to come by this train, will please telegraph to Lancaster the day previous, as the Coach can carry only nine passengers.

Any number of Passengers can be accommodated via Lancaster, as there is every facility from that point to carry them to the Springs. A daily Stage leaves Reading at nine o'clock, A. M., on the arrival of the Pottsville train; persons coming from Philadelphia will please telegraph to J. N. Miltimore, Reading, when the coach will be detained until the arrival of the Philadelphia cars—and arrive the same morning at the Springs. There is also a tri-weekly stage from Harrisburg via the Springs to Downingtown.

Telegraph despatches and packages sent by Adams' Express via Lancaster, will be delivered daily, and a daily mail during the season.

A stock of good livery horses for saddle and harness will be kept at the Springs. There are a great variety of pleasant drives over good roads, in the vicinity.

Experienced Assistants and good servants, an excellent Band of Music, a good Barber, and careful Hostlers are employed, so that every attention is paid to the comfort of visiters. The best refreshments — and every effort will be made to procure the delicacies of the Baltimore and Philadelphia Markets, in addition to a good Market at the place, will warrant us in promising a good table.

Visiters desirous of availing themselves of Hydropathic treatment, will find every convenience, and a competent physician to attend them, if required.

For further information, address

**JOSEPH KONIGMACHER,**

*Ephrata Post-Office, Lancaster county, Pa.*

PASCAL IRON WORKS
LOUDERBACK & HOFFMANN

www.ingramcontent.com/pod-product-compliance
Lightning Source LLC
LaVergne TN
LVHW021125110826
845150LV00005B/954